Palisades

PALISADES

The People's Park

Robert O. Binnewies

EMPIRE
STATE
EDITIONS

AN IMPRINT OF FORDHAM UNIVERSITY PRESS

NEW YORK 2021

Fordham University Press has no responsibility for the persistence or accuracy of URLs for external or third-party Internet websites referred to in this publication and does not guarantee that any content on such websites is, or will remain, accurate or appropriate.

Fordham University Press also publishes its books in a variety of electronic formats. Some content that appears in print may not be available in electronic books.

Visit us online at www.fordhampress.com.

Library of Congress Control Number: 2021902725

Printed in the United States of America

23 22 21   5 4 3 2 1

First edition

This book is dedicated to Aaron Gastellum,
who, for a moment, before a careless driver struck him down,
brought brightness and passion to our continuing search
for environmental wisdom.

CONTENTS

FOREWORD

Barnabas McHenry

BOB BINNEWIES HAS written an elegant chronicle of the battles, skir-mishes, and acts of valor that have given our nation one of its most esteemed and innovative park systems. The private citizens and professional admin-istrators who illuminate these pages are on the honor roll of American conservation. Here, they receive proper recognition for their passionate labor in solving the complex tasks associated with building and protecting an intricate park system in the graceful Hudson River Valley. *Palisades* is the tale of how green islands of nature and historic anchor points of our nation's birth are presented in an incalculable display that attracts millions of visitors each year.

Visitors discover multiple outdoor recreational opportunities. They may hike along miles of forest trails and cliffside pathways, swim in sparkling lakes, enjoy rustic camping, learn from nature educators and historians, follow George Washington and the Continental Army from the beginning to the end of the Revolutionary War, share space with wild birds, animals, insects, and plants, test climbing skills on vertical rock walls, enjoy spectac-ular scenic views, discover quiet wilderness vales, and relax at the historic Bear Mountain Inn. All this within the nation's most densely populated metropolitan region.

The challenge to protect these treasures began with a brutal blast of dynamite in the late nineteenth century and continues to this day, most recently when an industrial giant threatened to grievously wound natural scenery at the very heart of the park system. The story of getting from zero protected acres to the rich tapestry that is today's Palisades park system is one of huge determination, moments of crisis, caustic resistance to the very idea of conservation, glorious philanthropy, a steep learning curve, and responsibilities for guardianship passed with care from one generation to the next.

Assembling these public privileges has been "no walk in the park." The effort is immense and has attracted the allegiance of famous Roosevelts, Rockefellers, Harrimans, and thousands of not-so-famous like-minded citizens. All have given of their time, expertise, and money to build a park legacy of incomparable benefit. The pages of this book confirm that a conservation experiment begun at a rock quarry has influenced our nationwide conservation agenda, including the establishment of the National Park Service and the Environmental Protection Agency. Above all, *Palisades* is a celebration—a compelling statement that Planet Earth is our home and that we can maintain it wisely, if we try.

FOREWORD

Joshua Laird, Executive Director,
Palisades Interstate Park Commission

THE VISIONARIES WHO first protected what has grown into nearly 130,000 acres of parkland could hardly have predicted the extraordinary success of their multigenerational efforts. Over 100 years after its founding, the Palisades Interstate Park system has become a gateway and nurturing ground for the next generation of conservation leaders. Its landscape, a dazzling interplay of history and nature, spreads north up the Hudson River Valley on a scale that is hard to imagine in such a dense urban region. We are the fortunate heirs of this remarkable legacy but must remain dedicated to serving as responsible stewards. Our task, going forward, is to give nature a chance wherever we can find additional opportunity and to invite millions of future visitors along in this continuing journey of exploration, learning, and sheer enjoyment.

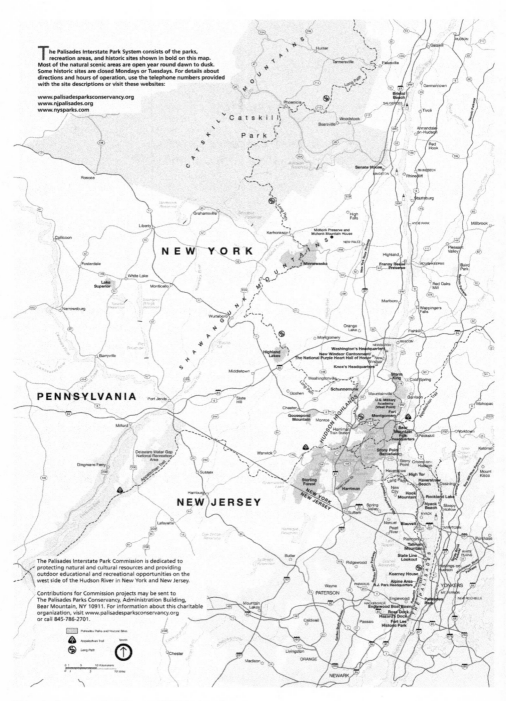

The Palisades Interstate Park System consists of the parks, recreation areas, and historic sites shown in bold on this map. Most of the natural scenic areas are open year round dawn to dusk. Some historic sites are closed Mondays or Tuesdays. For details about directions and hours of operation, use the telephone numbers provided with the site descriptions or visit these websites:

www.palisadesparksconservancy.org
www.njpalisades.org
www.nysparks.com

The Palisades Interstate Park Commission is dedicated to protecting natural and cultural resources and providing outdoor educational and recreational opportunities on the west side of the Hudson River in New York and New Jersey.

Contributions for Commission projects may be sent to The Palisades Parks Conservancy, Administration Building, Bear Mountain, NY 10911. For information about this charitable organization, visit www.palisadesparksconservancy.org or call 845-786-2701.

Palisades Interstate Park System (PIPC Archives)

Palisades

1

A Dynamite Park

The land was pleasant with grass and flowers and goodly trees as ever seen, and very sweet smells came from them. We were on land to walk on the west side of the river and found good ground for corn and other garden herbs, with a great store of goodly oaks, walnut trees, yew trees and trees of sweet wood in great abundance, and a great store of slate for houses and other good stones. Hard by was a cliff that looked of the color of white green and thought it were either copper or silver mine, and I think it to be one of them by the trees that grew upon it. For they were all burned, and the other places were as green as grass.
—Robert Juet, aboard Henry Hudson's *Half Moon*, 1609

ALICE HAGGERTY WAS the winner. According to a report in the March 5, 1898, edition of the *New York Times*, "This slip of an Irish girl had been chosen for the distinction of destroying one of the most widely known and splendid pieces of scenery in North America."

The piece of scenery was known as the Indian Head, a massive 200-foot vertical spire of diabase rock, a distinctive geological feature estimated to weigh 350,000 tons attached high on the Palisades cliffs overlooking the Hudson River. The Carpenter brothers, quarrymen by chosen profession, were dynamiting away more than a thousand cubic yards of the cliff each day. They "owned" the Indian Head, and it was their next target. Forty shafts, each two inches in diameter and twenty-five feet deep, had been drilled downward from the summit of the cliff into this celebrated rock formation. "Eighty feet below," reported the *Times*, "a tunnel was dug from the face of the precipice. It was five feet in diameter and ran back into the rock one-hundred-feet. From the inner end of this, two five-foot shafts descended almost to the river's level." The shafts and tunnels were packed with seven thousand pounds of dynamite.

During the Triassic period, thirty million years ago, semimolten igneous rock had been forced up through a fissure in the earth's crust. Overlying the fissure was a thick insulating layer of sandstone. The intruding rock slowly cooled, shrank, and solidified into giant vertical crystals of diabase lava consisting of silica, feldspar, magnetite, and pyroxene. Over eons, as the earth's crust shifted and the sandstone eroded, the flint-hard diabase was exposed, forming a thirty-mile cliff face ranging in height up to 550 feet along the western shoreline of the river named in honor of Henry Hudson. In 1524, when Giovanni de Verrazano sailed the French ship *La Dauphine* into what would become known as New York Harbor, he was so amazed by the cliff that he proclaimed the newly discovered land "La Terre de L'Anomee Berge," the Country of the Grand Scarp.

More than three centuries later, Alice Haggerty stood on the summit of Verrazano's "Grand Scarp." On a flat rock within arm's reach rested a small wooden box about twelve inches square. Two electric terminals and a T-shaped plunger protruded from the box. Wires from the terminals snaked off through grass and forest duff to the cliff's edge, then descended to connections with thousands of pounds of dynamite packed into the Indian Head. One quick pulse of hand-generated electricity from the box promised to deliver explosive power that would rattle buildings across the river in New York City.

By foot, on horseback, and in horse-drawn carriages, a large expectant crowd had gathered along the cliff summit. Other spectators, including the Carpenter brothers, floated in boats at a safe distance far out on the river. All waited for Alice to step forward and grasp the plunger, despite a whispered indictment among some of the spectators that she had won the contest with "pull." After all, went the whisper, Alice was a friend of Mrs. Hugh Reilly, wife of the Boss Blaster. Second only to the Carpenter brothers, the Boss Blaster was in command, and his word and influence were supreme.

The grand moment did not go as planned. According to news reports, Alice Haggerty "dabbed coquettishly" at the plunger, "then swept her skirts around with an involuntary movement of one hand and fled." A trickle of electricity reached one or two of the distant explosive caps that were embedded in sticks of dynamite. From a lower tunnel, a dull thud and small puff of smoke dislodged a few pieces of rock that clattered to the base of the cliff. Then silence. The Indian Head held firm, looming over the river as it had for fifty thousand years, ever since the last Ice Age glacier receded from the river valley.

Blasters describe a partial explosion as dangerously "hanging fire." Alice's meager blast likely had done some damage to the maze of connections that still awaited the electrical signal. These connections would require repair. At worst, a blasting cap somewhere in the network of tunnels and shafts might still go off unexpectedly. The Carpenter brothers, the Boss Blaster and his quarrymen, and the hundreds of spectators waited. Still more silence. Finally, Hugh Reilly and four quarrymen approached the edge of the cliff. The Carpenter brothers rowed close into shore for observation. With a rope around his waist, the Boss Blaster scaled down the cliff face and, for more than an hour, toiled to inspect connections and reset wires.

The mission was to bring the Indian Head down into a dust-swept, jumbled pile of broken rock. Italian stone workers, swinging sledgehammers and using more dynamite and machinery, would further pulverize the rock in preparation for its shipment on barges across the river to the city. Rock blasted from the Palisades was being used as base course and macadam for streets, building foundations, railroad beds, and other functional but unglamorous purposes. Diabase rock was too hard and brittle to be used for anything other than the most utilitarian projects. No diabase would

Palisades quarrying, circa 1898 (PIPC Archives)

decorate city skyscrapers or otherwise capture the creative attention of architects. Marble, granite, and sandstone were preferred. Palisades rock was "fill material," generally buried out of sight and forgotten.

When Reilly finally was hoisted back to the cliff summit, he walked resolutely to the plunger box. By then, the Carpenter brothers had rowed again to the safe haven of the far eastern shore of the Hudson. "The brawny arm of the Boss Blaster flew high as he drew the plunger out to its full length," reported the *Times*. "Then he forced it back with a quick, vicious thrust. The ground for hundreds of yards back from the brink of the cliff shook and trembled. There was an enormous all-pervading crash and roar. The solid face of the cliff bellied out at the middle and then the whole great surface collapsed and crumbled with a rush. Echoes of the explosion and fall reverberated along the cliffs and shores for six minutes. Where the Indian Head had been there was a huge, raw-looking concavity in the side of the Palisades, with a great pile of broken rock heaped at the bottom."

The crowd cheered. The Carpenter brothers rushed back across the river to congratulate the Boss Blaster. Alice Haggerty's moment of fame floated away in the dust.

2

###

The Commission

LONG BEFORE QUARRY operators started blasting away the Palisades, Native Americans had used the cliffs and summit for shelter, observation, and protection. When Verrazano stared at the cliffs from his ship, very likely his gaze was returned by Native American eyes. The Sanhikan, Hackensack, Raritan, and Tappan are said to have referred to the cliffs as "Wee-awaken," rocks-that-look-like-trees. The Palisades served as a citadel against threats from cross-river enemies, the Manhattans. But the real threat came from Dutch settlement of the lower Hudson River Valley in the 1600s. Dutch policy favored the establishment of large, manorial estates on the east side of the Hudson upriver from New Amsterdam (New York City) and more adventurous settlement probes on the west side in the dangerous territory of the looming Palisades. This caution proved prescient; in 1643, in response to the murder of Indians by Dutch soldiers, the Hackensack and Tappan attacked and destroyed the Vriessendael colony at present-day Edgewater, New Jersey, and an isolated farmhouse on the Hackensack River. Through much of the 1600s, what was eventually to become Bergen County, New Jersey, and Rockland County, New York, was described as a "howling wilderness with scarcely a single settler located within."

The Dutch gave up their claim of sovereignty in the Hudson River Valley to the British in 1664, and, slowly, with superior firepower, tools, and military protection, control shifted from Native American to Anglo settlers, but not until the early 1800s was the first house erected atop the Palisades. The house was framed in Boston, transported by ship to New York Harbor, floated by barge across the Hudson to the base of the cliffs, and hoisted, piece by piece, to the summit by its "active and enterprising" owner, Nathan T. Johnson. Johnson was followed by other settlers who scorned "anything as useless as aesthetics" and found a ready market for the forest of centuries-old hardwood trees on the summit that, until then, had escaped the axe man. Old-growth hardwood trees were felled, stripped of their branches, and "pitched" from the cliffs to the talus slopes below for shipment across the river. Railroad ties became the dominant market incentive, and, in time, the once heavily forested Palisades was "stripped practically clean."

Enterprising merchants also made use of the Palisades. The great vertical cliff faces became natural billboards. Letters twenty feet high, large enough to be read a mile away across the river, were painted on the cliff faces to advertise patent medicines and other products of the era. Fishermen settled at the base of the cliffs, where their livelihood depended on enormous catches of shad. They were joined by employees of "bone factories" where animal bones were ground into fertilizer and by others who worked at a steam-powered cereal mill.

The cliffs did not escape the attention of these early settlers. Small quarries began to appear along the cliffs, hinting at industrial-level rock crushing soon to come.

Wealthy men and women who could afford to cross the Hudson in private yachts sought out properties on the flat summit of the Palisades to escape the city's pungent congestion and to enjoy the spectacular views. They began erecting stately homes where simple cabins once stood. "Ownership" of the Palisades, from river's edge up and over the summit, was divided into scores of private holdings.

The nation was in a post–Civil War industrial boom. The iron, textile, shipping, mining, railroad, lumber, and oil industries were providing fortunes for adroit and skilled entrepreneurs. Except for a few solitary voices—John Burroughs, George Catlin, John Muir, Henry David Thoreau, Frederick Law Olmsted—these business adventurers paid little attention to environmental damage or even outright destruction. In California, three-thousand-year-old Giant Sequoia trees were being cut down to produce fence posts and other low-grade lumber products. On the Great Plains, remnant herds of bison were being shot for sport, sometimes from the windows of moving trains. Land, from the Mississippi to the Pacific Ocean, was up for grabs. Buffalo Bill had moved east to regularly present his circus-like Wild West Show in New York City, reminding audiences of the glamour and dangers of a frontier that had been breached almost to the point of disappearance.

But environmental glimmers of hope also were flashing in nineteenth-century America. The Yosemite Land Grant to California was proclaimed by President Abraham Lincoln in 1864, directing that the grant was made "upon the express conditions that the premises shall be held in public use, resort, and recreation and shall be inalienable for all time." In 1872, the establishment of Yellowstone National Park followed.

Predating these seminal land conservation actions, the Hudson River Valley School of art was begun in 1825 when two landscape paintings by Thomas Cole were spotted in a frame-maker's shop window. New York City mayor Phillip Home purchased the paintings, giving early market validation to artistic interpretations of "sublime natural beauty." The Hudson School of art, now renowned for highly valued art pieces, most in museums, attracted the talents of Albert Bierstadt, Asher Durand, William Bartlett, George Inness, Frederick Church, Victor Audubon, Winslow Homer, and Thomas Moran. They found timeless natural splendor in what others had referred to as "useless aesthetics." Jasper Francis Cropsey, especially, found inspiration in the charm of the Palisades and put his brush to canvas to capture indelible scenes.

But rolling explosions by quarrymen were predominant, causing Rudyard Kipling to grieve:

> We hear afar the sounds of war,
>   as rocks they rend and shiver;
> They blast and mine and rudely scar
>   the pleasant banks of the river.

Palisades Scene (Jasper Francis Cropsey)

Public outrage over destruction of the Palisades began to find voice in newspapers in the 1890s, but at the same time, in an example of dueling interests, the Carpenter brothers and other quarry operators, including Brown & Fleming, the Trevor brothers, and P. Gallagher were increasing production in response to market demand in New York City. The Carpenter brothers owned about one-half mile of cliff face and shoreline and could produce 1,500 cubic yards of crushed stone each day at a market price of $1 per yard. They had enough raw material to stay in business for decades.

New Jersey governor George T. Werts, serving in office from 1893 to 1896, became an early political champion for protection of the Palisades. Searching for an answer to the challenge, Werts turned to his state geologist, John Smock, who suggested passage of a "park condemnation act" that would allow the state to demand and take legal possession of river shoreline beneath the cliffs, thus blocking access to the river by the quarrymen. But the governor had a problem. While demolition of the cliffs could be seen easily from the New York side of the river, the view from New Jersey was of a bucolic summit landscape. Some members of the New Jersey legislature confessed no knowledge of the location of the Palisades cliffs.

To compensate, in 1894 the governor hosted a boat trip on the Hudson River to allow key members of the legislature a firsthand view of the problem. The glaring wounds of quarrying prompted New Jersey state senator H. D. Winton to suggest that the governor appoint a commission at the opening of the next legislative session to identify the speediest means of preventing further destruction. In January 1895, Werts appointed three members of the legislature to the task, but one of the members, J. J. Croes, seems not to have understood the mission. He said of the quarries that they "were capable of adding to the picturesqueness of the river shore."

In February, the *New York Times* reported that, on a "light legislative day," New York governor Levi P. Morton followed Werts's lead and appointed a three-member commission to work with its counterparts in New Jersey "for the purpose of securing action by the government of the United States of America in acquiring and setting apart the west shore of the Hudson, as designated, for the purpose of fortification and reservation in order that the brink of the precipice may be securely held and defended against attack and hostile occupation." One might argue that this resolution was a bit tardy, given that the British had stormed and taken the Palisades and its defending fort in 1776, thus forcing General George Washington and the Continental Army to retreat to Valley Forge. A reader of the *Times* commented that "it would take a pretty bold invader to threaten New York from the direction of northern New Jersey. It has been suggested that an enemy approaching New York from Boston by way of Albany might be checked by forts on the Palisades. It is certain, however, that such an emergency would greatly surprise the Naval War College at Newport, which has been engaged in the study of the defense of the eastern entrance of Long Island Sound under the conviction that an enemy from the east would surely try to get at New York that way."

From Washington, D.C., Secretary of War Daniel S. Lamont, a New Yorker, tersely reinforced this opinion, commenting that fortification of the Palisades "does not meet with favor." Officers down in the ranks, who were aware that Yellowstone and Yosemite National Parks were patrolled and protected by the U.S. Cavalry, were among the first to see park potential. One officer said, "I think the best course to pursue would be to secure from Congress authority and an appropriation for the purchase of the land and quarry plants and turn the whole thing over to the Interior Department." Another agreed: "I favor the idea of making a great public park out of the Palisades."

In 1895, a bill was introduced in the Fifty-Fourth Congress by New York congressman Benjamin Lewis Fairchild that would cede the Palisades to the federal government to become a national park/military reservation, but this legislative attempt never was brought to a vote. Three years later, with Spanish-American War paranoia gripping the nation, a second legislative attempt was made to transfer the Palisades to the War Department. This effort also failed.

As nascent attempts to save the Palisades sputtered along, voices in opposition by local citizens began to rise. A New Jersey property owner said, "If Bergen County prefers swamps and rattlesnakes and a dozen denizens on top of her miles of cliffs, and a half dozen dynamite fields along their face, she ought not be disturbed, and her rights infringed upon." The Carpenter brothers and other quarry operators wasted no time in seeking out politicians who would defend their rights to do as they pleased with their properties.

This was a man's world; men decided things, so little attention was paid in 1896 when seventeen women gathered at the home of Mrs. John A. Wells to form the Englewood Women's Club, a new chapter of the statewide New Jersey Federation of Women's Clubs. Led by Adaline Sterling, Elizabeth Vermilye, and Mrs. Chester Loomis, the club was formed "for the purpose of bringing together women interested in intellectual and cultural advancement, to stimulate inquiry concerning questions of public significance, to be a potent factor in the development of the community, and to promote its well-being through a program of philanthropy." Their "inquiry concerning questions of public significance" became sharply focused on the Palisades.

The women intended to find a way to stop the quarry operators. They had no voting rights and were expected to stay out of politics, but just before the first automobile rolled off the Ford assembly line and while the Klondike gold rush was underway in Alaska and Canada, the Englewood women formed a subcommittee on the Palisades, chaired by Vermilye. A retrospective, published years later in the *Rockland Journal News*, captured Vermilye's love for her task:

> In the 1860s, when she was a child, she would hike up Englewood's steep Palisades Avenue to the edge of the cliffs and roam the trails skirting the brink and the shoreline. She would clamber over the Indian Head and peer into the old tavern at Alpine where Cornwallis holed up as his British and Hessian troops scaled the cliffs to chase the Americans from Fort Lee. She marveled at the spectacular drops of Forest View, Bombay Hook, and the Giant Stairs. She watched the trains work their way up and down the opposite shoreline through Riverdale, Yonkers, and Hastings, silent as ships and barges at a distance.

Vermilye and her colleague Cecelia Gaines set out to enlist women from throughout New Jersey to the cause of the Palisades. They started writing letters to every newspaper in the state, accepted any invitation to speak before interested groups, and started lobbying legislators. In 1897, the Englewood women won the privilege of hosting the third annual gathering of the New Jersey Federation of Women's Clubs. Fifty women attended and were invited aboard the yacht *Marietta* made available by its owner, Harrison B. Morse.

Accompanying the women were several members of the newly formed American Scenic and Historic Preservation Society, chaired by Andrew H. Green. Green strongly shared the belief that quarrying of the Palisades must be stopped, and, as comptroller of Central Park and president of the New York City Board of Education, he brought strong credentials to the challenge. On board the yacht, Englewood club member Mrs. K. J. Sauzade stated that the practical duty of women was to "conserve the beauties of nature." She found no disagreement from Green, whose Preservation Society mission included the need to "promote public parks by private gift or the appropriation of public funds for the health, comfort, and pleasure of the people."

When the women debarked from the yacht, having seen the brutal evidence of quarrying, they arrived unannounced at the Carpenter brothers' operation just when midday blasting was scheduled. The scene only can be imagined: tough, gritty quarrymen who likely held strong opinions about the place in society of the weaker sex confronted by a group of determined, genteel ladies. The place for women was not at a quarry interfering with honest, purposeful work. Church sewing societies were fine places for disenfranchised women to chatter; politics and commerce should be left to men, who understood these complicated matters.

The resolute women obviously had a different opinion. They made no headway at the quarry trying to deliver a message to protect scenic beauty, but before the annual meeting of the Women's Clubs adjourned, a resolution was approved: "This famous Hudson River scenery which the citizens of New Jersey hold in trust for all the world will eventually become a thing of the past to their lasting shame and disgrace. As it is, the glorious heritage of the people of the state is being trampled under the foot of man and beasts in the streets of Gotham. Will the State Federation realize its power for good in this matter?"

In partial answer to that question, New Jersey governor Foster M. Voorhees, successor to Governor Werts, felt enough pressure from public opinion and the media to agree to meet with Elizabeth Vermilye and the Englewood women. His message was anything but encouraging: "Ladies, this is a hopeless task. I have tried for ten years to save the Palisades; it cannot be done." Voorhees said that the state had no money to buy the Palisades and that no support or interest in protecting the cliffs existed in "south Jersey."

But Voorhees had not counted on the 1899 arrival in the New York governor's chair of Theodore Roosevelt, the celebrated commander of the Rough Riders of Spanish-American War fame. No government leader would bring more energy and sense of mission to the cause of national conservation than this legendary man. Among his early public-service accomplishments, Roosevelt had served as police commissioner for New York City. Drawn, as always, to the outdoors, he had become an active member of the Englewood, New Jersey, rod and gun club and had roamed the summit of the Palisades in search of small game, gaining a detailed knowledge of the area and appreciation for its bold, natural beauty.

Soon after being elected governor, he met and conferred with Voorhees. The two governors agreed that preserving the Palisades would be to the benefit of both states. By then, the first study commission had gone by the wayside, having found no viable solution to the explosive problem.

Roosevelt and Voorhees appointed a second commission, five members representing New York and five New Jersey. In a female bounce against the glass ceiling, Vermilye and Gaines were among the New Jersey appointees. Meeting every week, this commission rushed to report back to the governors even as the Carpenter brothers accelerated their blasting schedule. By the end of the year, the study commission revisited a discarded idea, proposing to take a 737-acre strip of land along a fourteen-mile stretch of the Palisades between the cliffs and the low-tide mark of the river, thereby denying use of the shoreline for crushed rock commerce. The commission estimated the cost at $350 per acre; total price, $260,000.

Unknown to them, a resident across the Hudson was about to join their cause and would become crucial to the task. From his Glyndor estate on the eastern shoreline of the Hudson, George Walbridge Perkins was keenly aware of the Palisades. The cliffs were boldly within his view. He could hear the blasting, see the smoke and dust, and feel the tremors. The reverberating explosions would awaken Perkins's napping two-year-old child. These experiences would draw Perkins along a path of civic duty that would span two decades and involve his family for a century. He would establish precedent-setting momentum for the creation of parks and historic sites, actions that would bring energy and skill to the nation's fledgling conservation movement and place him squarely between two giant achievers whose opposing philosophies were legend: John Pierpont Morgan and Theodore Roosevelt.

According to biographer John Garraty, "George Perkins is a man little known today but was one of the most successful, controversial, and interesting Americans of the early twentieth century. The story of his rise from obscure beginnings to wealth and power would have strained the credulity of Horatio Alger's most devoted readers. Son of the warden of a boy's reformatory, he never went to high school but was invited to lecture at Columbia University. His father thought him slow in the head, but he revolutionized the insurance business, mastered the most complicated problems of corporate finance, developed the Palisades Interstate Park, guided creation of the International Harvester Corporation, made millions of dollars—and helped to organize and run Theodore Roosevelt's 'Bull Moose' Progressive Party."

So compelling was his personality that J. P. Morgan offered him a partnership, plus $125,000 for any worthy cause close to Perkins's heart, the first time they met. Jealous leaders of the hotly competitive farm-machinery industry, unable to agree about the value of their enterprises, asked Perkins to make an assessment and promised, in advance, to accept his decision.

The shrewd Russian finance minister Sergei Witte was so captivated that he refused to license American competitors of Perkins's life insurance company. Wall Streeters called him a socialist, and Western reformers considered him to be a tool of Wall Street.

Born in Chicago in 1862, Perkins died at age fifty-eight in Stamford, Connecticut. During his lifetime, and almost by accident, he became one of the nation's great environmental champions. A glimpse of Perkins's style occurred in 1911 when he appeared as a witness before a congressional committee investigating Morgan's alleged manipulation of the stock market. "What do you say to the statement that the panic was started to get rid of certain undesirable bankers, and that you gentlemen later were unable to manage it?" Congressman Charles L. Bartlett asked of Perkins. Perkins was reported to have stood up in "high theatrical dudgeon," slammed his fist on the table, and shouted, "I say that there never was a more infamous lie started than that! There is not a scintilla of truth in it! You might just as well say that a certain group of gentlemen made a contract with Mrs. O'Leary's cow to kick over the lamp that set Chicago afire!"

Perkins had wanted to be a missionary, but thin family finances prevented him from gaining the necessary training. Instead, he joined the New York Life Insurance Company as a clerk at age seventeen. In his mid-twenties, he was selling insurance from Wichita to Denver. By age thirty, Perkins placed his net worth at $51,000 and had arrived as third vice president at the company headquarters in New York City. He and his wife, Evelina, searched for a home in the metropolitan area and found a pleasant property in Riverdale-on-the-Hudson with a commanding view of the Palisades.

In 1900, when the second Palisades study commission was reporting to governors Roosevelt and Voorhees, Perkins had other issues on his mind. For him, the decades of the 1890s "had been one long triumph," according to biographer Garraty. Perkins reached an annual salary of $30,000 and, rising to second in command of the world's largest insurance company, enjoyed the allegiance of New York Life agents across the country. He instituted a pension plan for the agents, favored steady sales over high-pressure tactics, and believed in profit sharing. Reasoning that local agents could best serve their clients if they were given better financial incentives and less red tape, Perkins took steps to eliminate forty-four regional offices in favor of local shops. Old-guard middle managers were eliminated, and their traditional cut of profits went instead to local agents, all of whom reported directly to Perkins.

In the 1890s, when the massacre of Lakota Sioux Indians at Wounded Knee was juxtaposed against the founding of the *Wall Street Journal*, the construction of Carnegie Hall, and the opening of an immigration center on Ellis Island, the nation was still struggling to come of age and find its way in the international arena. American insurance companies were turning toward Europe and had established footholds in England, France, and Russia, but the lucrative German market, which held Swiss, Austrian, and Polish customers in its sway, was closed to American insurance companies. The financially conservative Germans were suspicious of upstart vendors from the United States. Competition among the American insurers was intense, and a breakthrough into Germany would be an impressive feat. Perkins took it upon himself to meet the challenge. On several transatlantic trips, he negotiated with German insurance regulators and invited them to visit New York Life and examine the books. The tough sell was to convince his own corporate colleagues to forsake more risky stock market investments in favor of the safe financial harbor of bonds. With these assurances, New York Life won the sole privilege of selling American insurance in Germany.

It was this success that brought Perkins face-to-face with Theodore Roosevelt. Competing insurance companies were seeking passage of a "Limitation Bill" that, if approved by the New York legislature and signed by the governor, would place a cap of $1.25 billion on the amount of insurance that any company could carry. Ostensibly, this legislation was said to encourage competition and reduce costs for working men and women, but the real purpose was to box in New York Life so that smaller insurance companies could gain more market share. Perkins said the Limitation Bill was "enough to make a man's hair turn white."

In March 1900, he gained a meeting with Roosevelt. The directness and clarity of Perkins's presentation convinced the governor that the Limitation Bill was a bad idea, prompting Roosevelt to signal to legislative leaders that it would not win his signature. The bill was withdrawn. Whether Perkins touched on the punishment being visited on the Palisades during this meeting is unrecorded, but the second study commission had completed its work. Two weeks after the Roosevelt-Perkins meeting, on March 22, 1900, legislation reached the governor's desk that would "establish the Commissioners of the Palisades Interstate Park." As authorized, the commissioners could "provide for the selection, location, appropriation, and management of certain lands along the Palisades of the Hudson River for an interstate park, and thereby preserve the scenery of the Palisades." The legislation allowed the governor of New York to appoint five commissioners from New

York and approve five from New Jersey. The legislation included $10,000 in operating funds.

Similar legislation was making its way through the New Jersey government, but Roosevelt's signature was the first on record, marking an initial step in what, for him, would become a national conservation legacy of immense proportion. Perkins was on vacation when the bill was signed. Roosevelt telephoned him to say that he wanted Perkins to serve on the newly established interstate park commission. When Perkins asked for time to "think about it," the governor responded, "I did not call you up, Mr. Perkins, to ask you to consider this thing; I called you up to tell you that you are President of the Commissioners of the Palisades Interstate Park."

In New Jersey, Governor Voorhees was contending with opposition to the Palisades initiative from landowners on the summit and by quarry operators who warned of government restrictions on business freedom and tax and job losses. In response, the New Jersey Federation of Women's Clubs swamped the legislature with letters, news articles, and personal visits. In various attempts to head off the women, the Palisades legislation was rewritten several times. A sticking point was the word *appropriation*, which had been included in the New York version of the legislation. This word meant that commissioners of the interstate park would have the authority to purchase privately owned land over the objections of landowners. This option to "condemn" land was considered essential if the commission were to have any real power. In an astonishing victory for Elizabeth Vermilye and her women's club partners, opposition to the New York legislative version was overcome, and Voorhees signed the legislation. New Jersey appropriated $5,000 for expenses.

The identical legislative acts cooperatively approved by the two states focused almost entirely on the New Jersey Palisades. The northern boundary of commission authority was set at Piermont Creek, a small tributary of the Hudson River barely across the state line in New York. Although membership on the commission was balanced, five and five, almost 100 percent of commission actions would be in New Jersey.

Vermilye and Cecelia Gaines hoped that their leadership and persistence, their service on the second study commission, and the hard-won legislative victory in New Jersey would result in appointments to the newly authorized interstate park commission. The response was that "some male members would be less free in their deliberations if women also served." Appointed to the commission were New Yorkers George H. Perkins, Nathan Barrett, D. McNeely Stauffer, Ralph Trautman, and J. DuPratt

White. The New Jersey contingent included W. A. Linn, Abram S. Hewitt, Col. Edwin A. Stevens, Franklin H. Hopkins, and Abram De Ronde.

Despite the wishes of Roosevelt, Col. Stevens was chosen as the first president of the commission, but soon deferred to Perkins. Technically, there were two ten-member commissions, but the membership was identical. The commissioners would meet on New Jersey matters, adjourn, and immediately be called back to order to work on New York issues. The agenda was self-evident. By the time the commission was authorized and began its work, quarry dynamite was booming daily, and the 3,000-acre summit of the Palisades had been cut into 127 separate privately owned parcels. Land prices were rising. Land on the summit was generally valued at $3,000 per acre; down along the shoreline, somewhat less. Perkins and his fellow commissioners had a total appropriation of $15,000. This meager sum was encased within the commission's broad authority to receive, control, and invest money, purchase land and property, construct facilities, maintain and operate parks, and take legal action as deemed necessary. The commissioners had no staff, no office, and no clear understanding of park stewardship. They knew only that they were expected somehow to conserve open space and natural beauty in the nation's most densely populated metropolitan region.

Ownership of the Palisades was constantly changing. Ulysses S. Grant had once owned a summit parcel, and Susan B. Anthony still did. In 1858, Joseph Lamb built Falcon Lodge, the first of many summer homes on the summit that eventually would be dubbed "millionaires' row." To reach his lodge on summer weekends, Lamb would ride up the east side of the Hudson River on the New York Central Railroad until opposite his property. At his request, the train would stop, allowing him, family members, and guests to be met as prearranged by the necessary contingent of shad fishermen. After being rowed across the river, Lamb and party would climb a rugged trail to his lodge.

At the northern end of the Palisades, near Piermont, New York, a terminus of the Civil War underground railroad, former slaves had established Skunk Hollow. At one time, seventy-five black farmers wrenched a marginal living from the rocky soil but had sold out and moved on. At the base of the cliffs, a few elderly widows lived in small shacks to which they had no legal claim. They were the last survivors of more than eight hundred fishermen, boat builders, and bone merchants, who, beginning in 1825, had settled along the Hudson River shoreline in the loosely defined community of Undercliff.

THE PALISADES MOUNTAIN HOUSE,
ENGLEWOOD, N. J.

This elegant new Hotel is beautifully situated on the Southeastern Slope of the elevated Plateau of the Palisades, 380 feet above the Hudson, opposite the upper end of New York, 11½ miles from the City Hall. It will be pleasantly and conveniently accessible at all times, night and day, by steamboat and cars. The views therefrom are magnificent, overlooking Manhattan Island, Central Park, New York Harbor and Bay, Long Island Sound, the Tappan Zee, Staten Island, &c.

Palisades Mountain House, circa 1860 (PIPC Archives)

A three-hundred-room hotel, the Palisades Mountain House, had been constructed in 1860 at a choice summit location, offering expansive views, landscaped grounds, bowling, billiards, fine dining, and music. The Mountain House was a favorite of city dwellers, who found respite there from crowded, pungent metropolitan environs.

William Dana, publisher and editor of the *New York World*, built the baronial Greycliff mansion in 1861. Cuban sugar baron Manuel Riondo arrived on the Palisades in 1904, four years after the interstate park commission was formed, and built an estate on his 200-acre Rio Vista property, complete with a one-hundred-foot-high water tower. Charles Nordhoff, editor of the *New York Herald*, who had sent Henry M. Stanley to Africa in search of Dr. David Livingston, had a home on millionaires' row, as did John Ringling of Ringling Brothers, Barnum & Bailey Circus. A much later arrival was Anthony Fokker, the Dutch engineer who had designed the tri-wing airplane flown with lethal skill in World War I by Manfred von Richthofen, the Red Baron.

George Perkins and his newly appointed colleagues on the Palisades Interstate Park Commission (PIPC) faced the daunting task of trying to decipher the ownership conundrum and decide how to proceed. They agreed to use $3,000 of their scanty financial war chest to retain the services of C. C. Vermeule, a reputable engineer from Englewood, N.J., who was charged with surveying the Palisades to untangle the question of ownership and property lines. But Perkins's primary target was the Carpenter brothers. He was determined to end the battering of the cliffs and the resulting assault on the sensibilities of residents in the lower Hudson River Valley, including his own.

In October 1900, he contacted the brothers George and Aaron Carpenter, beginning a negotiation that would span several weeks. Many successful, socially privileged, patrician businessmen of the day might not have been willing to sit down with hard-scrabble, blue collar, tough-minded quarrymen to try to hammer out a purchase deal, deferring instead to their lawyers. Not so Perkins: he sat across the table from the brothers.

The asking price started at $200,000 and then was "shaded" to $190,000. Perkins offered $100,000. In response, the brothers boosted the price back up to $200,000, offering to donate $25,000 to the park commission if the deal was struck. After several sessions, the Carpenters were flirting with a sales price of $145,000, and Perkins had come up to $125,000. In the meantime, the brothers were racing to produce as much crushed stone as possible. A compromise option-to-purchase finally was reached at $132,500. Perkins used the $10,000 appropriated to the commission by the state of New York for the down payment. No one bothered to ask whether spending New York money to buy land in New Jersey was legal. That detail aside, Perkins had to find $122,500 to close the deal, and quickly.

While the Carpenter negotiations were underway, James Stillman, president of the National City Bank of New York, had tapped Perkins to join the bank's board of directors. In a letter to Stillman years later, Perkins said, "You were the first man from the financial district to invite me to be connected with any of the important financial interests downtown, and I have always felt that this led to openings that have given me exceptional opportunities to broaden my education and better equip me for the work of the world." How right he was. One of the "financial interests downtown" was J. P. Morgan, who had started investing in New York Life Insurance bonds as a result of Perkins's agreement with Germany.

Morgan logically was on Perkins's list of potential donors, but there was a problem: Perkins did not know the legendary railroad, shipping, and steel tycoon. He asked Robert Bacon, a fellow member of the National City

Bank board and a Morgan partner, to arrange for an introduction. Reputed to be brusque, hardheaded, demanding, and distrustful, Morgan also was known, in contrast, to be a generous philanthropist. On his personal list of charitable projects were the American Museum of Natural History, the Metropolitan Museum of Art, Madison Square Garden, the Cathedral of Saint John the Devine, the American Academy of Rome, Harvard Medical School, and Morgan libraries scattered across America.

Bacon succeeded and introduced Perkins to Morgan at the tycoon's office, 23 Wall Street, then left the two men to private conversation. Biographer John Garraty reported, "Perkins launched at once into the story of the Palisades, but after a moment or two Morgan interrupted. I know all about that. You are Chairman of the Commission. What do you want?"

"I want to raise $125,000."

"All right put me down for $25,000. It is a good thing. Is that all?"

Perkins, delighted but probably off balance by the rapid exchange, asked Morgan who else might be approached for a contribution. Morgan suggested John D. Rockefeller. Expressing gratitude for the fundraising pledge and advice, Perkins rose to leave.

> Morgan: I will give you the whole $125,000 if you will do something for me.
> Perkins: Do something for you, what?
> Morgan: Take that desk over there.
> Perkins: I have a pretty good desk up at New York Life.
> Morgan: No, I mean come into the firm.

With lightning speed, and on first meeting J. P. Morgan, Perkins was being offered a breathless leap to the very pinnacle of the financial world.

> Perkins: I'll have to think about it.
> Morgan: Certainly. Let me know tomorrow if you can.

The decision was not easy for Perkins. Morgan was known as a notorious taskmaster. Family, friends, and professional colleagues gave Perkins mixed advice. New York Life responded by raising Perkins's annual salary from $30,000 to $75,000. Incredible as the conversation had been with Morgan, Perkins made an equally improbable decision—he turned down the Morgan offer.

Two months later, Perkins found himself again in the presence of the persistent Morgan, having breakfast at Morgan's Park Avenue mansion. Morgan appealed to the younger man by saying that Perkins's management skills were needed to respond to complex social and economic problems, that relationships between corporations, their employees, and the public must be improved, and that Perkins was the man for the job. On agreement that he could continue part-time at New York Life, Perkins acquiesced and joined the Morgan partnership.

Morgan wrote the check for $125,000 that would put the PIPC in position to close the deal with the Carpenter brothers, but the gift proved to be conditional. Morgan wanted to leverage the money. Years of looking for maximum benefit in any financial transaction had honed his instinct to find opportunity where others might not look. His gift would be anonymous, he told Perkins, but, if there ever were to be more, New York and New Jersey must first provide "sufficient funds," estimated at $450,000, to acquire various other properties on which the commissioners had obtained willing-seller options.

This demand by Morgan proved to come at an awkward moment. Perkins had momentarily lost his insider contact with the governor of New York. Only two years in office, Theodore Roosevelt had been chosen at the Republican National Convention to run as vice president, and the McKinley-Roosevelt ticket won a landslide victory in November 1900. Succeeding Roosevelt as governor was Benjamin Odell, a pragmatic politician who knew that the defacement of the Palisades had jangled the nerves of many voters in his state, but to him, the fact remained that the cliffs were in New Jersey. Governor Voorhees, approaching his last year in office, would have to take the lead in response to the Morgan financial challenge.

Carpenter Brothers Quarry (PIPC Archives)

Still, with check in hand, Perkins completed the purchase of the Carpenter brothers' quarry.

On Christmas Eve, 1900, the dynamiting ceased. Silence was a wonderful holiday gift for thousands. Boss Blaster Hugh Reilly, indicative of all that caused the interstate commission to be formed, was not witness to the transaction. Earlier in the year, while walking at night along a narrow trail at the edge of the Palisades precipice, he had slipped and fallen. His body was found the next morning at the base of the cliff.

3

Upriver

IN MID-JANUARY 1901, Perkins assisted in preparing an article about the Palisades that was published in the *New York Tribune*. He then purchased enough copies to deliver a *Tribune* to each member of the New Jersey and New York legislatures. Spreading around loads of newspapers with the Perkins article was good lobbying, but not enough to win prompt action on the Morgan financial challenge. On January 27, a special article in the *New York Times* was headlined, "Palisades Plans in Danger." Legislators from central and southern New Jersey who had resisted the original bill to create the PIPC were resisting again. "The prejudice that exists is as fixed as the rocks of the Palisades are now, and worse still, there is a widespread disinclination to be informed on the subject. The newspapers in the lower part of the state have treated the proposition flippantly and have prepared the legislators to resist any attempt that will be made to secure an appropriation for the Palisades scheme."

New Jersey legislators, always suspicious of New York motives, did not want to be drawn into an extended financial commitment that seemingly brought no benefit to many of their constituents. The two states once almost came to military blows over a boundary dispute, and, although closely tied by commercial and cultural ties, New York had the reputation either of bullying or ignoring its smaller, more rural neighbor. Of the eighty-one members of the New Jersey legislature, no more than twenty favored more money to buy land on the Palisades.

Elizabeth Vermilye, denied an appointment to the PIPC because of her gender, had largely directed her energies elsewhere, but on learning of resistance to expanded protection of the Palisades, she set about pulling another political rabbit out of the hat. She presided over the first meeting of a new organization, the League for the Preservation of the Palisades, formed only eight years after John Muir founded the Sierra Club. Among the activists joining Vermilye were Mrs. Ernest Thompson Seaton, wife of one of the founders of the Boy Scouts of America, Cecilia Gaines (now married to John Holland), Mrs. W. A. Roebling, whose husband oversaw construction of the Brooklyn Bridge, and clifftop property owners Mrs. Frederick Lamb and Mrs. Ralph Trautman. Frederick Lamb had served on the first Palisades study commission and Ralph Trautman on the second.

This group began pounding the halls in Trenton, attending hearings and buttonholing legislators. Palisades Park commissioner F. W. Hopkins, a wealthy investor, testified at the hearings. (The Reverend) Dr. Laidlaw, spokesperson for the Federation of Churches, participated as well, urging protection of almighty beauty. Vermilye learned from New York governor Odell that an appropriation of $400,000 would be forthcoming if New Jersey took the financial lead by approving $50,000. This was a sweet deal, and she and her colleagues made sure that elected officials in New Jersey knew it. On March 22, 1901, Governor Voorhees signed the necessary legislation, including approval of eminent domain authority for the Palisades Park Commission should the commissioners have to play hardball to acquire particularly crucial properties. State appropriations suddenly leaped from the small sum of $15,000, mostly spent, to a total of $465,000, and the Palisades Park Commission was in business well beyond the scars of the Carpenter brothers' quarry. Controversy followed immediately.

During the debate about financing in Albany, legislators claimed to have been left with the impression that the state of New Jersey would donate to the Palisades Park Commission riparian (underwater) land estimated to be worth $1 million. Acquisition of this watery real estate would strengthen the commission's control over access to the river shoreline. Based on assurance of "handsome" generosity by New Jersey, the New Yorkers had readily supported the $400,000 appropriation. After the fact, they discovered that any claim by the commission to the riparian lands in New Jersey was seriously in doubt. The commission's secretary, J. DuPratt White, took the heat. A New York assemblyman quoted in the *Hoboken Observer* said, "Now, Mr. Secretary White can take either horn of the dilemma he pleases—he is sure to be impaled. If his inter-state commission has a gift of $1,250,000 worth of riparian lands from the state, it has something the state can't give. If his inter-state commission has no gift of riparian lands, as he says, then New York has been what the sports call 'conned' out of $400,000."

New Jersey officials who served on the state's riparian board were caught off guard. They had not been involved in the Palisades maneuvers and were unaware that part of their submerged domain was being used as a bargaining chip. The secretary of the riparian board doubted whether legislation that directed the interstate commission to "improve the water's edge" applied to the drowned, muddy bottom of the river. But after due deliberation and pressure likely applied by Governor Voorhees, the board concluded that the New Jersey legislation must have "contemplated" that riparian lands running parallel to the Palisades cliffs should be excluded from dock and industrial site development. The underwater rights subsequently were transferred to the interstate commission, taking the bite out of this first cross-border park controversy, a hiccup promptly forgotten by politicians and the press.

J. DuPratt White was not a con man. A graduate of Cornell University Law School, he was admitted to the New York Bar in 1892 and, with a colleague, founded the prestigious law firm White & Case. All interstate commissioners were uncompensated and served at the pleasure of various governors. For many, this was a dutiful and honorable moment of public service, lasting for a few years. Not for White; he would serve on the commission for four decades regardless of shifting political winds, becoming a staunch defender of commission integrity and a key participant in the commission's many early achievements.

With muddy river tidal flats and blossomed funds in hand, the interstate commission hired its first employee in April 1901. Not surprisingly, Leonard Hull Smith was a lawyer with an annual salary from the commission of $1,200. Smith moved quickly, concentrating on the cliff face and shoreline of the Palisades, allowing Perkins to report at a meeting in mid-October 1901 that lands had been purchased from the estate of George Green, the Mahan heirs, Charles W. Opdyke, E. Ellen Anderson, John S. Lysle, the estate of George S. Cole, the estates of William B. Dana, Henry W. Banks, the Van Brunt properties, and, of course, the Carpenter brothers.

Soon after this meeting, a small article appeared in the *New York Times* under the interesting headline, "A Landscape Engineer Employed to Study and Preserve the Rocks." The commission had retained the services of Charles W. Leavitt, a civil engineer who, later in his career, would win kudos for landscape and design projects at the Saratoga and Belmont Park horse racetracks. Levitt's arrival signaled that the commissioners were not content with a preserved view of the Palisades from the New York side of the Hudson River, but wanted to develop public access to the summit of the cliffs and build a carriage road along the base. By early 1902, while Leavitt studied the rocks, the commission succeeded in acquiring thirty-four additional properties, spending funds as quickly as they arrived in the PIPC bank account.

By then, Perkins had organized a subsidiary Palisades Improvement Company intended to avoid most government red tape and move quickly when land buying opportunities beckoned. Part of the strategy was to enlist support from as many property owners as possible by encouraging donation of the vertical cliff faces while offering $500/acre for shoreline holdings. Many of these properties had been passed down by families through the generations and were held in small interests by widely scattered heirs. In one instance, the commission had to acquire a 1/240th interest in a 2.25-acre parcel from an heir who lived in the state of Washington. During this transaction, there was wording confusion when the proposed deed was reviewed by the heir. Back came the flawed deed via a railroad and carriage journey so that correction could be made, then the deed was dispatched again cross-country. The owner's signature was affixed; purchase price, $3.28, total mileage, 12,000.

Some of the properties acquired in this first bold conservation push included houses and cottages. The commission, searching for ways to build income, began renting them out. An exception was a carcass-filled, pungent fat-rendering plant on the river shoreline, which was torn down. With focus entirely on the New Jersey Palisades, enterprising quarry operators simply began moving upriver. Quarries sprouted at the New York river communities of Nyack and Haverstraw and on the slopes of Hook Mountain, a geologic feature almost as prominent as the Palisades. The kind of hue and cry that stopped the Carpenter brothers was heard again, with increasing citizen demand that PIPC authority be quickly expanded onto a new battlefront in New York.

Blasting reverberated across the river and reached the 3,500-acre Pocantico Hills estate owned by John D. Rockefeller. From Kykuit, his stately home, Rockefeller had a direct view of Nyack and Hook Mountain, about five miles distant. The view is so compelling that Alfred Bierstadt, among the most admired of the Hudson River School artists, captured the scene on canvas from a location near where Kykuit stands today. The scene is of a meadow decorated with ancient trees sloping down to the eastern edge of the broad Hudson River; the painting invites the viewer's eye

View across the Hudson River to Hook Mountain (Alfred Bierstadt)

westward across the water to hazy Hook Mountain, sheltered under a blue, cloud-touched sky.

Kykuit was constructed under the watchful eye of the elder Rockefeller's son, John D. Jr., but "Senior," known for his immense business success, personally designed the landscape plan to celebrate "bursting views of river, hill, cloud, and the great sweep of the country." This sense of design and appreciation for natural beauty would be passed on to Rockefeller Jr., who would leave a legacy of conservation from Maine to the Caribbean, from Virginia to California, and down through the Rocky Mountain region encompassing Yellowstone, the Grand Tetons, and Mesa Verde national parks. But the Palisades and the upriver scenes were right in the Rockefeller front yard. In his book *John D. Rockefeller, Jr.: A Portrait*, Raymond B. Fosdick says that Rockefeller Jr. "had known and loved the Palisades since his boyhood days when he often used to take his horse across the Hudson on the Fort Lee ferry and ride for hours beneath the cliff and through the woods to the top." Rockefeller Jr. was aware of the effort to stop the blasting in New Jersey. Now, his family could see, hear, and feel the explosions in what had been a bucolic Bierstadt scene.

In letters of March 18 and 21, 1902, to the New York governor and lieutenant governor, Rockefeller Jr. had taken up the interstate commission cause by supporting the extension of its activities into New York. He and George Perkins also were exchanging letters. Not everyone was on board. In a mid-March letter to Rockefeller Jr., a Nyack resident, James P.

Mcquaide, reported, "There has been strong opposition by the people who have stone crushers in Haverstraw, the principal being General Hedges, who, for many years, was the leading Republican politician in Rockland County." Hedges had clout. When a bill to expand the PIPC reached the desk of Governor Odell, it was vetoed.

The setback was made worse in early 1903 by the sudden death of one of the commission's founding members, Abram S. Hewitt (1822–1903). Hewitt's father had been a mechanic and his mother the daughter of a farmer. Hewitt attended public schools in New York City and successfully competed for a scholarship to Columbia University, where he earned a Doctor of Law degree. During his undergraduate days, he and his friend Edward Cooper toured Europe and, among their adventures, luckily survived a shipwreck. Hewitt returned to America in a borrowed sailor's suit with three silver dollars in his pocket.

The two friends went on to form the company Cooper & Hewitt, specializing in the manufacture of iron and laying the cornerstone for Cooper Union Institute in 1854. Their partnership surely must have been strengthened when Hewitt married Cooper's sister, Sarah. During the Civil War, Hewitt traveled to England and, in best cloak-and-dagger style, passed himself off as a copperhead, a Yankee who worked as an agent for the Confederacy. England was supportive of the rebellious southern cause, and Hewitt, playing to this sympathy, managed to purchase a large supply of guns and swords that he dutifully arranged for shipment to Union troops. Back in the U.S., and while visiting his wife's parents, he received a dispatch from President Abraham Lincoln. Lincoln said that General Ulysses S. Grant was in urgent need of twelve heavy mortars for an attack on Confederate fortifications, but that the U.S. Ordnance Department could promise delivery only in about a year. Could Hewitt & Cooper do better? The mortars were delivered in twenty-eight days.

After the war, Edward Cooper was elected mayor of New York City. Abram Hewitt also entered politics in 1874 and was elected to the U.S. House of Representatives. He was a skilled strategist and strong debater who won credit for establishing the Geological Society, a federal agency whose purpose was to measure the mineral wealth of the nation. Then another political goal beckoned. Following in the footsteps of his partner, Hewitt ran as a Democrat in 1886 for New York City mayor, competing against United Labor Party candidate Henry George, who came in second, and Republican Theodore Roosevelt, who came in third. After his term as mayor, Hewitt moved across the river to New Jersey and, in 1900, was

Park visitors with canoes (PIPC Archives)

appointed to the interstate commission by Governor Voorhees. By then, Hewitt was a well-known philanthropist and trustee of the American Museum of Natural History, the Carnegie Institute, and Barnard College.

His business reputation, scientific prestige, and seasoned political aptitude brought welcome strength to the commission as it searched to find footing in the tug-of-war between New York and New Jersey conservation challenges. With Governor Odell's veto on the table, Hewitt was needed more than ever. His death left the commission with a gaping organizational hole. Problems of property title, squatters, obscure owners, outright resistance, and slow action by New York to dispense funds continued to plague the land-acquisition initiatives in New Jersey. One of these owners was eighty-four-year-old Susan B. Anthony. In March 1905, the commission offered to purchase her land for $2,300. Anthony thought the price should be $2,400. The commission paid $2,400.

As land holdings accumulated, so, too, did interest by the public in camping, canoeing, picnicking, and hiking. The commission held properties, some disconnected from others, sprinkled along fourteen miles of river shoreline. Several key holdings remained beyond the commission's grasp, including an old, weatherworn structure that had been briefly commandeered 125 years earlier by Lord Cornwallis, whose troops struggled to the top of the cliffs in a vain effort to capture General George Washington. Some of the few fishermen or their widows who continued to cling to squatter's shacks were convinced that the ghost of Cornwallis returned every November to shout orders for king and country and beseech his phantom troops to climb fast and rush forward to capture the upstart rebel general.

Ghost or no ghost, public use of commission-owned lands was beginning to blossom. Members of the American Canoe Association applied for and received permission to camp under the cliffs. Four hundred canoeists paddled across from New York City to take advantage of the opportunity.

Hundreds of others, transporting themselves by foot, with equine help, or in sputtering vehicles, decided that this idea of camping was grand and availed themselves of free permits even as the commissioners were debating the merits and purposes of the park they seemed to be creating.

The need for this debate was underscored by a fast-developing impression that the Palisades was a "lawless" place on Sundays and holidays, when increasingly large crowds came to recreate outside of the usual police controls by which a metropolitan population usually lived. In 1905, the commission managed to retain the services of only one law enforcement marshal, who was stationed at a ferryboat landing. He was supposed to patrol miles of shoreline on foot on the assumption that a handful of well-behaved visitors might wish to stroll along the river's edge. In reality, and as word spread, this newly acquired public open space became a magnet for city dwellers. The lone marshal was overwhelmed. The Palisades was not going to be a static, protected natural scene; it was rapidly morphing into a people's park.

Perhaps the commissioners could be excused for being distracted by property transactions and a budding work-in-progress park that caused them to decline a proposal by Mrs. E. B. Miles of the New Jersey Federation of Women's Clubs to secure a site atop the Palisades for the purpose of memorializing the role women played in the fight to save the cliffs. In response, the commission advised Mrs. Miles that it "was not in a position to take official action in that direction." Subsequent commission minutes confirm a flow of "communications" on the matter, suggesting that the battle-tested women were unwilling to take "no" for an answer from the very commission they helped to create. In defense, the commissioners referred the matter to J. DuPratt White for further consideration.

The 1902 veto by Governor Odell of legislation to extend the jurisdiction of the commission into New York could not be forgotten because every morning and evening dynamite blasts at quarries on Hook Mountain would be heard for twenty miles up and down the Hudson River. John D. Rockefeller, John Speyer, and other wealthy river corridor residents were offering to pay for commission expenses in New York if its activities could be extended to Stony Point, fifteen miles upriver from the New Jersey/New York border.

Media attention, public outcry, and the influence of rich campaigners finally led to the introduction of the necessary legislation in January 1906, two years after Odell left office. Ironically, the legislation was introduced by a New York senator named Carpenter—no relation to the brothers Aaron and George.

Odell's successor as governor, Frank W. Higgins, suggested that written guarantees from private individuals to cover commission costs on the New York side of the border would be necessary to gain a favorable vote on the legislation. Despite informal expressions of philanthropic interest by the Rockefeller family, George Perkins, J. P. Morgan, and others in the highest financial orbit, the idea of a written guarantee did not go over well. Starr J. Murphy, a Rockefeller Jr. staff member, summarized the issue in a memo to his boss:

> This suggestion seems to indicate an entire misconception as to the nature of the pending bill to extend the limits of the Palisades Park. The beauty of the Hudson River is one of the great scenic attractions of the eastern part of the United States; it probably draws more visitors to this State than any other single feature outside of Niagara Falls, and the movement to preserve it is of the same public character as the movement to preserve the Falls, or the Yellowstone Park, or the Yosemite Valley. When the original Palisades Park was established no such assurances were required, and yet, outside of the original appropriation which I understand was spent entirely for acreage property and has not yet been exhausted, whatever funds were needed for acquiring business properties the Commission was able to procure. All that is asked in this bill is extension of the powers of the Commission so as to enable it to acquire additional lands. These lands are just as much needed for the preservation of the scenic beauty of the Hudson as the lands taken under the original act.

The Rockefellers were not without their contacts, including Timothy L. Woodruff, an influential businessman known to have close ties with Governor Higgins. In a follow-up memo, Murphy reported:

I met Mr. Woodruff (President of the Smith Premier Typewriter Company) at the Club today and had a little talk with him. He said he expected to attend a meeting of your Bible Class tonight and thinking that you might perhaps see him there I thought I would give you the situation up to date. He tells me that he thinks former Governor Odell is really back of the opposition. Odell is the member of the State Committee from Rockland County and the quarry owners are therefore his direct constituents.

Former governor Odell apparently was fanning a rumor that Rockefeller and Morgan had pledged to fully fund the 1902 bill to expand commission activities into New York but had reneged. The Hook Mountain quarry operators were advancing a parallel argument; they contended that if the commission's authority were to be extended into their territory, a "cloud" would be cast on the title to their land. They argued that the value of their holdings would decrease because of the commission's suspected intent to use the power of eminent domain to force them to sell. (This argument resonates to this day when property-rights activists claim that zoning controls and similar development restrictions reduce land values. The reverse is consistently true.)

Six quarries now were active along the riverside landscape in New York. This earthshattering reality was bringing more allies to the commission cause, including the American Preservation Society, the Albany Day Line Steamers, and the New York Central Railway. Among those joining in support of the commission was Cleveland H. Dodge, a founder of Phelps-Dodge, a company that, paradoxically, operated huge mines in the American West. Dodge lived in Riverdale on the eastern bank of the Hudson River. Like so many others, he had a low tolerance for the abusive tactics that he was witnessing across the river, so much so that he had purchased $25,000 worth of threatened rock outcroppings from one of the quarry operators, Brown & Fleming, and later would donate most of this holding to the commission. The rest was sold at a loss.

Political pressure marked by such a growing contingent of influential people and companies got the message across in Albany. In the same year that the New York legislature established Cornell University, Chapter 691 of the Laws of New York, 1906, extended the commission's jurisdiction to Hook Mountain, ten miles upriver from the New Jersey/New York state line, marking a milestone moment when "interstate" took on new force.

While looking northward, Perkins and his colleagues could take comfort in the fact that almost 80 percent of the fourteen-mile New Jersey shoreline had been acquired, even though New Jersey appropriations to cover commission expenses had dried up. In June 1906, he and his fellow commissioners took a trip up the Hudson on the yacht *Mermaid* to view the challenges ahead. But also, with a backward glance, Perkins was able to report that "almost all danger points" along the New Jersey Palisades "had been removed" with the recent acquisition of the last two rock-crusher plants. This was a grand statement, but the commission also dealt with the mundane. Among bills paid was $9.34 to Standard Oil of New York, a printing bill for $7.35, and $73.00 paid to John Jordan to take down the old fat-rendering plant. By the beginning of 1907, the commission had expended $470,534.80, chalking up a deficit of more than $5,000.

J. DuPratt White paid the bills. These included $1,500/year for a legal assistant and $15/month for the beleaguered law-enforcement marshal. In addition, most of the burden for issuing camping permits, dealing with the myriad details of land acquisition, evicting squatters, responding to complaints, preparing reports, and bringing definition to acceptable park uses fell on White's shoulders.

One man wanted to use the forests for random target practice. White thought that the unpredictable flight of rifle bullets through forests visited by park patrons was a bad idea. Two brothers, Obediah and John Older, held side-by-side leases on commission-owned houses. White ordered both brothers to vacate, explaining to one that "there has been so much trouble between your family and your brother's family that the commissioners have decided that this is the only way to settle the whole matter." In a letter to Perkins, White suggested that "the commission needs a receptacle for its files and papers. The commission has never bought any furniture and owns none." Lack of money was a persistent problem, with none coming from New Jersey, and the New York comptroller, an independently elected official, often responded slowly to vouchers sent forward for reimbursement.

Commissioner Abram De Ronde was not as involved in day-to-day management challenges as was White, but he certainly pulled his weight. His tenure would span thirty-seven years. De Ronde was a native of Teaneck, New Jersey, and had worked in New York City while attending night school. He won early business success in the chemical industry, then founded the Palisades Trust & Guarantee Company based in Englewood, New Jersey. While serving a term in the New Jersey State Assembly from 1889 to 1891, De Ronde proposed an idea whose time had not yet come: the construction of a bridge across the Hudson River that would connect New York City and New Jersey.

When De Ronde learned that the New Jersey women, steadfast in their hope that a memorial would be erected in recognition of their pathfinding conservation, had collected $3,000 with the intention of buying a suitable site on the summit of the Palisades and donating it to the commission, he joined their cause. He objected to the idea that the women would have to spend their own money to purchase a site for their memorial and convinced his colleagues to set aside commission-owned land for the purpose, thus allowing the women to reserve their hard-won money for "immediate" construction of the memorial. This was a victory for the New Jersey Federation of Women's Clubs, but it came at a time when the commission was struggling to find its way and keep financially afloat. A suitable site remained unidentified, and the women's money remained in a bank account. "Immediate" would stretch into years.

In July 1907, White posted a long memorandum to his fellow commissioners in which he summarized the many complex issues that needed careful attention. Property acquisition challenges in New Jersey remained, now primarily narrowed to entrenched owners who were not inclined to deal with the commission. White also mentioned the idea of "construction of a continuous driveway connecting the state roads of New Jersey with the view of making the outlet for the great system of good roads in New York down through the Palisades Park." This was not an entirely clear statement, but White obviously was floating the idea of a park driveway (parkway) that would meander through protected open space, giving travelers an opportunity to enjoy a linear park. A hint of the Palisades Interstate Parkway, now accommodating 65 million cars a year, cherished and sometimes cursed, was a glimmer in White's mind, even in an era when horse-drawn carriages still outnumbered mechanically fractious automobiles.

He also lamented the newly encountered dilemma at Hook Mountain, pointing to a lack of appropriations from New York: "It is exceedingly doubtful if it will ever be possible to accomplish the purpose of the act by an absolute purchase of the properties, and the destruction of the trap rock industry. Whatever is done should be formulated and undertaken without delay, as at the present time the commission is being severely criticized for inaction and neglect. It is not important that such criticism is unjust. The fact is that it exists and must be met."

Day-trippers and campers were being delivered to the commission shoreline holdings from ferryboat terminals in New York City and Yonkers. Where a handful of camping parties had appeared when the first river-edge lands were acquired, now two thousand tents would blossom on weekends. White sought advice about public use of parks from the faraway commander of Fort Yellowstone but found no help from an Army officer posted in a wilderness few people visited or even knew about.

In a letter to a commission patron, White returned $5.00 that had come with a request for a camping permit: "It would not be proper under any circumstances to accept any money." In the midst of overseeing the increasingly successful White & Case Law Firm and commuting from his home in Nyack to New York City, the indefatigable White found time to give advice to aspiring campers on the types of tents, equipment, and food to purchase to assure a favorable park experience. Public sanitation, accidents, the risk of forest fire, trespass by visitors onto private property, and rowdy visitor behavior were among the problems finding their way to White's desk. In the absence of policy guidelines, he invented park regulations to address the more persistent challenges.

As if this were not enough, White defended against an Alpine, New Jersey, tax collector who thought that adjacent commission lands should be a continuing source of property tax revenue even though the town provided no fire protection or law enforcement services. Seeking a compromise, White wrote, "Of course, if these taxes are paid by the commissioners, it is not to be in any way construed as a willingness on their part to pay any taxes on this property in the future." Compromise proved unnecessary; the commission refused to pay any taxes to the town.

White must have found irony in a communique from New Jersey state treasurer Daniel S. Voorhees, son of the former governor, who sought lease payments for the very riparian lands that had been ceded to the commission eight years earlier. The solicitation by Voorhees to receive payments on muddy land no longer owned by the state and regularly submerged by tides was declined.

Pressed by local residents and historians to build a battle monument on commission land at the Revolutionary War site Fort Lee, White wrote that "there is a dispute over the bills of H. A. Jaeger, who hauled several large boulders which were to be used for the monument. All agree that the bill is absolutely exorbitant and out of all reason, but the party is very stubborn and has declined to make any concessions." The bill was paid, and the commission owned a pile of boulders but had no funds for construction of a monument.

One day White "went tramping" along the river shoreline in search of a tent reported to be missing from a campsite. Along the way he recruited willing visitors as unpaid "park wardens," assigned them "districts," and asked that they make their own park warden signs to hang on their tents and to report unlawful incidents to the marshal. Two enterprising boys, law-abiding, but seeing opportunity for profit, started selling camping supplies from their rowboat. White decided that this was an acceptable commercial service, a very early hint of a huge contest in decades to come between those who value preserved open space and those who wish to cash in on the very popularity of parks. On the environmental front, the commissioners sent a letter to the Sisters of Peace advising that "the break in the sewer pipe runs over park property. This must be fixed, and in attending to it we would like to have the pipe continued further in the river so that the refuse will be carried out beyond the low water mark."

The commission found itself drawn to the idea that parks could be used for worthy social purposes, especially in cooperation with organizations that served underprivileged children. Working with representatives of the Hamilton House Settlement, a charitable organization located in New York City's notoriously crowded and impoverished Lower East Side, the commission began issuing long-term camping permits that allowed for the "relay of poor boys" to and from camps set up for that purpose.

Most of these park stewardship experiments went unnoticed, but on occasion the commission won praise. "I remember you very well in connection with the purchase of your property underneath the Palisades," White wrote to J. H. Magee. "The commissioners appreciate letters written to them in the spirit of your letter as they are glad to receive suggestions and assistance from outsiders, but we find that there are not so many public spirited citizens who will take the trouble to write about matters of this kind."

The commissioners were learning about the vagaries of park management as taught in the school of hard knocks. They were learning that people eagerly responded to the availability of preserved open space but did not necessarily appreciate how or why it was available or who must provide for its care. The commissioners were learning that in the public sector criticism gushed and praise was rare, that their investment in volunteer hours was unknown beyond a small circle of family, colleagues, and friends, and that politics could be even more fickle than assumed. They had no benchmarks to help them determine how best to be good stewards of wild and rugged parkland near the epicenter of the New York urban colossus. Theirs was no Central Park or Boston Common with manicured lawns, gardens, pruned trees, landscaped ponds and hills, constructed pathways, urban pace and mood, and high public watchfulness. Nor were they dealing with remote wilderness. They were somewhere in between, on their own, helping to invent a new type of park, finding their way as they went along.

4

⠿

Harriman

COMMISSION EYES UNDERSTANDABLY were focused on the vexing daily challenges of the New Jersey Palisades when, in November 1907, an article appeared in the *Outlook* magazine entitled, "The Preservation of the Highlands of the Hudson First Publicly Advocated by Edward Lasell Partridge, M.D." Partridge was looking beyond traprock quarries and toward the locally known "highlands," a spectacular segment of the Hudson River fifty miles upriver from New York City where the river cut through the northernly extension of the Appalachian Mountains. Rising near Mount Marcy in the Adirondack Mountains, the Hudson River cascades and descends for about a hundred miles to Fort Edwards, New York, then flows leisurely south for another hundred miles to the highlands bulwark at West Point. There, following a course bulldozed long ago by glaciers, the river narrows, deepens, and gains speed as it rushes through a sweeping twenty-mile S-turn past Bear Mountain before once again taking on a broad, stately flow down to its rendezvous with the Atlantic Ocean.

Early Dutch voyagers, tacking under sail against the current and wind in the "narrows," with threatening rocky bluffs abeam on port and starboard, referred to this section of the river as the "Devil's Horse Race." Henry Hudson managed to sail through the narrows to the upriver "gate" just above West Point, anchored, briefly explored, and then turned back, ending his quest for the fabled northwest passage to China. Continental Army defenders fortified the heights of this natural bastion and floated a massive chain across on barges from shore to shore in an attempt to stop shipborne assault by the British. And, decade upon decade, the narrows has attracted tourists who marvel at the scene and artists who try to capture its intricate beauty on canvas, paper, film, and microchips.

Partridge proposed that a sixty-five-square-mile section of the highlands, with the narrows as its centerpiece, be declared a national park. He noted that Civil War battlefields at Gettysburg, Chickamauga, and Shiloh had been preserved by the U.S. government in a manner that included the protection of scenery, forests, fields, and historic buildings and argued that the military installations at West Point and other Revolutionary War sites in the narrows deserved similar congressional favor. In his vision of a park, Partridge proposed nature-education programs, scenic roads, and preservation of "wild woodlands." He suggested tax incentives to encourage the protection of private land interests within the park boundary.

Partridge's conservation plea carried the weight of his reputation as an honored leader and teacher within New York City's medical community. His home was on the northern slope of Storm King Mountain at Cornwall-on-Hudson, just upriver from the narrows gate. The landscape design for the Partridge estate was courtesy of Frederick Law Olmsted, and his home was filled with art and historic objects. He served as the chair of Obstetrics at the College of Physicians and Surgeons and had an interest in philanthropic activities that included support for the New York Institution for the Education of the Blind, the Washington Square Home for Friendless Girls, the Huguenot Society, the American Scenic and Historic Preservation Society, the Society of Colonial Wars, the Garden Club of America, the Constitution Island Association, the Grant Monument Association, and the New England Society of New York. This robust man, who relished "light" lunches of two sandwiches, a thermos of coffee, and pie, took great delight in opportunities to pursue and influence public policy.

Among the landowners whom he was seeking to influence with his national park idea were Mr. and Mrs. Edward Henry Harriman. The Harrimans owned thousands of acres in the highlands and knew of Dr. Partridge as an acquaintance and respected neighbor. They took careful note of the *Outlook* article. Only two years later Mrs. Harriman would sweep the startled Interstate Park Commission forward to astonishing land stewardship responsibilities, influencing conservation initiatives across the nation.

Born in 1848, Edward H. Harriman, one of eleven children, "started with nothing" and became the "mightiest" railroad baron, according to biographer Rudy Abramson. Harriman built a railroad empire including the Union Pacific, Southern Pacific, and Illinois Central that controlled 75,000 miles of track and employed more men than the standing army of the United States. Named for a great uncle who was forced to walk the plank in a Caribbean pirate raid, E. H. was the son of a minister who settled in West Hoboken, New Jersey. His father became rector of St. John's Episcopal Parish, a position that paid $200/year, after returning from a misadventure in the California gold fields. A side benefit was that sons of the rector could attend the prestigious Trinity School in Manhattan. Harriman and his brothers would walk two miles to a ferry landing, cross the Hudson, and walk another mile to school. E. H. Harriman ranked first in his class at Trinity but dropped out at age fourteen to work as a copyboy at the brokerage firm of D. C. Hays. Promoted to pad boy, the fleet-footed Harriman raced from office to office, calling out the latest stock quotes. He favored his memory over the quotes scribbled on his pad and won admiration from the brokers for never making a mistake. Within eight years, Harriman had moved up in the Wall Street environment, so much so that he was able to buy a seat on the New York Stock Exchange. He gained a reputation as an outstanding trader, but was found to be a "loner, secretive and relentless" by some of his peers.

In 1879, he and Mary Averell were married. Biographer Abramson wrote:

> About the same time he was courting Mary, he struck up an equally fortuitous friendship with Stuyvesant Fish, who would eventually help him get his start in big-time railroading. It was, to put it mildly, an unlikely alliance. Fish's family had been pillars of New York commerce and society for two hundred years. Stuyvesant's father, Hamilton Fish, had been Governor of New York, a U. S. Senator, and more recently President Ulysses S. Grant's Secretary of State. Grandfather Nicholas Fish had served under General Washington from the Battle of Long Island to Yorktown.

Fish was physically large and easygoing. E. H. was "scrawny, near sighted, and cunning as a wolf," Abramson said. Stuyvesant Fish was on the board of directors of the Illinois Central Railroad, a company that controlled tracks extending through America's heartland from Lake Erie to the Gulf of Mexico. In 1881, a stock swoon caused by the assassination of President John Garfield prompted Harriman to buy heavily into Illinois Central and, soon after, join his friend Stuyvesant on the board. Together, they began going after more railroad routes. In the process, Harriman found himself in rivalry with the legendary J. P. Morgan for control of an obscure Iowa railroad, the Dubuque-and-Southern.

Morgan and a group of investors seemed to hold enough shares of stock to decide the fate of Dubuque-and-Southern at a forthcoming annual meeting by determining who would serve on the railroad's board of directors, but Harriman flooded the meeting with enough proxies from small investors to win election of a slate of directors of his own choosing. The board then approved sale of the railroad to Harriman's Illinois Central. Morgan threatened court action but finally acquiesced to the "little man" and sold his pool of shares at a price set by Harriman.

Harriman would not long remain a "little man" in terms of business muscle and wealth. By the beginning of the twentieth century, he controlled the Union Pacific and Southern Pacific railroads and stood among equals at the highest level of corporate power. This was confirmed when he and Morgan set aside their rivalry and joined with railroad tycoon James J. Hill in a transaction in 1901 to merge the Northern Pacific, Great Northern, and Chicago-Burlington-Quincy railroads. This merger sparked yet another high-profile contest, this time with President Theodore Roosevelt. Through the Department of Justice, the trust-busting president sought to dissolve the railroad merger on the grounds that Harriman, Morgan, and Hill would have a grip on national railroading verging on a monopoly. In a five-to-four vote by the Supreme Court in 1904, the president prevailed. But business was just business and did not preclude Harriman from raising $250,000, including a personal contribution of $50,000, in support of Roosevelt's 1904 presidential campaign.

Harriman, Roosevelt, and Morgan tripped over each other in a robust, competitive nation just beginning to feel its industrial might, but the natural treasure-trove along the New Jersey Palisades and in the New York Highlands brought them to an informal point of brotherhood.

The scenic values in the highlands attracted Morgan to acquire and maintain a grand weekend home in Highland Falls, just downriver from West Point. He and John D. Rockefeller, George Perkins, Dr. Partridge, and so many others were sharing the view and being influenced by one of the most superlative river ecosystems in America. Harriman's view was somewhat different. In 1885, he purchased land from iron-maker Peter Parrott, made famous during the Civil War for manufacturing the "Parrott Rifle," a cannon of improved range and accuracy that gave strategic advantage to Union troops. The Parrott iron mines and furnaces were in Sterling Forest, a splendid natural segment of the New York Highlands.

After the Civil War, when abundant iron ore was discovered in Minnesota and prices fell, Parrott's business collapsed. At a distress auction, Parrott placed his entire 7,863-acre land holding on the block, hoping that parcels would be divided among competing speculators, especially those interested in timber rights, to provide maximum financial return for him.

Harriman saw a different opportunity. He wanted to prevent the land from being subdivided and timbered and went for the entire acreage. His bid of $52,500 prevailed. In succeeding years, Harriman would add another 20,000 acres to the Parrott purchase by acquiring some forty surrounding properties. He named his estate "Arden" in honor of Parrott's wife. The Arden view was to the west over rolling woodland hills, the Hudson River miles away out of sight to the east, but all within an ecosystem that Dr. Partridge had proposed as a national park.

Harriman's conservation instincts were confirmed in 1898 when his doctor recommended a family vacation as treatment for business fatigue. Harriman decided to make the vacation into a scientific expedition to Alaska. When the day arrived in May 1899 for departure of the chartered ship *George W. Elder* from Seattle, Washington, Harriman welcomed aboard a traveling party of 126 people, including the eminent naturalists John Burroughs and John Muir. This was his idea of a restful excursion. The 9,000-mile voyage would take two months, reaching the coast of Siberia. Wild animal, bird, plant, and aquatic specimens were "collected" at the many shoreline stops along the way, as well as museum-quality art objects and artifacts found at the villages and campsites of indigenous peoples.

Muir had one later opportunity to visit Harriman, this time at the railroader's wilderness lodge at Pelican Bay, Klamath Lake, Oregon. Harriman had promised to "show you how to write a book" and made available a stenographer who spent many summer days recording Muir's ideas and musings. "To him I owe some of the most precious moments of my life," Muir said of his host. In subsequent writings, Muir followed this thought:

> Of all the great builders—the famous doers of things in this busy world—none that I know of more ably and manfully did his appointed work than my friend Edward Henry Harriman. He was always ready and able. The greater his burdens, the more formidable the obstacles looming ahead of him, the greater was his enjoyment. He fairly reveled in heavy dynamical work and went about it naturally and unweariedly like glaciers mining landscapes, cutting canyons through ridges, carrying off hills, laying rails and bridges over lakes and rivers, mountains, and plains, making the nation's ways straight and smooth and safe, bringing everybody nearer to one another. He seemed to regard the whole continent as his farm and all the people as partners, stirring millions of workers into useful action, plowing, sowing, irrigating, mining, building cities and factories, farms and home.
>
> In general appearance he was said to be under-sized, but though I knew him well I never noticed anything either short or tall in his

stature. His head made the rest of his body all but invisible. His magnificent brow, high and broad and finely finished, oftentimes called to mind well-known portraits of Napoleon. Every feature of his countenance manifested power, especially his wonderful eyes, deep and frank yet piercing, inspiring confidence, though likely at first sight to keep people at a distance.

In the first decade of the twentieth century, George Perkins and his commission colleagues had not yet joined in active alliance with Harriman, although they surely were aware of his immense landholding in the New York Highlands. They were struggling to close the final land deals in New Jersey and were on the defensive in New York for not having rid Hook Mountain of quarry dynamite blasts. J. DuPratt White wrote to one critic that "the reason the commission has not acquired the Hook Mountain, and the only reason, is that the commission is without funds—I can assure you that exhaustive efforts have been made to accomplish this end in every way that the ingenuity of the commission has been able to devise."

George Perkins saw opportunity in 1909 to raise the profile of the commission by joining in the 300th anniversary of Henry Hudson's "discovery" of the river named in his honor. At the same convenient time, the centennial of Robert Fulton's development of the steamship would be acclaimed— from sail to steam, so to speak. Perkins promoted the idea that the Hudson-Fulton celebration also should be combined with formal dedication of the Palisades Interstate Park in New Jersey, contending that protection of the scenery as it appeared during the voyage of the *Half Moon*, and for river travelers since, was a fitting tribute. From his upriver post, Dr. Partridge also wanted to participate to advance his national park idea. He succeeded in winning appointment to a Tercentenary subcommittee charged with studying the idea, but the subcommittee never met. Partridge, probably the most eager member of the subcommittee, was able to gain only a bit of attention in Albany and none in the U.S. Congress for his national park notion. The idea sputtered and flamed out.

Perkins won better results, especially fueled by a $20,000 donation that he and his employer, J. P. Morgan, made to help fund the celebration. Perkins was appointed to the Tercentenary Commission "so that there might be the fullest possible interchange of views between the two bodies, Interstate Commission and Tercentenary Commission." Even so, when the official announcement of the pending celebration appeared in newspapers on August 31, 1909, no mention was made that the commission dedication would be included. Perkins strongly objected. The chairman of the commission, General Stewart L. Woodford, had words with an army colonel who had arranged for the news release. More than 500,000 copies of an amended announcement were quickly distributed.

On September 27, 1909, a huge crowd gathered on the shoreline at the base of the Palisades cliffs to witness the dedication of the Palisades Interstate Park. Thirty-five musicians and many dignitaries were transported to the site on the *Waturus*, Perkins's private yacht. Governors Charles Evans Hughes of New York and J. Franklin Fort of New Jersey were in attendance. So, too, was Elizabeth Demarest, the official representative of the New Jersey Federation of Women's Clubs. The club members thought that completion of their memorial in conjunction with the dedication would be ideal. The commission's beleaguered messenger, White, responded that "everyone is overwhelmed with the work in connection with the celebration in general" and that the long-sought women's memorial must await another time. A group representing the Iroquois Nation performed a ceremonial dance, perhaps reminding some in attendance that not all had benefitted equally from voyages by Europeans.

Perkins announced to those assembled that a "Member of the Tercentenary Commission," meaning himself, had donated $12,000 to the commission, supplemented by a gift of land valued at $16,000 from Cleveland H. Dodge. He said that the commission privately raised a total of $284,000 in gifts of land and money, particularly crediting J. P. Morgan, Cleveland Dodge, Mrs. Lydia G. Lawrence, and Mr. and Mrs. Hamilton Twombly. Perkins noted that in the intervening nine years since the two states had provided the first $15,000 of appropriated funds to begin land purchases, an additional $20,000 had been received from New York and $17,500 from New Jersey.

He confirmed that almost all the 175 parcels, including twenty-one houses, needed to guard the integrity of the Palisades cliff face now were in commission ownership. "Here, within sight of our great, throbbing city is a little world of almost virgin nature which has been rescued for the people and now stands as a permanent monument to the discovery of the river by Henry Hudson," said Perkins. His statement was confirmed years later by Nancy Slowik in *A Nature's Guide to the Southern Palisades*. Her guide lists 11 species of amphibians, 12 of reptiles, 50 of butterflies, 29 of mammals, 232 of birds, 60 of trees, shrubs, and vines, and 51 of wildflowers, ferns, and grasses.

Perkins credited Elizabeth Vermilye, Cecelia Gaines Holland, Franklin W. Hopkins, William A. Linn, S. Wood McClave, Andrew H. Green, Frederick W. Devoe, Frederick S. Lamb, Abraham G. Mills, and Edward Payson Cone for assertive work in the citizen crusade to save the cliffs. "Man can do no more than preserve its natural grandeur and make the park accessible to one and all." The Navy warship *Gloucester* boomed a cannon salute from the river that echoed off the rocky wounds of silent quarries. The musicians struck up a lively rendering of the *Star-Spangled Banner*.

One person not mentioned at the celebration was Charles E. Howard, superintendent of New York Prisons, who was to become a highly unlikely and unwilling ally of the commission. Howard had been looking for a prison site and concluded that undeveloped wildland at Bear Mountain, just downriver from West Point in the Hudson River narrows, would do just fine. The state needed stone for public highway construction. While the commission was being applauded for the quarry demise in New Jersey and criticized for not doing enough at Hook Mountain, Howard was off on a different track, convincing the New York Prison Commission that there was plenty of stone for road-fill at Bear Mountain that could be mined by prison labor.

In 1908, just a year before the Hudson-Fulton celebration, the prison commission approved purchase of 700 acres at Bear Mountain from C. E. Lambert. Howard immediately set convicts to work clearing land and building a log stockade. The stockade would take on the appearance of a Wild West fort.

Bear Mountain Prison stockade (PIPC Archives)

The work force came from notorious Sing Sing Prison, and so many convicts took the opportunity to escape that nearby hamlets were said to be in a "state of terror." The intent was that the stockade would house 1,500 to 2,000 inmates, who, except on Sundays, would be marched up the slopes of Bear Mountain to mine rock, prybars, pickaxes, and shovels in hand. The commission was startled and anxious about this unexpected exploitive curveball. At the conclusion of the ceremony at Cornwallis Headquarters, Perkins hosted New York governor Hughes aboard the *Waturus* to express urgent concern.

Upriver, Dr. Partridge, checkmated on his national park idea, turned his considerable clout to forest protection. Partridge's new cause was prompted by Gifford Pinchot, former dean of the State School of Forestry at Cornell University, founder of the Society of American Foresters, and chief of the U.S. Forest Service in the Theodore Roosevelt Administration. Pinchot had sent out the alarming message that all mature trees in the United States would be cut down within twenty-five years if the present pace of out-of-control logging were allowed to continue. Added to this threat, deadly chestnut blight (*Cryphonectria parasitica*) had reached the New York Highlands. The grand and ancient chestnut trees had no defense. Most of the iron mines in the highlands had been shut down, but the Forest-of-Deane mine still was in operation, consuming large quantities of wood for its furnaces. Brick manufacturing also was common, the red-hot kilns fueled each year by thousands of cords of hardwood.

In response to the forest threat, on May 22, 1909, the New York legislature approved the *Highlands of the Hudson Preservation Act*. The act declared that "perpetuation and improvement of forest growth is declared to be in the public interest." To secure forest protection, the act called for designation of a resident forester whose daunting tasks would be to develop regulatory principles for public and private lands, see to construction of a highway to provide access, and prevent forest fires. Partridge provided data to help along the process, pointing out that private holdings in the highlands, excluding the Harriman estate, ranged from fifty to 5,000 acres, and that 500- to 1,000-acre plots were "common." The average per-acre cost was between $25 and $50. Property taxes ranged from $4 to $8 per acre.

The good news was that careful forest practices already were being implemented at the 16,000-acre U.S. Military Academy, West Point, and on the 27,000-acre Harriman estate. But the very bad news was that E. H. Harriman was extremely ill with cancer. The man whom John Muir described as going about work like "glaciers mining landscapes" was doomed. He died in September 1909 at age sixty-one, leaving his $70 million estate to his wife, Mary.

Before his death, Harriman confided in Mary a strong interest in the park concept proposed by Dr. Partridge and made this interest known to

Governor Hughes. The prison project at Bear Mountain, just beyond the eastern boundary of the Harriman estate, caused his interest to rise to a level of resolve that would pull the commission into park stewardship at a level not imagined. Harriman wanted to make an offer to the state of New York that could not be refused: if the prison board would move the prison somewhere else, Harriman would donate thousands of acres to the commission and donate $1 million to help with stewardship costs. Mary Harriman was determined to see his wishes fulfilled.

George Perkins scheduled a meeting of the commission on December 16, 1909, accompanied by the minimalist message that he "particularly asks that every member of the commission will endeavor to be present, as matters of great importance to the commission will be brought before the meeting."

The minutes of the meeting include reference to a letter from Elizabeth Vermilye urging acquisition of a property referred to as "Clinton Point," but the commission had declined because of lack of funds. A bill for $294.17 was submitted for approval to cover the costs of purchasing and mailing 357 copies of the book *The Palisades of the Hudson,* by Arthur C. Mack, to every member of the New Jersey legislature. Perkins submitted personal receipts for $45.00, representing expenses for a recent trip to Albany, then added that he would consider the expenses a contribution to the commission.

Then, the bland minutes confirm that "Mr. Perkins announced that Mrs. E. H. Harriman had signified her willingness to deed 10,000 acres at Arden, New York, to the commission, together with $1,000,000 in cash, provided others would contribute to the extent of $1,500,000, and all of the contributions to be conditioned upon the State of New York contributing an equal amount of $2,500,000."

Perkins added that he had completed negotiations "for a plan which would permit extension of this commission to West Point." The plan also would "permit construction of the boulevard and the acquisition of such lands between the present jurisdiction and the proposed jurisdiction as might be considered necessary." He then added that anonymous private pledges, already in hand, had reached the $1 million mark. The minutes said nothing about the magnitude of such a breathtaking leap in purpose and obligation. Perhaps the commissioners, most of whom were accustomed to big business deals, were completely comfortable with this giant leap in purpose and responsibility.

In a follow-up meeting on December 23, the details of the Harriman proposal were spelled out:

1. That in order that the Palisades Park Commission may carry out the proposed plan and receive and hold the land and money offered the

State by Mrs. Harriman, its jurisdiction shall be extended to the northward along the west bank of the Hudson river to Newburgh, and to the westward as far as and to include the Ramapo mountains, giving the Commission the same powers granted to it at the time it was created and at the time its jurisdiction was extended in 1906, including the right to condemn land for roadway and park purposes.

2. That the State of New York appropriate $2,500,000 to the use of the Commission for the acquiring of land and the building of roads and general park purposes.

3. That the State discontinue the work on the new State prison located in Rockland county, and relocate the prison where, in the judgment of the Palisades Park Commission, it will not interfere with the plans and purposes of the Commission.

4. That in addition to the aforesaid appropriation from the State, a further sum of $2,500,000, including Mrs. Harriman's pledge of a million dollars, be secured on or before January 1, 1910.

5. That in addition to the above $5,000,000, the State of New Jersey appropriate such an amount as the Palisades Park Commission shall deem to be its fair share.

On motion, the meeting was adjourned. Perkins and White already had scheduled a meeting the following day with Governor Hughes. They presented an imposing list: fundraising pledges that exceeded the Harriman challenge by $150,000.

John D. Rockefeller	$500,000
J. Pierpont Morgan	$500,000
Margaret Olivia Sage	$50,000
William K. Vanderbilt	$50,000
George F. Baker	$50,000
James Stillman	$50,000
John D. Archbold	$50,000
William Rockefeller	$50,000
Frank A. Munsey	$50,000
Henry Phipps	$50,000
E. T. Stotesbury	$50,000
E. H. Gray	$50,000
George W. Perkins	$50,000
Cleveland H. Dodge & James McLean	$25,000
Helen Miller Gould	$25,000
Eileen F. & Arthur Curtiss James	$25,000
V. Everitt Macy	$25,000

Perkins had been joined by John D. Rockefeller to tap into high levels of New York society through personal contact to achieve the fundraising goal. Among those contacted was Andrew Carnegie, who declined, explaining to Rockefeller that he was raising $5.5 million for libraries.

On January 6, 1910, Governor Hughes delivered a message to the legislature urging positive response to the proffered Harriman gift, but the governor hedged his bet. Instead of seeking an outright appropriation, Hughes used as excuses lack of tax revenue and other state financial obligations and, instead, recommended that a $2.5 million bond be issued for the purpose. The caveat was that a bond would have to be approved by the entire New York electorate. This was a novel political maneuver. Never before had New York voters been asked to approve borrowing by the state in the form of a bond to protect nature's handiwork. Bonds had been used to fund construction projects, not prevent them. But the governor was willing, the legislature was supportive, the heavily influential names of Rockefeller, Harriman, Morgan, and Perkins were associated, and the scenic magnificence of the Hudson River Narrows and the New York Highlands was at stake.

On October 29, 1910, an eighteen-year-old nervous Yale college student named William Averell Harriman, acting on behalf of his mother and family and speaking publicly for the first time, handed a deed for 10,000 acres to George Perkins, along with three checks totaling $1 million. In ceremonial coordination, the 700-acre prison tract was transferred from the state of New York to the commission. Mrs. J. Pierpont Morgan and her daughter, Mrs. Herbert Satterlee, sat with Mrs. Harriman to witness the ceremony. Mr. and Mrs. Henry Phipps and Mrs. Perkins sat nearby.

Among other attendees were Dr. Partridge and ex-Governor Odell. Partridge's Forest Preservation Act had been rescinded as part of the political strategy to win approval of the bond issue, but the doctor was happy. His vision of a highlands park was being fulfilled, and, unknown to him at the time, he soon would be appointed to the interstate commission and would serve for fourteen years, until 1929. Odell, who had first supported the commission as Theodore Roosevelt's successor but then opposed its expansion when he became a lobbyist for quarry operators, must have been amazed by the sweep of purpose and financial clout being conferred on the commission that October day. Commission authority now extended from Fort Lee, New Jersey, for forty-five miles upriver to West Point, New York.

Ironically, the opening speaker at the ceremony was William J. McKay, chairman of the New York Prison Commission. He alluded to a retreat of Hessian soldiers from Bear Mountain during the Revolutionary War, adding that the prison commission also welcomed the opportunity to retreat. (McKay had his facts backward; an overwhelming British and Hessian force attacked forts Clinton and Montgomery at Bear Mountain in 1777, forcing the retreat of 600 Continental Army troops.) But the prison commission retreat was not without a price tag. The commission found itself purchasing prison assets including a railroad bridge, fifty thousand bricks, the guard barracks, stockade, and warden's quarters, furniture, plumbing fixtures, and seventy railroad rails, each thirty feet long, weighing a total of 42 thousand pounds. The prison commission provided without charge a flagpole, barn, old sheds, and one set of ice tools.

The next speaker was J. DuPratt White, who, on acceptance of the 700-acre prison tract, said, "I shall avail myself of this, my first opportunity to speak of the work of George Perkins as Commissioner. His time, his thought, his advice, and his energy have been unstintedly devoted toward the accomplishment of what has been done." White then read letters from governors Hughes and Fort. Governor Hughes referred to a slight technical detail; the $2.5 million bond was not yet in hand, but the governor was confident that New York voters would approve of it in the November election, a prediction that proved to be accurate only because the vote count in New York City and nearby counties outnumbered the universally negative upstate response. Governor Fort confirmed in his letter that a $500,000 New Jersey appropriation to the commission had been made to fund construction of a road along the base of the Palisades cliffs.

After the governors' letters were read, Sargent H. C. Lieb, in command of a West Point field cannon, ordered nineteen rapid-fire salutes to the ear-splitting delight of those assembled. Then it was young Averell Harriman's turn:

> In accordance with the long-established plan of my father to give to the State of New York, for the use of the people, a portion of the Arden estate, and acting on behalf of my mother, I now present to the Commissioners of the Palisades Park the land comprising the gift. I also hand you my mother's contribution to the expense of future development of the Harriman Park. It is her hope and mine that through all the years to come the health and happiness of future generations will be advanced by these gifts.

On acceptance, George Perkins predicted that the day marked "the beginning of what certainly will become one of the largest, most beautiful, and practical recreation grounds in all the world."

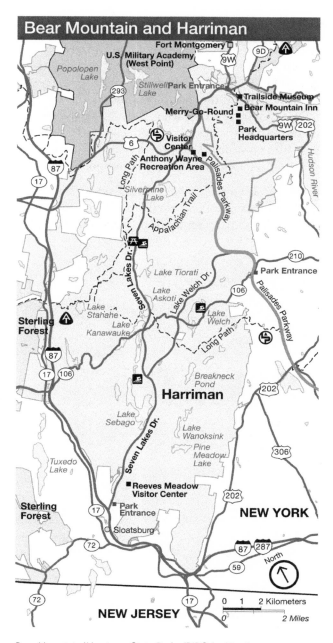

Bear Mountain / Harriman State Parks (PIPC Archives)

A follow-up report in the *New York Times* stated:

Nature itself had provided the setting. On three sides of the little plateau where the ceremony was held were the hills which form the rugged group clustering around Bear Mountain, just across the Hudson towered Anthony's Nose, jutting out into the river, and over the foothills shone the top of Storm King, the highest and most rugged of the mountains. All were colored red and gold by the Autumn foliage, a natural picture on which the group of men and women on the plateau gazed in admiration during the hour or more of the ceremony.

Mr. Kunz of the American Scenic and Historic Preservation Society collected the brass casings of five shells fired from the West Point field cannon, polished them, and sent them to Mary Harriman to be used as flower vases.

5

···

Legend and War

THOSE IN ATTENDANCE at Bear Mountain to celebrate the Harriman donations were on historic ground. In the eighteenth century, the highlands had been described as a "wild and shaggy wilderness, densely covered with forests of white pine and hemlock, the haunt of bears and wolves." Native Americans of the Algonquin nation were known to have found good hunting in the highlands and had established campsites there, but almost all were gone, victims of conflict with their Iroquois enemies, smallpox, slowly advancing European settlement, and deadly entanglements in the struggle between Britain and France for control of the colonies and Canada. In their place, a few rustic cabins and a primitive tavern were clustered at the base of Dunderberg Mountain on the west bank of the Hudson. The tavern, known as Caldwell's, named for the first family that settled in the tiny hamlet, was the "last oasis," convenient for fortifying courage before sailors attempted upriver navigation against the current and winds of the most dangerous part of the Devil's Horse Race. According to Rockland County, New York, historical records, it was at Caldwell's in 1720 that a ship with alien rigging and lines arrived offshore.

> From it landed a party of dark, bearded, fierce looking men. The landing party carried picks, shovels, and sacks, told no one of their business, and asked no questions. They started west along an old Indian trail that led through the pass between Bear Mountain and West Mountain and seemed to know where they were going. No one

saw them during the summer, but in the fall, they came back, each staggering under a heavy load in his sack. Before boarding the ship, they stopped at Caldwell's for some cheer, and one of them, in his cups, showed the contents of his sack, loaded with rich silver ore.

They made sail and disappeared down river. During the winter, hunters and trappers searched the woods and on the north slope of Black Mountain, found a rude log cabin. It was a two-room, saddle back affair, and the porch in the middle was oriented toward the summit of the mountain, as if the occupants had wanted to keep the heights under observation. Those who discovered the cabin suspected a mine, but further searching revealed not the slightest trace of any opening.

A year later, the Spaniards came back again in the same ship and made their way to Black Mountain. Their fierce appearance discouraged interference. They worked unmolested, again coming out in the fall to sail away, sacks full.

According to legend, the ship returned again a third year. Six of the mystery men disappeared into the highlands:

But in the fall, only two of the six came out and returned to their ship and sailed away, never to return. Soon after, a search party explored Black Mountain. They found the cabin and the bodies of two men inside. In the ribs of one was a Spanish dagger. The other had a broken skull. There was no sign of the other two. It was getting dark, but the searchers climbed higher on the mountain, looking for the opening of the shaft. Late the next day, the haggard and terror-stricken search party staggered into Caldwell's with a fearful story. As they approached the summit of Black Mountain, they were met by the ghosts of two men, sheeted in a light of phosphorescence. When they tried to flee, they couldn't move. They huddled under a great tree and shivered in fright until dawn, with the luminous ghosts whirling madly about. At the first light, the spirits disappeared, the searchers recovered command of their arms and legs and hurried down the mountain.

Hidden treasure in the highlands guarded by ghosts is legend, but the presence in July 1776 of Viscount Sir William Howe, Knight of the Bath, member of Her Majesty's Privy Council, and commander in chief of red-coated British forces, green-coated Hessian mercenaries, and a fleet of warships in New York Harbor was undeniably real, and Bear Mountain would prove to be a target. Howe had withdrawn his forces from Boston in the face of stubborn American resistance, but this time, he assumed, the result would be different. His 22,000 troops landed ashore and fought in several skirmishes and battles with the Continental Army and by November 1776 were in military possession of New York City. Strategically, Howe also wanted control of the Hudson River. If British warships could maneuver up the river to rendezvous with British troops marching down from Canada, the rebellious colonies of New England would be cut off from New York and Pennsylvania and the other colonies to the south. Collapse of the uprising might soon follow.

This fact was not lost on General George Washington. In his attempt to defend the city and river, Washington had ordered the fortification of the Palisades. He named this defensive location Fort Constitution, then renamed it Fort Lee in recognition of Charles Lee, a British general who switched sides and joined the Continental Army. An abatis of felled trees sharpened to spearpoint was positioned across the vulnerable northern approach to the site. Out in the river, just beneath the surface, a *chevaux-de-frise* made up of wooden cribs and sharp-pointed iron poles designed to rip at the hull of any passing ship had been floated and anchored from shore to shore. Log huts were built in the interior of the fortress to house hundreds of troops. From the high cliffs, cannoneers aimed their weapons downward toward the river to interdict hostile ships. According to historian Adrian C. Leiby,

> The site was almost a natural fortification. A clove in the Palisades, where a farm road from English Neighborhood twisting its way from the heights down to the river landing, left a high promontory standing out from the Palisades, inaccessible from three sides because of precipitous rocks, which fell off hundreds of feet to the river on one side and far enough on the other to discourage any assault. Ten acres were cleared, partly on this promontory and partly on the high land to the west, all rocky and heavily wooded wasteland on the farm of Peter Bourdet, whose farmhouse and cultivated land lay to the west, along the road to Bergen.

In October 1776, while battles for control of Manhattan Island and White Plains (present-day Westchester County) were being fought, Fort Lee also was tested. The British warships *Phoenix* and *Roebuck* maneuvered around the sunken log chain and sailed upriver despite repeated attempts by American cannoneers positioned on both sides of the river to sink them. In November, General Washington arrived at Fort Lee with 2,000 troops after commanding a vital tactical retreat from White Plains and a river crossing at King's Ferry, well above the position of the *Phoenix* and *Roebuck*. Once across the river, the troops marched twenty-five miles back downriver on the west side of the Hudson to join General Nathanael Greene and 2,400 Continental Army soldiers stationed at Fort Lee and in nearby Hackensack, New Jersey.

The last point of American resistance on Manhattan Island under the command of Colonel Robert Magaw was Fort Washington (where the modern-day George Washington Bridge connects to the east shore of the Hudson). On November 16, 1776, superior British forces overwhelmed Fort Washington, killing scores of defenders and capturing almost 3,000 Continental Army troops.

The next step for the British was to capture Fort Lee and, with luck, George Washington himself. Under overcast night skies on November 19–20, the "portly and awkward" British general Charles Cornwallis oversaw the clandestine transport of British and Hessian troops on flatboats across the Hudson River. The troops landed under the darkened Palisades cliffs near the "Blackledge House," ten miles upriver from Fort Lee. Cornwallis claimed the house as his headquarters and ordered his troops to climb up a precipitous half-mile trail to the summit of the Palisades.

While British and Hessian soldiers laboriously dragged cannons up the trail, delaying their assault on Fort Lee, watchful Continental Army sentries signaled the alarm. When Cornwallis's troops finally reached Fort Lee, its defenders and generals were gone, escaping to a miserable winter

Attack on Fort Lee, Sketch by British Officer Thomas Davis (New York Public Library)

encampment at Valley Forge. With the capture of Fort Lee, the great river avenue that could divide the colonies now seemed open for the taking.

In June 1777 the British strategy was put in motion. From Montreal, the British general John Burgoyne moved south with his army of 7,000 men. With little difficulty, his troops retook Fort Ticonderoga, a remote outpost captured early in the Revolutionary War by Ethan Allen's Green Mountain Boys and Colonel Benedict Arnold. From Fort Ticonderoga, Burgoyne faced a trek through wilderness that sapped the strength of his men and slowed progress almost to a standstill, but he doggedly pushed on. A smaller British force under the command of Lieutenant Colonel Barry St. Leger had been transported by boat from Montreal to Lake Ontario with the intent of following the Mohawk River for more than 150 miles through wild country to its confluence with the Hudson and to a rendezvous with Burgoyne. General Sir William Howe, the British commander in New York City, would send ships and forces up the Hudson River to meet the British troops descending from the north, completing the dismemberment of the colonies.

Standing against this strategy were Continental Army forces gathering in the vicinity of Albany/Saratoga under the command of General Philip Schuyler, a Continental Army group under the command of General Israel Putnam holding a defensive position on the east shore of the Hudson at Fort Independence (Peekskill), and the small, rustic fortifications at Bear Mountain manned primarily by tenant and freehold farmers from surrounding counties. The principal fortifications were forts Clinton and Montgomery located in proximity on the west side of the river. Using boulders, logs, abatis, and earth to erect the defenses, the militiamen had done their best to impose military control on the river narrows.

Preparations also included the Herculean task of floating a massive chain on logs across the river and securing it firmly shore to shore. Iron for the chain had been mined and forged in the highlands. Each link in the chain weighed upwards of 100 pounds, and the fully completed chain weighed many tons. The gambit was that any British ship trying to bypass the forts would be blocked within point-blank range of shore-based Continental Army cannons.

The defenses were placed under the command of the governor of New York, George Clinton. He was a prominent politician and lawyer-farmer from Ulster County, New York, who had gained military experience in the French and Indian War and held the rank of brigadier general in the Continental Army. Clinton was elected in June 1777 just as Burgoyne was slowly moving southward from Canada. The high urgency of defending the

Hudson River narrows prompted the governor to join his brother, Col. James Clinton, to oversee construction of the defenses at Bear Mountain. The larger fort was named in honor of Major General Richard Montgomery, a casualty of an ill-fated Continental Army attempt in December 1775 to capture Quebec. The smaller fort probably was named by James Clinton for himself.

Far downriver in New York City, General Howe, seeing a chance to take Philadelphia, divided his forces and sailed for Chesapeake Bay. Left in charge of the Hudson River strategy was Sir Henry Clinton, Canadian-born, a veteran of battles at Boston and New York, and perhaps a distant relative of the brothers Clinton. The river assault was delayed pending arrival of 1,700 reinforcing troops from England who reached New York Harbor in late September.

In early October 1777, Sir Henry launched the river offensive. Signals from Continental Army sentinels posted on high elevations along the river corridor were quickly passed to the waiting defenders. On learning of the British threat, Governor Clinton, who had returned to the provisional seat of New York government at Kingston, rushed back to Bear Mountain to rejoin his brother and 600 militiamen. The "battle of the Clintons" was about to begin.

Sir Henry had his own informers and sentinels and determined not to confront the currents of the narrows, its forts, and chain head-on. Covered by an opportune fog, British and Hessian soldiers landed at Stony Point on the western shore of the Hudson, well downriver from Bear Mountain. These forces were strengthened by the Loyal American Regiment commanded by Colonel Beverly Robinson, a wealthy landowner opposed to the revolution. Robinson's home was just across the river from West Point, and he had roamed the forests on both sides of the river on many hunting trips. He knew the terrain in the vicinity of Bear Mountain and helped Sir Henry plan a surprise flank attack against the forts.

Robinson identified a narrow trail that led through an unguarded pass on Dunderberg Mountain to Mountville (later renamed Doodletown), an isolated hamlet tucked in a small valley within two miles of forts Clinton and Montgomery. The hamlet had been settled by woodcutters, most of whom now were among the militiamen at Bear Mountain. So, too, among the defenders, were tenant farmers from across the river who, ironically, paid rent to Robinson.

At Mountville, the attacking British/Hessian troops split into two columns, one looping around the western slopes of Bear Mountain well out of sight of the forts. The other contingent was held in reserve at Mountville.

Attack on Fort Clinton Redoubt. Diorama (Palisade Interstate Park Commission)

The signal for the troops at Mountville to advance at fast march directly toward the forts would be the sound of gunfire when the first column attacked its intended target, Fort Montgomery.

The movement of so many hostile troops was detected by the defenders at Bear Mountain, although the size of the force being brought against them was unknown. A message asking for reinforcements was sent by Governor Clinton to his commanding officer, General Putnam. Putnam, though, was confronted by a British raid in a trick intended to distract from the larger British mission unfolding on the west side of the river. The deception worked; no reinforcements would arrive at Bear Mountain.

The battle at Bear Mountain began with a skirmish at about 10:00 A.M. on October 6, 1777, when militiamen on scouting patrol discovered the hostile troops waiting at Mountville. The farmers, woodcutters, and regular Continental Army soldiers soon would learn that they faced attackers of far superior strength in numbers. The quickly developing battle was fierce. British attempts to breech the fortresses were beaten back, and blood flowed. Warships on the river closed within range and began to bombard the forts. As the day faded the stubborn Continental forces appeared trapped, suffering, and low on ammunition, prompting the British regimental commander, Lieutenant Colonel Mungo Campbell, to demand surrender. No white flag was raised.

In the final attack, Campbell lost his life even as George Turnbull of the Loyal American Regiment came first over the wall at Fort Montgomery. Count Grabouski, a Polish nobleman and aide-de-camp to Sir Henry Clinton, was killed. British soldiers, pushing one another up and over the works at the two forts, were met hand-to-hand by defenders with bayonets, clubs, rocks, and a few remaining bullets from muskets. The struggle turned from defense of the doomed forts to attempted escape by the defenders.

At dusk, Governor Clinton and his wounded brother, James, managed to reach the river and cross by boat to the opposite shore. Militiamen, covering the escape, fired last volleys and threw down their muskets in surrender. Forty-one British-led combatants, mostly Hessians and loyalists, lay dead, with 142 wounded. An estimated 70 of the fort's defenders died, 40 were wounded, and 240 were taken prisoner. Two Continental frigates anchored above the chain as part of the river defense were unable to sail from harm's way against the strong current and were set on fire and abandoned by order of their captains. The battleground at Bear Mountain lay quiet. Several months after the battle, a chaplain visiting the site reported that bodies still could be seen where they had been dumped into a nearby pond.

The Hudson River narrows was in British hands, but the huge, ponderous chain still blocked the river. British forces were delayed for three days while the chain and a less formidable upriver *chevaux-de-frise* were removed. By October 16, Sir Henry Clinton's forces, led by Major General John Vaughn, were in Kingston, still more than forty miles from Albany/Saratoga. The delay at Bear Mountain had proven strategically costly to the British.

Lieutenant Colonel St. Leger, attempting the Mohawk River route to the Hudson, had been driven into retreat in mid-August by Continental Army forces at Fort Stanwix, New York. Messengers brought the more urgent news to Vaughn that Burgoyne was facing unexpected disaster. In fact, Burgoyne, without reinforcements and after suffering losses in two sharp battles, surrendered his army on October 17, 1777, to General Horatio Gates, successor in the Albany/Saratoga region to Philip Schuyler.

The British strategy to divide the colonies through control of the upper Hudson River Valley was in shambles. Vaughn did not wait for final confirmation of the Saratoga defeat. Kingston, New York, considered by the British to be "a nursery for every villain in the country" and a "hotbed of perfidy and sedulous disloyalty to King George the Third and His Majesty's

Parliament," was put to the torch the day the Vaughn forces arrived. By the end of October, British ships and troops were back in New York City.

More than a century later, George Perkins and his fellow commissioners would find themselves holding title to and managing priceless Revolutionary War sites far different in stewardship terms from the mission of protecting natural beauty, but of highly complementary public benefit. Rescuing the Palisades and acquisition of a prison site at Bear Mountain put the commission squarely in the business of preserving, presenting, and interpreting historic events at the very doorstep of United States nation-building. Among these historic gems, the commission would, over time, assume responsibility for the:

Blackledge-Kearney House where Lord Cornwallis set up headquarters in 1776 beneath the cliffs of the Palisades during the British assault on Fort Lee,

Fort Lee, taken by the British in November 1776, forcing George Washington and the Continental to escape to Valley Forge,

Forts Montgomery and Clinton, standing in defense of the Hudson River narrows and taken by the British in October 1777,

Senate House, Kingston, New York, first capital of New York, built in 1676, partially destroyed by fire in October 1777 in the aborted British assault on the upper Hudson River Valley,

Stony Point, the "Gibraltar of the Hudson," taken by force from the British in July 1779 by Continental Army light-infantry soldiers under the command of General (Mad) Anthony Wayne,

Knox's Headquarters, a home owned by the Ellison family that was used during the Revolutionary War as headquarters by Continental Army generals Henry Knox, Nathanael Greene, and Horatio Gates,

New Windsor Cantonment, New York, the encampment in 1782–83 of 7,000 Continental Army troops restively awaiting the end of the Revolutionary War,

Washington's Headquarters, Newburgh, New York, recognized in 1850 as the nation's first publicly designated historic site. By stature and persuasion, General George Washington held the Continental Army together in 1782–83, awaiting the Treaty of Paris and the end of the Revolutionary War.

The distance from Fort Lee to Washington's Headquarters is about forty miles. The distance in history is immense, from submissive colonies, erased, to proud states, united.

6

⦚⦚⦚

Welch

April 15, 1910
Mr. Frank E. Lutz
Assistant Curator
American Museum of Natural History

Dear Sir:

Replying to your favor of the 14th instant, I beg to advise you that the Commissioners gladly extend permission to you to collect insects in the Interstate Park.

Yours very truly,
J. DuPratt White

SCIENTIFIC STUDY OF moths, beetles, ants, mosquitoes, ticks, fireflies, crickets, and insects in general seems so esoteric to most people that it usually wins nothing more than amused and fleeting curiosity; an eccentric person in a pith helmet, romping through fields with a butterfly net is the popular image of the entomologist. Dr. Frank Lutz very likely did use a butterfly net, and his work with the American Museum of Natural History established him as a major presence in the nation's scientific and educational community. Before his career ended in 1943, Lutz developed a museum collection of more than two million specimens of insects, one of the great baseline scientific collections in the world.

Lutz was drawn to the open space being protected by the commission for its potential as an outdoor laboratory. Where others saw recreational opportunity, he saw in the highlands a rich source of scientific knowledge. Like almost all environmental scientists, Lutz needed access to protected lands that were rent-free, or nearly so. Budgets for his type of work ranged from meager to nonexistent. In addition to cost considerations, he was lured to the commission parklands because of proximity to the Museum of Natural History in New York City. Lutz could develop baseline data knowing that lands on which he roamed with his butterfly net could be revisited time and again, secure from the risk of urban or suburban burial.

He enjoyed delivering messages in a manner that had staying power. Once, he tried to make a wager with the director of the Museum of Natural History that he could find more than 500 species of insects in his 75-by-100-foot backyard in Ramsey, New Jersey. When the director declined the wager, Lutz countered by proposing that his salary be raised by $10 per year for every species above 500 that he could find. Still the director declined, but Lutz made the count anyway. He documented 1,402 insect species.

In 1925, through a grant provided by W. Averell Harriman, Lutz was able to establish the *Station for the Study of Insects* on a forty-acre commission-owned plot in the highlands. The station became an outdoor classroom, allowing young students to work and stay in the camplike setting during summer months. One of these students was ten-year-old David Rockefeller, youngest son of John D. Rockefeller Jr. Insects are not for everyone, but David was so fascinated that he became an expert in beetles. Of his later travels as banker, statesman, and business leader, Rockefeller said, "I can go anywhere in the world and will know with certainty about one thing, beetles." He personally assembled a spectacular, comprehensive, and scientifically meaningful collection of more than 150,000 beetle specimens now available for study at the Harvard University Museum of Natural History.

To encourage his students, Lutz developed a trail around the station and posted small signs to identify interesting plants, insect haunts, and other natural features. In so doing, he created the first nature trail in the United States. At the beginning of the trail, the first sign read,

> The spirit of the training trail: a friend somewhat versed in natural history is taking a walk with you and calling your attention to interesting things.

In 1926 Lutz expanded his educational and scientific interests to Bear Mountain by establishing the Trailside Museum in cooperation with the American Museum of Natural History. The Trailside Museum and Wildlife Center remains a significant, highly popular asset for park visitors.

By bringing the prestige and integrity of the Museum of Natural History to the woods, meadows, marshes, lakes, and streams of the highlands, Lutz confirmed the scientific and educational benefits of parks. He recognized the potential of these lands to engage the intellect as well as the senses. Research and educational activities aimed at a better understanding of the environment, so widespread in parks throughout the nation today, can be traced to a man with a butterfly net.

George Perkins likely was aware of Lutz's initiative to bring practical science under the commission's banner, but he had other concerns on his mind. At the end of 1910 Perkins retired from J. P. Morgan & Company. He had worked hard, won significant business victories for Morgan & Company, and gained great wealth. Perkins judged that the time was right to enjoy the independence that wealth provided. He freed himself from corporate structure to pursue personal interests, most especially the Palisades project.

The Harriman gift, too, was a motivating factor. Perkins had been continually at the helm of the commission but had not been immersed in the many day-to-day details that were rapidly transforming its assumptions and needs. In a memorandum that accompanied his retirement announcement, Perkins explained, "I have long felt that it is not wise to leave all our public affairs to the politicians, and that business men of sufficient leisure and means should for patriotic reasons give their attention to great public problems." His retirement from active business life also reflected growing stress in his relationship with J. P. Morgan. Perkins's biographer, John A. Garraty, wrote that Morgan's "increased irascibility and distrust" had found its way to Perkins himself. The break was stormy but not irreparable. Although a professional coolness thereafter existed between the two men, Morgan continued to respond to commission fundraising pleas, and Perkins always used the quaint phrase "dear Senior" in a respectful and affectionate manner when referring to the elder financier, who had catapulted him to the top of the business world.

Perkins's more focused attention on commission matters came none too soon. The commissioners were under criticism for not responding to the demand that Henry Hudson Drive, envisioned for the base of the Palisades in New Jersey, be extended northward along the river shore, perhaps all the way to Bear Mountain, thirty miles north of the New Jersey–New York border. J. DuPratt White again served as the lightning rod. In a lengthy letter to the editor that appeared in the *New York Times*, White defended the commission and explained the practical difficulties of the road project, adding that "as is usually the case with writers of letters to newspapers for the purpose of correcting supposed public wrongs, your said correspondents are hopelessly ignorant of their subject matter. They show no knowledge whatsoever of the laws creating or governing the great Interstate Park, of the plans of this Commission, or of the topography or geography of the region."

Controversy swirled around the question of roads. Perkins was accused of bowing to the wishes of Mrs. Lydia G. Lawrence by refusing to condemn her property on the border of the two states to make way for the road project. A New Jersey assemblyman wrongly denounced commissioners William A. Linn and Abram De Ronde, both appointed in 1900, for cashing in on commission projects through the National Bank of Hackensack and the Palisades Title and Guarantee Company of Englewood, New Jersey, controlled, respectively, by Linn and De Ronde.

Despite these criticisms, on March 11, 1911, a headline in the Rockland Journal News proclaimed, "COMMISSION BUYS THE HOOK."

> What a few months ago seemed to many like an impossibility, or a difficult proposition at the least, has been accomplished really in a few weeks by the Palisades Park Commission.

Five years after the commission's authority was extended north of the New York/New Jersey state line to include Hook Mountain, the Barber Asphalt Company sold its subsidiary, the Manhattan Trap Rock Company, to the commission for $415,000, including a large concrete powerhouse at Nyack Beach. Attorney Irving Hopper of Nyack, New York, assisted in settling the details of the purchase. Only two quarries, the Rockland Lake and Clinton Point Trap Rock companies, remained in operation. The hard-fought contest with the quarry operators was coming to an end.

The heightened profile and success of the commission led landowners to approach the commission in the hope of selling their properties at inflated values. Proposals were advanced, only to collide with the time-tested business and legal skills of the commissioners. Publicity caused by the Harriman gifts of land and money and the associated increase in private and government funding for the commission did not change the basic style of Perkins, White, and their colleagues. Each dollar was squeezed for maximum benefit. Those looking for quick windfall profits, including Addison Johnson, who owned 500 acres at Bear Mountain, were firmly discouraged.

Johnson proposed to sell his land to the commission for $100 per acre, total price $50,000. The commissioners learned that Johnson had purchased the property two years earlier for $7,500 and refused further contact with him for several months. When Perkins did reopen the negotiation, he offered Johnson $10,000 and gave the owner two weeks to accept. Johnson accepted.

Steven Rowe Bradley of Nyack, New York, made a much more attractive offer. Bradley proposed to donate 212 acres on South Mountain, to be designated "Rockland Park." Bradley was a community leader who was instrumental in forming the Nyack National Bank in 1878, the Nyack Library in 1879, and the Nyack Hospital in 1895. At their May 1, 1911, meeting, the commissioners accepted Bradley's generous offer to begin the creation of a park on South Mountain. But the donor was in ill health. Before the transaction could be completed, he passed away at the age of seventy-five. His children immediately took up the cause, and title to the Bradley property was transferred to the commission in October.

In a foresighted letter to the commissioners, the Bradley heirs expressed their collective expectation that the land "shall be deeded for a natural park, to be held for the benefit and enjoyment of the public at large, open at all times for their use—to secure the perpetuation of the birds, animals, plants, trees and other natural features—and to restrict and govern the admission of motor vehicles." On behalf of their father, and so early in the century, S. R. Bradley, Mary T. Bradley, Augusta B. Chapman, and William C. Bradley were recommending guidelines for park stewardship that would become the focus of constant debate between preservation advocates and proponents for the exploitation of parks. Their concern about "motor vehicles," at a time when horse carriages still outnumbered automobiles, is reflected in today's many hotly contested park debates. Popular parks suffer traffic gridlock; off-road vehicles scar fragile desert habitat; snowmobiles shatter winter quiet. (One example: in 1980, a plan was approved with strong public support that called for restricted automobile access to Yosemite Valley, California. Since then, because of commercial pressure, traffic congestion has increased.)

The Bradley heirs clearly expressed their hope for the perpetuation of natural features but may not have known about incoming flying bullets. A state rifle range existed on the summit of South Mountain adjacent to the Bradley property. Samuel Broadbent, president of the Board of Health, Village of Grand View-on-Hudson, New York, wrote to Perkins to urge the commission to acquire the range. J. DuPratt White responded:

In the form of affidavits, I suggest that you gather as much evidence as you can of actual trespass of bullets upon the properties of residents in the villages. It is common talk that bullets have entered houses through the walls or roofs and the windows, and that bullets have also been seen to strike the ground within the limits of the several villages. These circumstances, it seems to me, should be run down and proper evidence embodied in affidavits before any Legislative committee is asked to pass upon a bill.

In other words, the commission was interested and willing to help if the residents of the village could provide convincing proof that bullet holes in their houses had come from the rifle range. With facts in hand, and by an act of the legislature two years later, the 500-acre rifle range was transferred to the commission.

South Mountain is a high ridge above Piermont, New York, offering a sweeping, forty-mile view of the Hudson River. The ridge accommodates Tweed Boulevard, named for the infamous "Boss" William M. Tweed, master of New York City machine politics. Tweed sponsored a road connection from Hoboken, New Jersey, to Nyack, New York, and the narrow road on the summit of the ridge is testimony to a powerful politician who could make things happen. Tweed ultimately fell from his commanding position in the rough-and-tumble of his political world when he was convicted of fraud and sent "up the river" to Sing Sing Prison to spend his last days behind bars. But Tweed Boulevard, providing access to land donated by the Bradley heirs and the adjacent commission-acquired rifle range (present-day Blauvelt State Park), remains a fixture on the ridge summit.

Almost in concert with the Bradley gift, Dr. James Douglas of New York City offered to donate to the commission several tracts of land atop the New Jersey Palisades. This proposal, too, was accepted, drawing the commissioners ever closer to involvement with the rolling lands and elegant estates that capped part of the cliffs. These properties had escaped the attention of the commission during its battle with the quarry operators, but the spectacular river views from the summit lands, their inherent and exquisite wildness, and the need to improve access to increasingly popular park holdings were proving to be irresistible to the commissioners.

Acquiring manor-like properties was running parallel with growing commission interest in the idea of a grand boulevard that would allow for leisure motor travel from New York City to Bear Mountain. Charles W. Leavitt Jr., chief consulting engineer for the commission, recommended that the services of consulting engineer Alfred Nobel (famed as the inventor of dynamite and later to endow the Nobel Prize) be retained to study the boulevard concept. The obvious challenge was that no bridge existed that connected the city with the Palisades. Automobiles would arrive via ferryboat. Among the options suggested by Leavitt and Nobel was to drill an ascending road tunnel through the cliff face from the base of the Carpenter brothers' old quarry site to the summit of the Palisades. The idea was not discarded out of hand. At the invitation of Perkins, the commissioners would discuss boulevard options at their next meeting to be held aboard his yacht, *Thendara*. A meeting on the yacht was timely because another concern much on the minds of the commissioners was the obvious and increasing pollution of the Hudson River.

Raw sewage was being dumped into the river from every village and municipality along its shores. Ships and boats plying the river commonly jettisoned whatever they wished. Driftwood was a menace to small craft and was clogging the New York Harbor shoreline. The river, popular with swimmers, canoeists, sailors, yacht owners, and anglers, was a handy, untreated Hudson River Valley disposal system. The commission began petitioning governors John A. Dix of New York and newly elected Woodrow Wilson of New Jersey to clean up the river.

Problems were mounting, and one, in particular, persisted. White responded to a resolution critical of the commission that had been approved at the 1911 annual convention of the New Jersey Federation of Women's Clubs. White wrote in a letter to Mrs. Joseph M. Middleton:

> The resolution is a protest against the present condition of the park, and a demand by the Federation that the original idea of the Federation be given consideration, but I do not know what the Federation's original idea was. The resolution further provides that measures shall be adopted to make the tract a true park, with a fitting memorial approved by the women of New Jersey. Will you kindly inform me whether or not the Federation is under the impression that this Commission has ever undertaken to expend any money on said Memorial Park?

After more than a decade, commission institutional memory and White's personal memory apparently had faded, and the Federation of New Jersey Women's Clubs, still without its monument, would have to wait still longer.

Another vexing challenge was about to surface. In 1912, people who wished to establish a summer camp to be supervised by the National League of Urban Conditions Among Negroes contacted the commission.

The commissioners promptly approved the camp in concept, finding logic and appeal in the idea that parklands could benefit any child, regardless of race. But they were to confront the racial attitudes of the early twentieth century and would struggle to open the park door to people of all races.

Other matters large and small came to the table for commission decisions. In one meeting, the commissioners focused their discussion on choices for the width of a steep road planned from the Palisades summit at Englewood, New Jersey, down to the shoreline below. Apparently, the idea of tunneling through the old quarry site had been discarded. Facing the commissioners and their chief consulting engineer was a precipitous drop down the cliff face that would be daunting for any road builder. The middle-aged and elderly bankers, lawyers, and businessmen of the commission scheduled a field visit to the site, where they scrambled up and down the proposed route, huffing and puffing, to determine the road width needed to allow automobiles to get through the hairpin turns.

In November 1912, Perkins wrote to John D. Rockefeller Jr. to report the following developments: the quarries at Hook Mountain were being acquired either through willing-seller transactions or by condemnation; the Henry Hudson Drive in New Jersey was under construction; docks had been built along the river to improve public access, including a very large dock at Bear Mountain; the Bear Mountain site itself would be opened to the public the following year; water supplies, sanitation facilities, and picnic and camping areas were being installed to meet increasing visitor demands; and forests were being cleared of deadwood. Perkins concluded the letter by reminding Rockefeller Jr. of his financial pledge to the commission: "The amount of your subscription is $500,000; you have paid $200,000, leaving a balance of $300,000. The Commission will very much appreciate receiving a check for this amount on or before December first of this year." Rockefeller willingly complied.

While adjusting to his retirement from Morgan & Company and grappling with the increasingly complex commission agenda, Perkins took on another demanding extracurricular task, as described by Garraty:

It is the evening of June 20, 1912; the scene, a large room in the Congress Hotel in Chicago. About twenty men are present. Perhaps a dozen of them are seated around a large table. Others sprawl wearily in armchairs or lean against walls. One, a solid, determined-looking fellow with thick glasses and a bristling mustache, paces firmly back and forth in silence, like a caged grizzly. He is Theodore Roosevelt, and these are his closest political advisers. All of them are angry, very angry.

In a nearby auditorium, the Republican National Convention is moving with the ponderous certainty of a steamroller toward the nomination of well-fed William Howard Taft for a second term as President of the United States. All of the men in the hotel room believe that this nomination rightfully belongs to Roosevelt.

It is growing late, and everyone is weary. Conversation lags. But gradually attention is centered on two men who have withdrawn to a corner. They are talking excitedly in rapid whispers. One is the publisher Frank Munsey; the other, George W. Perkins. Neither has had much political experience, but both are very rich and very fond of Theodore Roosevelt. Now everyone senses their subject and realizes its importance. All eyes are focused in their direction. Suddenly the two millionaires reach a decision. They straighten up and stride across the room to Roosevelt. Each places a hand on one of his shoulders. "Colonel" they say simply, "we will see you through." Thus, the Progressive party, "Bull Moose" some call it, is born.

Munsey's involvement in the Progressive Party was brief, but Perkins made a commitment and intended to do exactly as he said, see it through. He became campaign manager for Roosevelt's bid for another presidential term. Perkins's ideals and philanthropic activities were suggested to reporters in those rare moments when he nudged open the door to his personal thoughts. "Shall I go on and pile up a few more millions on top of those I have already acquired and make a big money pile the monument to my memory? A man should ask himself, what is this all about? Where is my work going to lead me?" By his actions, Perkins, at age fifty, was answering these questions for himself.

He and Roosevelt seemed the odd couple, Roosevelt the trustbuster joining Perkins the monopolist. But there was no contradiction in Perkins's' alignment with the Progressive movement. The interstate commission had formed a Roosevelt-Perkins bond, and both the commission and this new venture into politics provided avenues that promised social benefits for working men and women, values long championed by Perkins. From the very beginning of his business career, he had maintained a strong interest in socially beneficial government policies and regulation. His involvement with the Progressive Party was an extension of this long-held interest.

The candidates in the 1912 election were President Taft, former President Roosevelt, and New Jersey governor Woodrow Wilson. The popular Wilson prevailed in the election. Roosevelt carried six states and finished in second place. Even after the loss, Perkins remained involved with reform politics in the Progressive and Republican Parties, although his duties as president of the commission were an increasingly central focus of his public work.

That work was not going well. Within the commission, controversy developed over the performance of chief consulting engineer Leavitt. The New Jersey commissioners sought Leavitt's removal from direct supervision of the Englewood approach-road project. As a result, Leavitt remained in charge of general engineering work for the commission, except for the Englewood project. Commissioner De Ronde resigned as chair of the roads subcommittee in favor of Col. Edwin A. Stevens, who had raised objections about Leavitt's performance. During his career, Leavitt would claim many laudable accomplishments, including construction of the Yale Bowl, but his relationship with the commission was becoming untenable.

On Leavitt's staff was young engineer William Addams Welch. Welch caught the eye of Perkins, who recruited him to the commission as an assistant engineer. Before Welch concluded his career forty years later, he would blaze a path in park development and management second to none in the United States.

Welch's family connection to the Hudson River Valley was represented by "Welch Island" in the headwaters of the Hackensack River, named for an ancestor who had settled in the region in 1695. The Welch roots extended even further back to Plymouth, Massachusetts, where John Welch landed soon after the Mayflower colonists. The Welch family migrated west to New Jersey, then westward again before the Revolutionary War. The family settled in Kentucky, where William Welch was born in 1868. Welch's father had ridden with Morgan's Raiders during the Civil War. His mother, Priscilla Addams, was descended from John Adams, second president of the United States. An extra "d" had been added to the name when the Southern branch of the family disagreed with the policies of John Quincy Adams, the sixth U.S. president.

Welch attended college in Colorado and Virginia. On graduation, he worked on engineering projects in Alaska for six years, then in the western states, Mexico, and South America, where he specialized in railroad construction. Yellow fever caused him to return to the United States, where, in private practice, he designed the beautiful Havre de Grace racetrack in Maryland and built the boardwalk at Long Beach, Long Island. When he was tapped by Perkins to join the commission, Welch, with no park experience, stepped into untested territory. But long before his career ended, park advocates from all over the United States and Europe would be seeking him out for advice and guidance at his small, secluded Bear Mountain cabin. Welch, who granted no interviews with the press and sought no personal accolades, became the "father" of the state-park movement and greatly influenced the creation of the National Park Service.

One of his passions was the construction of camps for children. During his tenure, 103 children's camps would become available within the park system capable of accommodating a total of 65,000 children each summer season. Eleanor Roosevelt attended the dedication of one of these camps. Given the honor of breaking a bottle of champagne over a boulder, Roosevelt swung hard and missed, hitting Welch in the head. As reported in the *New York Times*, "After he came to, the ceremonies were resumed."

7

Bear Mountain

March 1, 1913
Hon. William Sulzer
Governor of New York
Albany, New York

My dear Sir:

The situation existing in New Jersey, relative to the appointment of Commissioners of the Palisades Inter-State Park is such that I feel it incumbent on me to offer you my resignation.

I hereby resign as a Commissioner of the Palisades Inter-State Park of the State of New York to take effect at your pleasure or upon the appointment of my successor.

Very truly yours,
Abram De Ronde

WILLIAM WELCH WAS joining the commission staff at a time of upheaval and an almost head-spinning surge forward in the park stewardship experiment. Within two years of his employment, he was promoted to chief engineer and general manager, leaping from a career that had been focused primarily on building railroads to hosting a growing wave of park visitors rapidly increasing from the hundreds to the thousands, and finally into the millions. Welch was not in on-the-job training; he was in self-training with no prototype as a guide.

Politics and mortality were taking a toll on the commission. The signal from Democratic governor James F. Fiedler of New Jersey, who succeeded Woodrow Wilson, echoed the former governor's opinion that Abram De Ronde, an active participant in conservation activities since the inception of the commission thirteen years previously, was no longer politically acceptable. De Ronde, a member of the New Jersey Democratic Committee, apparently stepped on the wrong toes within the power structure of his own party. With his resignation, De Ronde's name disappeared from the commissioners' roster, ironically to resurface nineteen years later when he was reappointed to the commission by Democratic New Jersey governor A. Harry Moore. De Ronde thereafter served until his death in 1937.

Two other charter members departed the commission in 1913: William A. Linn, a second political casualty in New Jersey, who, like De Ronde, submitted his resignation, and D. McNeely Stauffer, who passed away. Four years earlier, death also had taken William B. Dana. Succeeding Dana was Richard V. Lindabury, a founding partner of Lindabury, Depue, & Faulks in Newark, New Jersey. Through various legal proceedings, including a successful challenge directed at the American Tobacco Company for the restraint of trade and a defense of the United States Steel Corporation against trust-busting, Lindabury gained a distinguished reputation in his chosen profession. He served as a director and counsel for the Prudential Life Insurance Company and, as a lifelong Democrat, was urged by party leaders to run for governor. Lindabury declined, preferring, instead, to return to his childhood farming roots by creating and devoting attention to "Meadowland," his 600-acre estate near Bernardsville, New Jersey. Lindabury brought sharp legal talent to the commission and would serve for fourteen years until a tragic equestrian accident took his life.

Succeeding De Ronde was Frederick C. Sutro, a Harvard University graduate and campaigner for Woodrow Wilson who helped found Sutro Brothers Braid Company, a textile manufacturer. Sutro's avocation was history and park conservation. He was elected president of the New Jersey Parks and Recreation Association, a post he held for twenty years. His service with the commission extended from 1912 to 1940. Within the PIPC, Sutro performed the impossible in 1931 by stepping down from his appointment as a commissioner to assume the task of executive director for the next nine years, answering as a paid manager to the very men with whom he had shared authority as a volunteer and peer.

Appointed to succeed Stauffer was Charles W. Baker, an engineer with ties to Theodore Roosevelt and William Howard Taft. Baker played a significant role as a consultant on the Panama Canal project and was a champion for municipal ownership of utility companies. Baker's tenure with the commission would extend for thirty-five years. He joined his colleagues at just the right moment to provide important guidance in the construction of the Henry Hudson Driveway at the base of the Palisades cliffs.

Filling a fourth vacant slot was Dr. Edward L. Partridge, who was promptly elected treasurer. Partridge would hold that office and serve with the commission until 1929. The remaining commissioners were George Perkins, Edwin Stevens, J. DuPratt White, Franklin Hopkins, Nathan Barrett, and William Porter. These men and their predecessors, successors, colleagues, and employees seemed to share a common affliction: once bitten by the park bug, always bitten. The idea of serving for two or three years in dutiful public service was alien to most commissioners. Several of them would be carried feet-first from their commission responsibilities.

March 6, 1913
Mr. George F. Perkins
Hotel Raymond
Pasadena, California

Dear Mr. Perkins:

I enclose copy of a letter just received from Senator George A. Blauvelt in regard to certain proposed legislation hostile to Park interests. Hagar has plenty of nerve to attempt legislation of this kind. The exception of the Conger property would be a fearful breach of good faith on the part of the State with the people and the large private contributions.

Yours very truly,
Leonard Hull Smith
Assistant Secretary, PIPC

New York state senator Hagar, otherwise unknown to the commission, appeared to be representing a holdout quarry operator at Hook Mountain, New York. The senator was attempting by legislative maneuver to exclude the quarry site from the commission's acquisition power. Hagar attempted a clever legislative tactic by seeking to legally mandate expansion to twenty commissioners, with the obvious intent of diluting the influence of Perkins, White, and their eight fellow commissioners. Fortunately for the commission, it had a staunch, watchful, and influential champion in Albany in the person of New York state senator George A. Blauvelt. Senator Blauvelt, a senior and respected Republican in the state legislature from Rockland County, New York, quickly smothered this bit of political smoke and mirrors.

While the commission had its guard up in Albany, the tax collector for the borough of Alpine, New Jersey, put two commission properties on the block for sale, claiming tax delinquency. At about the same time the New Jersey comptroller was holding up $200,000 of the $500,000 appropriation made for construction of the Henry Hudson Drive. The comptroller claimed that the funds should not be made available until the commission purchased the "Lawrence" property so that the drive could be extended to the village of Piermont, New York, ignoring the fact that he had no authority to mandate such a requirement across the border in another state.

Then on the New York side of the border the commissioners encountered reluctance on the part of Governor William Sulzer to approve legislation intended to clarify the commission's prerogative to establish fish and game regulations, including a restriction against hunting on parklands. Sulzer expressed concern that the legislation would undercut the authority of the New York Conservation Department.

These challenges were met and overcome, only to be replaced by new challenges. After years of negotiation, Perkins succeeded in acquiring the E. I. DuPont de Nemours Powder Company property on the shoreline in New Jersey near the Carpenter brothers' old quarry site. Despite earlier claims that all necessary properties had been acquired to protect the cliffs, the DuPont property, including a narrow, 3,400-foot strip of land and riparian rights on the river shoreline and eight acres on the summit of the Palisades, was an exception. The $142,500 purchase price was tucked out of sight into the paper shell of the commission's subsidiary corporation, the Palisades Improvement Company, to ensure that the high per-acre cost would not come back to haunt future acquisition initiatives.

One of these initiatives was taking place in New York. Charles T. Ford, the estate manager for Mary Harriman, managed to win agreement from a "Mr. Cunningham," who boasted that he once stood gallantly alone and "blocked E. H. Harriman" from acquiring even more land in the highlands. Ford took it upon himself to negotiate with Cunningham, setting a $15 per acre price for 3,000 acres and $30 per acre for an additional 400-acre parcel. The proposed route for a road intended to connect the commission-owned Bear Mountain property with land donated by Harriman went right through the Cunningham holdings. Ford's offer was accepted and promptly paid for with commission funds, opening the way for the construction of a road later to be named Seven Lakes Drive, one of the most delightfully scenic byways in New York.

May 2, 1913
Mr. S. Gilchrist
Upper Nyack, New York

Dear Sir:

I have your note of the 29th about Dr. Helm's cows. I did not know of any such arrangement which you state was in force last season. I really feel that this coming season it would hardly be just the thing to allow this because the land will be more and more used by the general public and such action might be criticized, unless, of course, Dr. Helm was willing to pay some small amount for the privilege. It may be when Mr. White gets back we can take the matter up with him again.

Yours very truly,
Leonard Hull Smith
Assistant Secretary

May 9th, 1913
Mr. A. M. Herbert
Iona Island, New York

Dear Mr. Herbert:

Mr. Legien tells me of the manner in which he has been paying off some of the laborers. I don't think the arrangement is desirable, from many standpoints. In the first place, he tells me that it is really dangerous to take up the amount of money that we now have each month and pay off possibly two hundred laborers in the small shed which is now in use. In other words, some arrangement must be made at once whereby the man paying off can sit behind either a railing or at a door, so that they cannot be crowded or jostled. Furthermore, I think each man should receive his envelope personally, open it and count it, and if he cannot write, at least make his mark, and the name can be filled in by someone else, as is now the practice.

Yours very truly
Leonard Hull Smith
Assistant Secretary

May 10, 1913
Mr. Chester C. Platt
Secretary to the Governor
Albany, New York

My dear Mr. Platt:

I am exceedingly anxious that the Governor sign a bill that is now before him and about which I telephoned him the other day, that gives the Palisades Commission the right to accept or reject any award which may be made to purchase property which the Commission is condemning and may condemn along the west bank of the Hudson. This bill was introduced by Senator Blauvelt and I have just heard that the Governor may be inclined to hold it up or veto it because it is a Blauvelt bill. Of course, I am sure the Governor would not be actuated by any motive of this kind if the bill is all right. If there is the slightest doubt remaining in the Governor's mind as to whether he should sign it, won't you please telegraph me at once, because in that event I will come to Albany on the first train possible and see the Governor about it.

Sincerely yours,
George W. Perkins

Under the guidance of Welch, work at Bear Mountain was in full swing, accomplished by laborers being paid a daily wage of $2.00. Under the supervision of Andrew M. Herbert, the commission's superintendent, the labor force came primarily from New York City. Within months, the force expanded to 563 foremen, firemen, mechanics, drill runners, carpenters, helpers, and laborers. An additional eighty-six woodcutters were clearing the forest of dead trees and underbrush. Each day, the men would embark on the ferry to Weehawken, New Jersey, then board a West Shore Railroad Company train for the forty-five-mile ride to a special Bear Mountain stop arranged by the commission with the railroad owners. Ticket costs came from the pockets of the workers. Upon arrival at 7:00 A.M., this small army set to its tasks. Land was cleared, and roads were being built. Baseball diamonds, lawn tennis courts, a running track, and a huge lawn were under construction.

The commissioners and Welch also were grappling with the question of how to deal with "refreshment concessions." The question was how many commercial enterprises and for what purpose. Welch wanted to encourage competition and bids, but control prices and maintain quality.

Any concessionaire who failed to live up to the commission's standards risked summary eviction, with no right of appeal. One vendor proposed to provide chewing-gum machines and weighing scales for visitors. "No thanks," said Welch. Sandwiches being sold by F. D. Lockwood for ten cents were judged to be "too thin and lacked ham," and glasses of milk, sold for five cents, were "too small." The PIPC was quick to learn that some park concessionaires, assuming that the customers they served were day-visitors likely never to be seen again, put high priority on profit and less attention on service. Correction was demanded. Lockwood's sandwiches gained weight and substance, and his milk glasses increased in size.

The McAllister Steamboat Company was planning summer excursions from the Battery in New York City to Bear Mountain for $.50 per passenger, round-trip, with a rule imposed by the commission that no intoxicating beverages could be taken ashore. This restriction did not include intoxicated passengers. To accommodate the steamboat passengers, a tunnel was dug under the railroad tracks, allowing easy pedestrian access from the dock to the playfields and refreshment stands on the plateau above. Beginning with the first trip, the steamboat carried 600 to 800 people daily on excursions up the Hudson River from the city. In the minds of the commissioners and Welch, what some people had called a "worthless wilderness" at Bear Mountain was being transformed into a park. But not everyone approved.

Newburgh Daily News,

Governor Sulver's object in appointing a "commission" to investigate the Palisades Interstate Park Commission is tolerably plain. He wants to "reorganize" the commission, the members of which have been reluctant to rise up and declare him the anointed one, the fount of wisdom and the oracle of the people. Sulzer is convinced a lot of perfectly good money is not being distributed in patronage in a commission that should be responsible to political exigencies. Fully half the funding the commission possesses was subscribed by individuals, some of whom, including Chairman George W. Perkins, are members of the commission. No "investigation" is required to establish the fact that commissioners who are spending funds which they in part subscribed are not squandering them.

Publicly, Governor Sulver explained that his call for an investigation centered on his impression that the commissioners, "who were very, very busy men of large private interests," had not been giving enough personal attention to commission activities, but, privately, land owners had complained to the governor that the commissioners were less than generous in their offers to buy land. Perkins hotly countered in the *New York Times* that the commission did not turn over its executive responsibilities to subordinates, and that "no lands have been bought or condemned that are not absolutely necessary to the preservation of the Palisades."

While the commissioners were struggling with Sulzer's reproach, an even more complex quandary faced them. Miss May Milne, an unmarried, unescorted woman, wanted to camp at Bear Mountain! Then, on top of this startling request, Mrs. Albert W. Staub, wife of the headmaster of Riverdale Country School, proposed a camping trip for herself and ten Campfire girls! The commissioners had an answer for these unprecedented requests: women and girls could camp at the old rifle range on South Mountain, twenty-five miles downriver from Bear Mountain. To their credit, however, the commissioners did take a tiny step toward gender equality. When campgrounds first opened in Harriman Park, the commission rule was that wives could accompany their spouses, so long as they separated from their husbands on arrival and stayed in a women-only section of the campground. Single women need not apply.

Group camping activities reflected the gender dilemma. The Young Women's Christian Association (YWCA) was given primary use of the old rifle range at Blauvelt Park, and a large central camp for Boy Scouts was established in 1913 at "Car Pond" in Harriman Park. As with other initiatives, Perkins, his colleagues, and Welch could only guess at the eventual scale, diversity, and benefits of the group-camp program. Most rural parks are accessible to a mobile population with money to spend, private transportation, and time for leisurely travel. City parks protect small patches of green surrounded by the machines and structures of humankind. The transition between city and rural parks was and still is rare for lower-income urban residents. However, the commission sat on the doorstep of the nation's most densely populated metropolitan region while controlling an increasingly vast swath of wildland. Welch was determined to make a functioning connection between the two.

For many inner-city children, the commission's parklands afforded a first opportunity to step away, even for a moment, from the crash and contest of the city to hike along forest paths, climb to scenic vistas, learn about trees, plants, wild birds and animals, swim in lakes, invent toyless games, watch campfires blaze, listen to strange night sounds, and soak in the experience of nature. The YWCA and Boy Scout camps were a start, definitely not a finish line.

But one contest did end. Perkins was able to write to John D. Rockefeller Jr. that "after a very long hard struggle," and thirteen years after the Carpenter brothers' operation was shut down, the Clinton Point and Rockland Lake Trap Rock quarries at Hook Mountain were acquired, resulting in final victory in the quarry war.

As the dynamite smoke and rock dust disappeared, the commissioners found themselves explaining and justifying purposes and legal prerogatives to auditors, government appointees, and elected officials from the two states who frowned on the quasi-independence of the commission. Government officials in Albany and Trenton much preferred a more familiar style of pursuing the public's business, one in which political power, patronage, and budget priorities were usually molded to the next election and against the "other" party. The nonpartisan commission, a private-public hybrid that did not fit easily within preconceived notions of political privilege and leverage, was grudgingly tolerated, but not applauded.

The obvious antidote for criticism was success, and the first summer of visitation at Bear Mountain confirmed public enthusiasm at such a high level that the commissioners began to consider the purchase of a steamship to provide even more access, an idea especially appealing to the nautically minded Perkins.

One vessel that caught Perkins's eye was the steamship *United States*, moored in Michigan City, Indiana, and offered for sale for $175,000. The sidewheeler could carry at least 2,000 passengers. At the waterline, the vessel measured 208 feet in length. Its beam was forty-eight feet and its draft sixteen feet. A 2,500-horsepower engine provided propulsion. But the commissioners hesitated, not quite ready to begin creating a commission navy. At their annual meeting in September 1914, hosted by Perkins at his Wave Hill Estate in Riverdale, New York, they agreed, instead, to begin the construction at Bear Mountain of a "large restaurant" patterned after the Old Faithful Inn in Yellowstone National Park.

At the same time, a sense of neighborly cooperation developed between the U.S. Army and the commission. This loose alliance started on a shaky note when the commissioners pressed to gain a right-of-way through adjoining land at the Military Academy, West Point, to allow for road access along the base of dramatically scenic Storm King Mountain. Military officers were not keen on the idea, but while the right-of-way issue was under debate, J. DuPratt White was reminded by H. Percy Silver, the West Point chaplain, that while officers had their privileges, enlisted men and their families at the academy had few options for summer holidays. In response, the commission arranged for a special summer camp that accommodated 300 soldiers, their wives, and children. Agreement for the road project was reached soon after. The result is the Storm King Highway, a three-mile road connection between the villages of Highlands and Cornwall-on-Hudson that is an engineering and scenic marvel. William Welch oversaw construction. Today, the Storm King Highway is listed on the National Register of Historic Places.

In October 1914, with development of the Bear Mountain area accelerating, the commissioners agreed to accept an offer from Henry V. Gilbert, representing the heirs of John S. Gilbert, to buy the "Fort Lot" at the village of Fort Montgomery for $1,500. The site had quietly slipped into obscurity, shrouded by a canopy of trees and dense brush that made the old fortification all but invisible. One hundred thirty-six years after the "battle of the Clintons," few people were aware of or gave thought to the "twin forts" or to the militiamen, soldiers, and mercenaries who died there. Fort Clinton had come into commission ownership with the acquisition of the prison site at Bear Mountain, but the Fort Montgomery site, lost in the thick woods on the northern side of Popolopen Creek and held in private ownership for so many years, might have slipped away entirely from public attention except for the generous offer made by Henry Gilbert. The commissioners promptly accepted.

Architects from the firm Tooker and Marsh were retained to design the "large restaurant." Welch would supervise construction. The cost for construction was estimated at $95,000, but Perkins and Welch thought the cost could be scaled back to about $65,000 if certain efficiencies could be achieved, chief among them the avoidance of government-contract procedures. Perkins was leery of asking for state funds to cover the cost, knowing that delays and cost increases would probably result. He wanted the building up in a matter of months, each dollar squeezed to the maximum. Perkins turned to his fellow commissioners and won their unanimous approval for the use of privately donated gift funds, largely his own, to construct the Bear Mountain Inn.

There is no better symbol of Perkins's commitment than the Bear Mountain Inn, ultimately described by the commission as "a rugged heap of boulders and huge chestnut logs assembled by the hand of man, and yet following lines of such natural proportions as to resemble the eternal hills themselves." In late 1914 and early 1915, Welch led the commission's work force in a rapid construction effort that resulted in a three-story art piece

Bear Mountain Inn (PIPC Archives)

of more than 50,000 square feet completed at a bargain price of about $100,000. Perkins took criticism for avoiding contracts and union wages, but the result is magnificent. The Bear Mountain Inn was officially opened on June 3, 1915, and accommodated thousands of park visitors during its first summer season of operation, returning a profit to the commission of $600. The stone-and-log appearance is today recognized and celebrated as classic "park architecture," presenting design elements that have influenced similar projects in many state and national parks.

Positive news at Bear Mountain was not necessarily echoed on the New Jersey side of the state line. Challenged by the Appalachian Mountain Club about the pending construction of the Henry Hudson Drive under the Palisades cliffs, feared by club members to become an "ugly gash in the landscape," Perkins was quoted in the *New York Times* as saying, "We are not building any automobile boulevard along the shore." The *Times* headline proclaimed, "No Road for Palisades." This came as a surprise to Commissioner Sutro, who had been working hard for two years to convince the New Jersey legislature that the Henry Hudson Drive was a good idea.

Perkins quickly corrected the record by explaining that "a great many people seem to think that the commission has been determined to make out of this drive a great broad boulevard near the water and constructed in such a way as to almost obliterate many of the trees and make a long disfiguring scar through the Palisades." Referring to the commission's plan to carefully avoid visible roadway impact on the shoreline, Perkins added, "We have among ourselves been referring lately to the Henry Hudson Drive as an automobile trail." Perkins and Sutro promptly settled their misunderstanding. In a follow-up letter to Sutro, the commission's president said, "Whatever criticisms that can be made of those of us who are on the commission, there is one thing certain, that no halfway fair man could say we are not all desperately in earnest in our work and take everything that occurs in connection with it very much to heart."

A witness to the activity at Bear Mountain was twenty-three-year-old W. Averell Harriman, recently graduated from Yale, who officially attended his first meeting on April 7, 1915, as a newly appointed commissioner. The gift of land and money provided by his mother five years earlier had prompted the young man to pay attention to the commission's activities. Now, as a duly appointed commissioner, he would remain consistently involved for the next fifty-nine years. Harriman also would become known as the "durable Democrat." His credentials eventually would include service as ambassador to Russia in the Franklin Delano Roosevelt administration and ambassador to Great Britain in the Harry S. Truman administration. Through the years, Harriman came to know Joseph Stalin, Winston Churchill, Charles de Gaulle, and a host of world leaders. He was elected governor of New York in 1954 and twice contended unsuccessfully for the Democratic nomination to run for president of the United States.

Throughout his career, Harriman always viewed himself as a "volunteer," using the railroad wealth left by his parents to maintain personal independence and pursue goals of interest to him. No other allegiance matched his commitment to the commission. When he was elected New York governor, Harriman appointed his brother, Roland, to take his place on the commission. In the 1958 election, the Republican candidate, Nelson A. Rockefeller, prevailed over Harriman, prompting a bit of family interlacing at the commission. Roland Harriman graciously resigned so that the newly elected governor could reappoint Averell. In so doing, Averell Harriman rejoined his fellow commissioner and the governor's brother, Laurance S. Rockefeller. Both served together collegially for many years.

Before Harriman completed his work with the commission, he would donate an additional 12,000 acres to Harriman State Park. His constant allegiance and diplomacy brought stature and strength to the commission for more than five decades.

But that was in the future. In 1915, Harriman was sworn in with another new appointee, John J. Voorhees, son of the former governor of New Jersey. Harriman and Voorhees arrived just in time to hear a proposal from Mr. R. G. Hazard, president of the New England Bridge and Railway Company, who urged the commission to accept a seventy-five-acre gift of land in exchange for a right-of-way that would allow a railroad tunnel to be dug through Bear Mountain. The railroad would be connected to a bridge across the Hudson River, requiring a massive concrete anchor point in the park. The commissioners, including the two new appointees, were quick to decline this offer.

In New Jersey, the increasing popularity of the river shoreline was providing another challenge. Visitors were flocking to the small beaches, docks, and riverside campsites in increasing numbers. More law-enforcement patrols were scheduled, but few benchmarks existed to guide the unarmed patrolmen. Like so many of its activities, the PIPC found itself inventing more rules and regulations to control park use. The patrolmen watched for unattended campfires, directed that camping permits be visibly affixed to tents, and ordered that rubbish be either burned or buried before campers left. Camping permits were issued for a minimum of one week for a fee of $1.00. The commissioners reminded patrolmen that "severely injured persons must be taken to the nearest hospital as soon as possible." A commission regulation directed that "bathers in bathing suits shall not be allowed to parade around amongst camping and picnicking parties, but must keep to the beaches, roads, and paths." Patrolmen were advised to keep freshwater springs from becoming polluted, to prevent boulders from being rolled from the summit of the cliffs onto the crowds below, and to enforce the rule that "all boy scout parties shall be required to leave their axes with patrolmen on entering the park." "Patrolmen will be allowed one day off every two weeks, the day to be designated by the Captain," said a personnel rule.

The commissioners discovered, too, that the New Jersey Palisades were attracting "moving picture companies." Requiring that the filmmakers have permits, the commissioners gave patrolmen discretion to stop or delay any filmmaking that might interfere with the public's enjoyment of the park. Where public use was not encumbered, cameras cranked while cars plunged from the Palisades summit, and damsels clung to the cliffs by their fingertips while villains stalked and sneered. Thus, introduced into the national lexicon was the phrase "cliff hanger."

The commission was branching out in many directions, all aimed at creating "the right sort of healthful playground for the people," as Perkins explained in a letter to Rockefeller Jr. In response to this letter, on February 19, 1916, the commission received a $1 million donation from the Rockefeller Foundation. Perkins and his commissioner colleagues could take credit for raising more than $4.7 million in private funds over the first years of activity. In today's dollars, the amount raised would exceed $100 million.

Momentum was everywhere. "If you love me," Perkins pleaded with Welch, "do get the steamboat business straightened out next week. I am literally snowed under with work and it is going to be next to impossible for me to go into the differences between the McAllisters and the Albany Day Line. Please get after this and let me know as soon as possible in the fewest possible words what can be done to put us in the best possible shape for this year's travel." In 1914 almost 115,000 people arrived at Bear Mountain aboard regularly scheduled dayliner trips, and an additional 15,500 came by special chartered excursions. Palisades-bound passengers coming over from the city on the Dyckman Street Ferry numbered 199,973 in 1915. The inadequate docking facilities at Bear Mountain were already congested, and more boats would soon be underway. One of these was the *Palisades*, built for the commission in 1916 by the Mathis Yacht Building Company of Camden, New Jersey. This new 120-foot-long vessel was intended to provide free outings to Bear Mountain for poor mothers and children in care of social agencies. "So far as we know," reported Welch to Perkins, "the *Palisades* is the first passenger boat in America using diesel-type engines."

The commission continued its initiative to bring underprivileged children to Harriman State Park. A camp was established to provide for about 425 "poor boys," one hundred of whom were segregated for a nutritional experiment. Under the direction of the New York Association for Improving the Condition of the Poor, the study resulted in an average weight gain for each well-fed child of 3.5 pounds. Welch also reported that 2,369 young women had camped at Blauvelt Park, adding that, otherwise, "no record" of women was to be found in the general camping population, but by 1916 a camp that could accommodate twenty-four girls and two chaperones was established at Bear Mountain, shattering the tent ceiling.

Presaging demands made on park managers everywhere was the insistence of hunters that they be granted access to the commission's lands in New Jersey to kill foxes. Their claim was that these predators were a menace to birds and other small animals. Setting an early standard on this troubling question, the commissioners rejected the request. Another use of parkland was welcomed. Herbert Whyte, representing the Outing Publishing Company, asked for permission in January 1916 to take a two-week "walking trip" through commission lands. The vanguard of "trampers" represented by Whyte and his peers would help create a generally firm and friendly alliance between hikers and the commission.

Unlike the hikers, landscape architect Nathan T. Barrett, a charter appointee to the commission who served until 1914, saw the natural terrain through different eyes. After he left his post as a commissioner, Barrett recommended to Perkins that equipment acquired with the quarries be used to drill a huge tunnel from Rockland Lake through Hook Mountain. His concept was that water flowing through the tunnel from the lake would tumble down a precipice into the Hudson River, creating an artificial 160-foot-tall waterfall. The commissioners declined.

But in 1916 other commissioner-approved projects were blossoming. One example was the sale of "soft drinks" of sugar-flavored Bear Mountain spring water, produced and bottled by the commission. In Harriman State Park, several dams were under construction to "enhance" existing lakes and ponds. Just north of Bear Mountain, a six-hundred-foot-long steel bridge was constructed across the deep Popolopen Creek gorge to greatly improve access for automobiles and horse carriages to the military academy at West Point and destinations farther upriver. The Bear Mountain Inn was expanded, reaching a size of 81,000 square feet. Far out in Harriman State Park, group camps continued to expand, but one camp was quarantined when two members of a Boy Scout troop who had been swimming in a nearby lake were diagnosed with infantile paralysis (polio). (Years later, Franklin Delano Roosevelt would contract polio at age thirty-nine, likely caused by swimming in a Harriman State Park lake.)

Land acquisition remained high on the commissioners' priority list. With the acquisition of the "Queensboro" property between Bear Mountain and Harriman parks, the commission added 5,000 more acres to its holdings. In New Jersey, $100,300 was spent to acquire the 113-acre Anderson Avenue Realty Company's property known as Greenbrook Park on top of the cliffs, confirming that the commission was taking a strong, yet publicly undeclared, initiative to control the heights as well as the shoreline of the Palisades. The commission now owned almost 20,000 acres.

While most parks around the nation were counting visitation in the hundreds, more than 500,000 annual visitors were journeying to the commission's park system. This fact did not go unnoticed by Stephen Tyng Mather in Washington, D.C. Mather was just succeeding with the formation of the National Park Service and invited William Welch to speak at the first "National Parks Conference" in 1917. Enos Mills from Colorado, who introduced Welch to those in attendance, said, "The subject of discussion this morning is the recreational use of the national parks. The first speaker on this subject is Mr. W. A. Welch, who is Chief Engineer in charge of the Palisades Interstate Park. As some of you know, we will say that this park embraces 30,000 acres of land. Altogether, $8 million have been expended in this park. More than one-half of this entire sum has been donated privately."

Referring more specifically to Welch's work, Mills added, "The work that this man is doing is really evolutionary and revolutionary." Welch responded by describing himself as a "little kitten" among the "lions" and went on to describe the array of opportunities and challenges encountered by the commission. He offered real-world examples of solutions to park-management problems learned in the school of hard knocks. At the conclusion of the event, the "Supervisors of the National Parks" traveled to New York at the invitation of Welch and Henry Fairfield Osborn, trustee of the American Museum of Natural History, to tour the Palisades Park system.

In his conference remarks, Welch gave strong credit to the New Jersey State Federation of Women's Clubs for its dauntless role in establishing the Interstate Park Commission, but still lost within the pace of the activities was the women's memorial project. Only a small plot of land had been set aside for the memorial at a scenic location atop the cliffs. Almost two decades had passed. "Thus, alas, for the best laid plans of women, the park is a reality, the Palisades are preserved for ever and ever, but the tall tower to commemorate the work of the women is still in the air," stated a 1917 Women's Club report. "According to Miss Demarest, the money is still in the Knickerbocker Trust Company of New York and amounts to $3,500. So, although the club women of New Jersey . . . virtually saved the Palisades by bringing about the first interstate movement, their own project of a permanent monument is unfulfilled."

Perkins and the commissioners were meeting every month, communicating with one another and Welch in between, and dealing with a growing public clientele and critics and partisans alike. Somehow, the women's memorial kept being pushed down the agenda. This was due in part to the constant struggle for state funding. Perkins was quick to point out to anyone who would listen that the breakdown of funding consisted of 54 percent from private sources, 40 percent from New York, and 6 percent from New Jersey. Welch was distracted from the memorial project by letters of inquiry, now arriving by the sacksful, from park enthusiasts from all over the country asking for information ranging from what type of toilets to use in campgrounds to why parks were even necessary.

The biggest distraction was just around the corner. The *New York Times* announced that Norman Selby, better known as the popular prizefighter and role model "Kid McCoy," had enlisted in Company K of the Seventy-First Regiment to begin military training. In 1917, England, France, and

Italy were already at war with Germany. The commissioners encouraged the staff to enlist for military training and agreed, as a contingency, that female waiters could replace male waiters at the Bear Mountain Inn, if it came to that.

Rockefeller Jr., impressed by the continuing success of the commission, wrote to his father in April 1917, proposing that a fund of $15 million to $25 million be established for the purpose of "developing parks and playgrounds in the City of New York and also in the State." He also suggested that a portion of the fund be used to "build a bridge across the Hudson River connecting the Interstate Palisades Park with Riverside Drive" at an estimated cost of $10 million. "I should like to see you build this bridge yourself," Rockefeller Jr. said to his father. The senior Rockefeller was attentive to his son and seemingly not startled by the proposition that he personally bankroll the construction of a huge bridge where the present-day George Washington Bridge exists. War clouds, though, put this initiative on hold.

From his distant perch at Long's Peak Inn in Colorado, naturalist and author Enos Mills was continuing to push his national park agenda. Writing to Welch, he said, "It is probable that a vigorous five-thousand-word account of your Palisades Inter-state Park, by me, will appear in the *Saturday Evening Post*. As I told yourself and, I think, Mr. Perkins last winter, our National Park Service is not a model by any means. The inclination of those in power is to play politics and no one more than yourself knows what this means."

By midsummer 1918, Perkins was able to report to Rockefeller Jr. that the interstate-park model that Mills found so appealing was increasingly popular. "More people visited Bear Mountain than ever before." On one day, Perkins said, "six large steamers landed passengers at our docks, and a number of people came by way of the West Shore Railroad. Something over a thousand automobiles stopped at Bear Mountain." So many people were coming that the American Scenic and Historic Preservation Society speculated that the PIPC was having a difficult problem "reconciling its desire to preserve as nearly as possible the natural beauty of the tract and its desire to make the park practically useful to the people."

Problem or not, the inn and facilities at Bear Mountain were being given over to increasingly urgent preparations for the United States entry into World War I. The 102nd Company, U.S. Marines, was housed in a dormitory formerly occupied by park workers. Their mission was to guard an ammunition depot developed by the U.S. Navy on Iona Island. This glorious island is only about one-half mile downriver from the Bear Mountain dock. Hikers who reach a point just below the summit of Bear Mountain are rewarded with one of the most picture-perfect river scenes anywhere, letting their eyes glide southward through the river narrows from West Point, past geologically prominent Anthony's Nose on the opposite side of the river, past Iona Island nestled between the flanks of Bear Mountain and Dunderberg Mountain, then farther downriver to the Tappan Zee, where the Hudson widens again on its southern course to the Atlantic. On clear days, the classic New York City skyline is visible from the Bear Mountain summit.

Iona Island is the centerpiece of this enchanting scene. On its western shoreline, the island is edged by a large 200-acre marsh that traces the path of an ancient horseshoe bend, bypassed when the last glacier straightened and deepened the river channel on the island's east side. Native Americans called the 120-acre island "Manakawaghkin" and left behind archaeological evidence attesting to their use of it for more than 5,000 years.

Through succeeding ownerships and name changes, title to Iona Island eventually came into the hands of John Beveridge in 1849. He promptly sold a part-interest to his son-in-law, Dr. C. W. Grant, a self-styled fruit grower and winemaker. Grant planted orchards and vineyards on the island, taking advantage of the glacial soil and sea breezes from the distant Atlantic that tempered the climate. For two decades, he struggled to develop the "Iona" grape with the intent of successfully competing against French imports. The resultant mediocre taste of the wine did not match Grant's enthusiasm. His business venture failed, but he is credited with the first attempt in the United States to produce marketable wine. New owners transformed the island into a resort and picnic ground, presaging the popularity of Bear Mountain. World heavyweight boxing champion John L. Sullivan was a frequent visitor.

In 1899 the navy arrived. By Act of Congress, the island was purchased for $160,000, primarily because the navy had run out of storage space in New York Harbor. World War I transformed the comparatively small navy depot on the island, and storage capacity reached more than 2 million pounds of explosive black powder, which was being packed into cannon shells and ordinance of all shapes and sizes for the U.S. and British fleets. Submarines sailed up the Hudson, tied up against the Iona Island shore, and were loaded with torpedoes. Just next to the submarines, antisubmarine depth charges intended for the "other guy" were loaded with other ordinance on barges and seagoing transports. This activity was within watery yards of the passage of dayliners full of park visitors traveling to and from the Bear Mountain dock.

Iona Island U.S. Naval Ammunition Depot (PIPC Archives)

One hundred forty-six buildings were constructed on the island, most of them made of brick, some including architectural flourishes a step above military square-and-basic. Ammunition bunkers designed to direct any accidental explosion upward and away were chiseled into the island's topography. This beautiful island carried the official designation U.S. Naval Ammunition Depot.

Wartime demands caused Perkins, while retaining his position as president of the commission, to assume new responsibilities as president of New York's Food Control Commission to prevent war profiteering, encourage bulk purchases of food commodities from upstate farmers, maintain price supports, and urge backyard gardening. His food commission consisted of 130 members. Welch joined the Aviation Section of the Signal Corps and was posted to the Pacific Northwest to oversee the production of spruce-tree lumber for the construction of "aeroplanes." Perkins proudly reported to Rockefeller Sr. that Welch, given the rank of major, had been assigned to such a crucial wartime task. While engaged in what turned out to be a mammoth duty involving legions of men, Welch was visited by National

Park Service director Stephen Mather, who, on a swing west, stopped by to further discuss park management and development. Neither man probably considered that the old rifle range at Blauvelt Park was again being used for military training with the same result: bullets whizzing through nearby residential neighborhoods.

A few successes were being achieved even during the war years of 1917–18. Perkins, his hand always on the commission's tiller, continued to negotiate purchases of land, including the purchase of the Dana property on the summit of the Palisades for $175,000. One observer wrote to Perkins, commenting that "few will ever express their appreciation of what your Commission is doing; one man wants to do so." The letter was signed by Rockefeller Jr.

8

Perkins

GEORGE PERKINS SAILED for Europe in December 1918 even as the Allied forces struggled with the remnants of a destructive war-to-end-all-wars that killed 5 million combatants. During an extended stay in Europe, Perkins involved himself in the general effort to rebuild postwar France and Germany, bringing his special skills and political access to the need to rehabilitate the German economy to ensure that its war debts were paid.

Faraway, in the Pacific Northwest, Maj. William Welch was completing his wartime assignment and was anxious to return to the Palisades Interstate Park, but the director of the Bureau of Aircraft Production was in no hurry to release him. Welch finally returned to the Hudson River Valley in late February 1919, having been recommended by General Brice P. Disque for the Distinguished Service Medal. Welch was cited for directing the construction of thirteen mainline and spur railroads through mountainous terrain "in record time," for building dozens of truck roads and sawmills, for training and supervising thousands of raw recruits and untrained officers, and for increasing the production of airplane lumber by 1,500 percent.

But even as Perkins and Welch focused on the war effort, separated by oceans and continents, the work of the commission remained in their thoughts. Signaling the strain of events, Perkins wrote to Welch from Paris in January 1919, saying, "It is quite evident now that I shall not be able to sail for home nearly as soon as I had expected when I left New York. My stay in the hospital, where I still am, will have consumed two and more nearly three weeks. In addition, I find more to do over here than anticipated." Perkins went on to provide a lengthy list of concerns and ideas for the spring opening of the parks, ranging from the need to improve sanitary conditions and increase drinking-water supplies to the prospect of upgrading rather than enlarging the Harriman Park group camps. Funding, always a concern, would have to depend for the moment, at least, solely on New York and New Jersey appropriations. Perkins, stating the obvious, confirmed to Welch that "this is a bad time to raise new funds."

In a long letter to Commissioner Richard Lindabury, Perkins expressed great concern over continuing political and state funding problems, particularly those being caused by New Jersey's acting governor, Walter E. Edge, who seemed unsympathetic to the commission. Perkins stated to Lindabury that "the commission should not be subject to senatorial patronage or political influence but should be filled by joint action of the two Governors." He was referring to Edge's resistance to the reappointment of Franklin W. Hopkins, who had been serving with the commission since 1900. "I think it should be pointed out to Governor Edge that the Commissioners consider continuity in office of vital importance in the management of this interstate enterprise, especially because it constitutes a great private as well as public trust. Frequent changes in the personnel, especially if those changes would seem to be distracted by partisan or factional considerations, would tend to weaken that confidence."

Theodore Roosevelt had vigorously affirmed this belief, but the great champion of conservation was gone. He died on January 6, 1919. The "lofty ideals and freedom from political influence" that Roosevelt expected of the founding class of commissioners he had appointed in 1900 were proving to be under constant trial. New York governor Al Smith was more responsive to the commission and used a meeting with Governor Edge concerning a New York–New Jersey tunnel project to make the point. Welch, newly returned to his position as chief engineer and general manager, reported to Perkins, still in Paris, that "Governor Edge said (to Governor Smith) that New Jersey was ready to put up her share of the cost. Governor Smith replies, 'of course you will also put up the $600,000 New Jersey owes the Commission which you understand is a debt you owe to New York.' This took Edge's breath away, and then Smith rubbed it in by adding, 'You might begin by reducing your action of a few days ago when you cut out of their appropriation $100,000 for the Henry Hudson Drive, which is several years overdue.' That is the best thing I have heard since my return."

Inspired by the Smith-Edge meeting, commissioners Hopkins (who survived the reappointment challenge and would continue to serve until 1929), Baker, and Sutro took up the challenge. They scheduled a field trip to the

Palisades and Bear Mountain for New Jersey legislators. Message delivered: New Jersey appropriations were forthcoming in the amount of $42,500 for operations and $100,000 for the Henry Hudson Drive. Partial credit for the fieldtrip success was given to Mrs. Camille Welch, who, like many park spouses to follow, provided what all participants agreed was a "very fine lunch."

Dr. Charles C. Adams, of the Syracuse College of Forestry, was unaware of the dialogue and concerns at the top of the commission's organization but was anxious to become more involved in the still novel and expanding park system. In the 1919 edition of the *Empire Forester*, he reported that the commission was interested in an "inventory of natural history resources." He added that it "has not been customary for park officials to see the value of such studies." According to Adams, the study would bring several benefits, including:

The possibility of interesting visitors and campers in birds, fish, game, and other wild animals.

The capability for better protecting visitors from "intrusive" plants and animals.

The possible development of a balance between people and nature.

The opportunity to issue a publication for students of park administration.

Adams pointed out that "miles of rocky, wooded wasteland" were in existence along the lower Hudson River, primarily because of "lagging economic and social ideals." He saw the PIPC's holdings as a fitting testing ground for the budding theories and concepts of park management.

Not all the land along the lower Hudson River was "wild and wasted," as Adams put it. Property on the summit of the Palisades owned by former commissioner William B. Dana had been acquired from his heirs through Perkins's persistent efforts. The acquisition brought with it Greycliff, a massive twenty-six-room mansion built in 1861 of blue trap rock and pink sandstone. Proposals for the use of this elegant Victorian structure quickly followed, including a recommendation by the Federation for the Support of Jewish Philanthropic Societies that the mansion be used as a vacation resort for working girls.

Edward F. Brown, manager of the commission's camp department, was trying to juggle requests for use of the Dana mansion with other camp proposals, including one from Mrs. Victoria Dyke for the establishment of a camp for undernourished girls. At the beginning of 1919, Brown was overseeing nineteen camps, including two for military training, eight for Boy Scouts, two for Girl Scouts, a camp for the Children's Aid Society, one for the Harlem & Heights Business Girls Athletic League, another listed as "The Negro Fresh Air Camp," and a YWCA camp. Children from Brooklyn, the Bronx, and Queens were transported to the scout camps, usually for ten-day adventures. By year's end, thirty-seven camps were in operation, seven more were under construction, and more than 50,000 campers were in attendance, a tenfold increase over prewar numbers.

At Upper Twin Lake in Harriman Park, "Miss Lewis" was in charge of one of the Girl Scout camps. Her demands of Brown and the commission's staff were unending. She wanted twenty-four feet added to the length of the mess hall and thought that the kitchen wall should be moved outward by three feet. Roofs needed attention, the camp needed more cabins, the tents were not satisfactory, and the grounds required better landscaping. "If Miss Lewis is so insistent upon our pruning the grape vines around her camp," commented an exasperated Welch, "she will probably want someone to pick the grapes next year."

In Perkins's absence, J. DuPratt White stepped forward to again help deal with the day-to-day challenges. There were many. White focused significant effort on developing Hook Mountain, prompted by a letter from King suggesting the construction of a bathhouse and lunchroom. White thought the development of Hook Mountain would be a wonderful asset for nearby Nyack, New York, and proceeded to win concurrence from the town's leaders.

Along the rugged shoreline of this mountain, British major John André had met secretly with Benedict Arnold during the Revolutionary War to receive the plans for West Point. André was captured with the plans just before reaching the British lines, convicted as a spy, and forfeited his life on a hanging tree in Tappan, New York. Arnold, until then a hero of the Revolutionary War, escaped to England, leaving behind his sullied name and reputation, ever after associated with the word "traitor." White and his fellow commissioners wanted this story told in a way that would enfold visitors in the event itself by encouraging them to seek out and sense the very place on the river shoreline where Arnold and André met. Educational opportunities for park visitors, both to honor history and introduce city dwellers to the natural world, were beginning to emerge as integral to development and management of the parks.

In the meantime, Welch grappled with other issues. He turned down a request from Abram De Ronde, commissioner from 1900 to 1912, who asked that the commission hire his son. Welch also rejected a suggestion from a taxidermist who thought that stuffed park animals and birds would be a wonderful way to educate the public and make a profit for the commission. Postwar costs for labor and materials were soaring, labor by 60 percent and materials by 100 percent. A copper sulfate treatment of some of the lakes in Harriman Park to get rid of aquatic plants and algae killed fish instead. Welch and Dr. Hugh Baker, dean of the Syracuse College of Forestry, collaborated further to find the right chemical dosage for the lakes.

Problems kept coming. Maurice M. Lefkowitz wrote, "I have been to Bear Mountain twice this season, and at both times was pushed off the dance floor by one of the patrolmen for doing absolutely nothing else but proper and decent dancing. This is not my case alone, but the same incident occurs at least ten or fifteen times during each dance, and the patrolmen do not bother persons dancing the "shimmy" but seem to pick out certain people. They handle the girls as though they were common street walkers. I say this because at both times my partner was a respectable refined girl who felt deeply hurt and ashamed at being roughly pushed off the dance floor."

Welch investigated and learned that Lefkowitz was a ten-time dance champion who could make moves that the patrolmen deemed "improper." Generally, the young dancers and patrolmen maintained an uneasy truce, but a visitor to the Bear Mountain Inn was not so sure they would or should. Writing to Welch, she complained, "As to morals, I realize that it is beyond your power to pacify and check the sexual impulses brought into play in the park. Still, something should be done. When the so-called 'shimmy' proved to be fatal to the morals of the young, the government saw it to be proper and necessary to interfere and stop it. And so, it should be—any place where vice and immorality will flourish should be uprooted. I have seen young boys and girls of fifteen shame those of twenty in the making of love."

Welch's engineering books had not covered this particular subject, and he was getting an extra dose of on-the-job training. Dancing and the shimmy continued at Bear Mountain. There is no record of how Welch handled the issue of lovemaking. In a more serene and appropriate setting, camp manager Brown was able to report to Perkins and Welch the first successful "musical evening in the park," provided by Miss Ruth Linrud, who played a large harp and sang to 500 children at Carr Pond. (The name of this pond changed from Car to Carr and would change again, to Lake Stahahe.)

At the various road entrances, "intelligence officers" were posted to "tell occupants of automobiles that they were entering the public park and that the Commission is trying to preserve its beauty by prohibiting the picking of flowers and destroying shrubs." Steamboat companies were vying for the river traffic, including the Hudson River Day Line, the Central Hudson Steamboat Company, Steamboat Agent C. T. Mallory, and the McAllister Steamboat Company.

Intertwined with daily problems, two threatening political missiles were fired at the commission. The first came from residents of the town of Fort Montgomery who heard rumors that the commission intended to condemn and wipe out their whole village. They sent an angry petition to Albany. The commissioners hastily reported to Governor Smith's counsel that only four properties in the town were being acquired, two through eminent-domain settlements agreed to by the owners and two from willing sellers.

Just as this matter was resolved, the second missile, in the form of the "Peck Bill," came whizzing in. Somehow, New York assemblyman Peck managed to move through the legislature a bill that directed the commission to transfer to Conger family heirs a rock-quarry property acquired from the erstwhile senior Conger for $368,500. Within the commission, shock and surprise were understatements. On its face, the bill seemed illogical and averse to any rational financial doctrine of public trust, yet it had won majority votes in both the New York Assembly and Senate. It also carried with it a potentially damaging precedent. If the acquisition of the Conger property could be overturned, so, too, could scores of other acquisitions. The whole effort to build a park system would come tumbling down. Led by White, the commissioners scrambled to make their case directly to Governor Smith, who promptly vetoed the bill.

In New Jersey, the dangling matter of the women's memorial surfaced once more.

Miss S. Elizabeth Demarest
Passaic, New Jersey

Dear Miss Demarest:

Your letter is very interesting and after reading it, I feel quite sure that you and your associates will be able to raise sufficient money to carry out the monument project. Of course, it should be approached in a rather business-like way—You will understand that all this is very tentative and I am not in any way presuming to force any particular sketch for the acceptance of the women who are to erect the monument, and I do not believe that the Commission will, but simply render such assistance as we can.

Yours sincerely,
J. DuPratt White

Elizabeth Demarest had been secretary for the New Jersey Federation of Women's Clubs since the mid-1890s. Representing many colleagues, some now gone, who fought so skillfully to save the Palisades, she was still trying to win one more victory: the realization of the long-sought women's memorial. The $3,000 memorial donation made to the commission by the women in 1909 had accumulated interest and stood at about $3,800, but inflation had boosted the projected cost for the memorial to $10,000.

Despite his largely noncommittal letter to Demarest, White was a strong behind-the-scenes advocate for the women. He asked a personal friend, architect Henry G. Emery, to sketch a design for the memorial. White posted the sketch on the wall of the commission's meeting room at 61 Broadway in New York City to ensure that the matter would remain visible, figuratively and literally. A committee consisting of White, Charles W. Baker, and Edward L. Partridge was appointed to meet with the Ladies of the Memorial Park Association to pursue the monument.

Perkins returned from France in April 1919. Two months later, the *State Service Magazine* published his article "Sight-Seeing Buses in the Palisades." The lead paragraph captured a slice of the obvious:

> The so-called sight-seeing trip is a by-product of the early development of the automobile industry. Formerly these trips consisted chiefly of excursions into the slum districts, where misery was the object of morbid curiosity, just as the wretched cripples of the eighteenth century were exhibited at circuses for the delectation of the mob. Not infrequently in the metropolitan cities, sight-seeing explorations degenerated into trips to questionable resorts where vice careered unafraid. Gradually there came into existence here and throughout the country the sight-seeing trips which constitute a source of enjoyment and education.

Perkins went on to describe the delights of a sight-seeing trip to the park system, assuring safety for the travelers and emphasizing the social values that the system provided. The cover photo for the article showed three women standing on the Palisades cliffs, with the Englewood boat basin far below.

Learning, too, of the demands and growing tensions between various group-camp directors on the commission's staff, Perkins, who referred to the staff as the "managing force," summoned the combatants to a dinner (in business dress) in a little log cabin perched incongruously on the roof of the Abercrombie & Fitch Building in midtown New York City. The cabin was symbolic of the outdoor equipment and clothing business in which

Abercrombie & Fitch specialized. This unusual setting produced the desired result. The camp directors became more understanding and tolerant of the workload carried by Welch and his men, and vice versa.

The cabin was symbolic for another reason. Welch, who continued to search for the right equation in balancing park uses, turned down a request to build a large number of rental cabins in the parklands. His rationale was that the commissioners "always agreed that such a procedure would not be in line with the general ideals which we are following." Welch contended that the parks should be "enjoyed by all people" and favored rustic camps over more exclusive resort-style developments.

Welch's belief was justified, but naïve. The national parks in particular would quickly become vulnerable to commercial resort development, a predictable outcome driven by what Perkins termed "the profit motive concessionaires." He and Welch knew that parks would rapidly become hot commercial properties, confirmed by visitation that leaped from a handful, to hundreds, to hundreds of thousands, and, in the commission's park system in 1919, to more than 1 million for the first time.

The commission's experiences and crystal ball were increasingly in demand. Welch was invited to be a principal speaker at a national City Planning Conference, held in Niagara Falls, New York. Following his presentation, he received inquiries from officials in Cleveland and Boston who were eager to receive additional information about park management and to arrange for field trips to the commission's parks. Another attentive listener at the conference was Frederick Law Olmsted, son of the famed landscape architect. On Olmsted Brothers letterhead, he wrote, "I did not have opportunity after your talk at Niagara to express my thanks for your presentation of the importance, scale, and great value of the undertakings of the Commission." Olmsted, too, proposed a field visit to the parks.

John D. Rockefeller Jr. continued to keep a careful eye on the commission, and he liked what he saw. In a letter to his father in June 1919, Rockefeller Jr. proposed that the Laura Spelman Rockefeller Memorial Fund provide $500,000 for the purchase and refurbishment of the steamboats *Clermont* and *Onteora*. Making the financial argument, he pointed out that the two vessels were available and could be purchased for $400,000. Another $100,000 would be required to refit them. By comparison, he estimated that the construction of new steamboats of similar size would cost $2 million. The commission's operation of the steamboats "would be a service rendered directly to the people of the city in large numbers and would have an intimate and personal aspect which would peculiarly commend it."

In Perkins's tradition of hard bargaining, the *Clermont* and *Onteora* were acquired for $340,000.

On shore, the PIPC was busy improving its park services. The *Messenger of Haverstraw* reported, "Mothers especially throughout the state will be interested to learn about the provisions made for their comfort and enjoyment at Bear Mountain. One of the features is the mothers' rest stations maintained by the Red Cross. Here, numerous comfortable baskets are provided with covers, where babies may rest in cool, quiet spots under the kindly eye of women matrons. There is no charge, and the theory is that the mother derives a double benefit of having her children cared for in pleasant surroundings, while she herself may enjoy a brief respite from the duties of motherhood."

A tourist suggested that the commission should post multilingual signs to inform non-English-speaking visitors of park rules. He had witnessed a visitor "wearing an undershirt, and nothing else." The visitor was "at the edge of the water rinsing a bathing suit. Every time he leaned over to put the suit in the water, he exposed the main parts of the back side of his body, and when he turned around, the front part was not hidden. I believe some visitors may not understand our customs or be able to read English." Welch referred the matter to the sign committee.

Welch continued to field questions from distant places. The Texas state highway commissioner asked how the commission acquired its parks. "Did you have to buy them? If so, where did the money come from?" Welch responded by confirming that the commission now owned more than 35,000 acres of land and explained the details of the financial public-private partnership that fueled the park engine.

Through his friend Enos Mills, Welch was alerted to a pending visit by Mrs. D. A. "Mother" Curry, owner of Camp Curry in Yosemite National Park. She decided to make the long journey from California specifically to learn about the Palisades parks. Welch was also making plans at the invitation of Stephen Mather to attend the Superintendents of the National Parks Conference, scheduled to be held in Denver in November 1919. Perkins, too, had been invited but could not attend. At the conference, Welch spoke and answered questions covering many topics, including patrol services and enforcement of regulations, publicity, maps, education, museums, preservation of forests and flowers, concessions, signs, sanitation, and individual and group camping. In his report about the conference to Perkins, Welch said, "We have gone so far ahead of all the other parks in our development that they so fully recognized this that they kept me continually answering questions and discussing these problems."

Following the conference, Mather helped arrange for the governor of South Dakota to visit Perkins and Welch for the purpose of gathering information to help manage the newly created Custer State Park near Rapid City.

The PIPC's reputation may have been gaining stature on the national stage, but close to home the trip was bumpy.

August 30, 1919
My dear Mr. Perkins: The terrible Mike McCabe has been vanquished. Between Judge Arnold of the Knickerbocker Press and Mike McCabe of the Haverstraw Times, it has always been difficult for me to know who was the worst enemy of the park.

The letter was signed by camp manager Brown. McCabe had been editorially in favor of the quarry operators during the fight to save Hook Mountain and had been a persistent critic ever since. Brown took a leap of faith and invited McCabe for a tour of Harriman Park, something the opinionated newsman had not bothered to undertake on his own. The tour included lunch at the Bear Mountain Inn, where Brown and McCabe were when Welch and the inn manager walked in and sat at a nearby table. McCabe commented about the "desperate characters" of the two men, then left with Brown for the rest of the tour.

At the end of the day, he said goodbye to Brown, not giving a hint about his reactions to the many features of the park he had seen, including several of the camps. Two days later, under the headline "Making Others Happy," McCabe published a long, four-column, frontpage article in the *Rockland County Times* singing the praises of the commission, Perkins, Welch, Brown, the camp program, and park activities in general. "While from a literary standpoint the thing is atrocious, the fact is that the terrible Mike is vanquished, and we have one more friend, and a powerful one," Brown claimed.

The Ramapo Mountains Water, Power, & Service Company, a utility company operating along the western border of commission holdings, was not a friend. Its directors had decided to condemn lands owned by the commission and take them into water-company ownership. The commissioners took the case to the Court of Appeals, winning unanimous judgment that the water company had no legal right to condemn commission land. This was only one of the many instances when the commission would find itself in court defending its purposes and legal prerogatives.

As the postwar year 1919 came to a close, the commissioners could look back over their brief history and count $13,096,903 in expenditures, including $4,712,644 in cash donations and $1,862,765 in donations of land. Fifty-one charitable organizations were active in the park system. Despite surprises, demands, and crises, the commission was beginning to take on a certain maturity. It had no equal in service to park visitors anywhere in the nation.

The National Geographic Society, just getting into the publishing business, was planning to devote an early issue to the national parks and asked for photographs from Welch, who willingly complied. The commission/national park connection became apparent when the commissioners agreed to accept seventy elk from Yellowstone National Park in an effort to help save the remnants of a herd of 15,000 that had been gunned down by hunters and poachers until only 750 animals remained. Disease conceivably could wipe out these few remaining survivors. To guard against this potential calamity, a sanctuary for some of the herd had to be found far enough away to guarantee that an epidemic in one place would not reach the other.

The possibility of extinction of the Yellowstone herd prompted the commissioners to act on Welch's recommendation. A 500-acre fenced field was established near Arden, the Harriman estate, to accommodate the elk. Historically, elk had ranged from Canada down through the Adirondacks into the Hudson River Valley. Arrangements for the shipment were made through Horace M. Albright, Yellowstone's superintendent. The elk, exhausted and petrified, were released into their new fenced home the day after Christmas, 1919.

As the year ended, winter snows slowed the pace of business. Perkins planned to leave at the end of January for a well-deserved rest in Florida. He planned, too, to immediately begin preparations for a new $5 million fundraising campaign, to be launched on his return to New York in the spring. Before leaving for Florida, Perkins found himself once again in a familiar posture, fighting to restore a funding cut by New Jersey. Acting governor Edge had stepped down at the conclusion of the war in favor of Governor William Runyan. Runyan served only one year but managed during that brief time to cut the commission budget, including the funds for the on-again, off-again Henry Hudson Drive.

Now Perkins and Commissioner Sutro were meeting with Runyan's successor, Governor Edward Edwards, to explain, justify, and defend the commission's purposes and track record. The pattern had become all too familiar. Every time a political change occurred in Trenton or Albany the commission had to reinvent itself. An interstate agency with a private bank account and ownership of thousands of acres of land did not readily fit on the typical government organization chart. New governors bring with them a bevy of new and eager political appointees who are in a hurry to establish their own priorities and styles of control. New arrivals are often suspicious, even resentful, of those who might have had allegiance to a previous political administration, are deemed too independent, or enjoy legal authority that is not automatically overturned by election results.

The legal structure of the commission was not easily understood and stood beyond the immediate grasp of newly arrived occupants in the hallways of government. Perkins and his commissioner colleagues always maintained a stature that allowed them direct access to the governors, regardless of partisan instincts lower down in the ranks, but, increasingly, it seemed that the purpose of these visits with the governors was to seek a cure for the latest bruises of the budget hammer rather than to discuss the commission's social and environmental purposes and achievements. If appointees and elected officials could not corral the commission, they could hamper it with the budget. They did and would.

As the park movement across the nation gathered momentum, an ironic budget pattern developed. The creation of parks is euphoric, but once they are established, the excitement fades. Even the greatest of parks can drift along for years, taken for granted and always perceived to be there, almost as though parks were ordained from the beginning. Parks have amazing staying power. Closing or dismantling an established park for lack of funding virtually never happens. To make the attempt would be to risk unleashing the political clout of a vast but usually silent constituency of park users who plan family vacations, weekend jaunts, daily exercise, and personal explorations on the assumption that the parks are permanent fixtures, always available, woven securely into society's fabric.

So long as the gates are open and some semblance of maintenance is evident, park users usually do not complain. They may expect sturdy and freshly painted facilities, clean recreation grounds, picnic areas, and campgrounds, well-maintained roads and trails, and a friendly and capable management presence, but if the facilities are shabby and neglected, the roads are falling apart, trash is overflowing, and park rangers are nowhere to be seen, most park users just cope.

This curious relationship between parks and their users is usually manifested in government's budget process. Parks are a low priority. As elections approach, appropriations for parks often increase momentarily, with accompanying claims of good deeds by those seeking reelection, but otherwise, the operating budgets for parks are usually pegged just above the poverty line. The irony is that a false economy then takes over. The parks drift along, sometimes for years, as facilities deteriorate. Eventually, a big dose of funding is required to avoid a real management crisis. The commissioners were on this budget ride. Fortunately, the inflow of private funding softened the ups and downs of government budget cycles.

On the fundraising front, Perkins was networking in anticipation of creating the third major gift fund. His personal relationship with Rockefeller Jr. was reflected in a brief tongue-in-cheek note that Perkins received just before departing for Florida: "Thanks for your friendly note. When the six-hour day and the five-day week come into effect generally, you and I will grow fat, and probably our wives will be tormented to death by our so constant presence at home."

Rockefeller Jr. was interested in the new gift fund and pledged further contact on the subject with Perkins. The fundraising effort would be aided on many fronts, among them a major article, with photographs, scheduled for a summer edition of the monthly magazine *The House Beautiful*. Mary Alden Hopkins, who was putting the article together, exclaimed to Welch, "If the thousands of park visitors enjoy the park as much as I do, there is a tremendous amount of treasure laid up in heaven or somewhere for all you who have worked so hard to make it possible for others to play."

The strategy for the fundraising campaign came sharply into focus in a February 11, 1920, letter from Rockefeller Jr. to Perkins. Rockefeller Jr. reconfirmed the $500,000 commitment made by the Trustees of the Laura Spelman Rockefeller Memorial that allowed the commissioners to purchase the two river steamboats *Clermont* and *Onteora*. This commitment assumed matching amounts from state and private sources, as already agreed with Perkins. Rockefeller Jr. then added, "The Memorial will contribute one dollar for every two dollars contributed, one by the State, one by other contributors, up to a total additional contribution from the Memorial of $500,000. Under this pledge, therefore, the possible gift from the Memorial would be increased from the present $500,000 to $1,000,000, in the event that $1,000,000 is received from the State and $1,000,000 from other contributors."

If all proceeded according to plan, the Rockefeller grants, laid down as a challenge to Governor Smith and Perkins's loyal group of private donors, would generate an additional $2 million, plus $1 million from the Rockefeller Memorial. From Florida, Perkins sent a handwritten letter to Rockefeller Jr. gratefully acknowledging the "wonderful" news and enthusiastically agreeing to the terms of the challenge. Perkins added a short comment about why the letter was handwritten: "I am writing in this way as I am flat on my back in bed for a week under Doctor's orders to, if possible, get rid of a cold and some digestive disturbances. The fact is, I have been overdoing for some time and a good rest is required—am confident I'll be myself again shortly."

Health problems that had sent Perkins to a hospital in Paris during the war years were with him again. Even so, his spirits were high. The news from Rockefeller Jr. came just before Perkins learned that the Supreme Court had ruled in favor of the United States Steel Corporation after a lengthy process of antitrust litigation. Perkins was proud of the role he had played years earlier with J. P. Morgan in creating the giant steel corporation. Despite his well-known support for Roosevelt and the Progressive Party, Perkins had always been leery of antitrust actions seemingly intended to dismantle the industrial might of the United States. In a second handwritten note to Rockefeller Jr. in which other PIPC fundraising matters were the primary topic, Perkins said, "I am of course very happy today over the Steel Corporation victory. I believe it may point the way for better things in our larger industrial affairs and it makes me even more keenly alive."

Almost from the moment the Rockefeller challenge grant was confirmed, the wheels were in motion in Albany to win an appropriation of $1 million for the commission, aided by the personal involvement and advocacy of New York Assembly Speaker T. C. Sweet. By coincidence, Stephen Mather wrote to Welch in mid-February 1920 seeking up-to-date information about the commission's funding initiatives, explaining, "I am very anxious to get for the national parks the kind of authority that will enable me to develop the parks under my charge along the lines of the methods used by the Palisades Interstate Park Commission." Welch responded with the latest details, providing Mather with important facts that worked to his advantage in testimony before the House of Representatives Appropriations Committee.

At the operating level, Welch continued to encounter rough situations. Heavy winter snows collapsed buildings at Bear Mountain. Many of the camp directors were asking for extra assistance because of severe weather that was hampering efforts to get the camps ready for the coming spring and summer seasons. Welch and his staff still were operating from office space in New York City after a fruitless search to find better facilities near Bear Mountain and in New Jersey.

Strong consideration was given to purchasing the United States Hotel in Haverstraw, New York, just to move the staff closer to the action. Perkins commented to Commissioner Lindabury that "regarding office room, it is a vexatious question, but one thing I am convinced—we cannot go to Bear

Mountain. In the winter, the location would be impossible and at any time the question of obtaining clerical help and getting them to and from work would be a very hard matter."

The office remained in the city, but city and rural cultures and styles were not easily mixed. Welch experienced a hard management bump when a crew from New York City was brought to the park to labor alongside the commission's workers on a camp project for the Boy Scouts of America. In a follow-up letter to B.S.A. camp director H. A. Gordon, Welch described the problem. "Our men, as you know, are all mountain men who have been with us for a long, long time and they have not associated with the City labor nor the City people and it is very easy for a few radicals to stir up a tremendous amount of trouble. I would very much rather get along with just the men we have than take any chances of introducing trouble-makers among them."

On March 21, 1920, the *New York Times* proclaimed, "Palisades Park Has 20th Birthday Today." Over two decades, the small experiment in landscape protection had blossomed into a full-grown conservation mandate that would forever change the character of the New York–New Jersey metropolitan region. Without fanfare, almost invisibly, Perkins and Welch had decided on a strategic move that would strengthen commission capabilities even more. Thousands of acres were being added to the commission's holdings in New York. Theoretically, these lands should have been removed from the tax rolls of local towns when acquired. For some towns, the tax loss would have been significant. In Rockland County, and later in Orange County, New York, legislative steps were taken to retain the commission's lands on the local tax rolls.

Perkins and the commissioners could have challenged the legislation and probably prevailed, but they chose not to. A letter from Welch to Jay Downer of the Bronx Parkway Commission explained the rationale: "As you know, we have in some instances taken practically all of the property in a number of school districts." Welch confirmed that the commissioners "considered it wisest" not to oppose the taxation bills. To do otherwise would financially wound school districts and local governments.

In time, property-tax payments for almost all commission lands in New York became a routine part of the annual state budget allocations. Had Perkins decided to oppose the tax legislation, future acquisitions would have been placed in jeopardy because of the always controversial matter of lost property-tax revenue. Instead, the commission remained a property-taxpayer. Local towns so benefitted continue to receive tax revenue, but the commission pays for many services that otherwise would be the obligation of the towns, such as law enforcement and facilities maintenance. This early decision by the PIPC is reflected across the nation. Many local government entities receive in-lieu tax payments for lands in national parks, forests, and other agencies. For example, Yosemite National Park is the largest single taxpayer in Mariposa County, California.

The commission's first twenty years spanned a sensational time in the history of the United States. The Wright brothers flew at Kitty Hawk in 1903; an assembly line for Model "T" Fords cranked up in 1908; Standard Oil was declared a monopoly in 1911; the *Titanic* went down in 1912; the League of Nations was founded after World War I; Prohibition arrived in 1920. Perkins may have dwelt for a moment on the events of the era and the commission's anniversary, but other matters were much on his mind, including a personal battle to regain his health and strength. In a letter to Rockefeller Jr. after his mid-April 1920 return to New York from Florida, Perkins affirmed continuing health problems:

> I am a trifle better than when I wrote you from Florida but still under the doctor's care, with a strict diet and regulation as to my hours of rest, and am facing the proposition of having to go away again just as quickly as I get the Palisades Park campaign out of the way—Everything in connection with the campaign just now is at the touch-and-go point. I have been soliciting a number of people here in town and, so far, have succeeded in obtaining pledges for $350,000. I must, if possible, get this up to at least half a million before the end of next week, because at about that time I think the Governor will act on the bill which I talked over with you and which has been introduced in Albany.

Perkins made a personal donation of $50,000 to the "third gift fund." (Jack) Pierpont Morgan, heir to the Morgan empire, made a donation of $25,000. Perkins's prediction of favorable action in Albany was confirmed when both houses of the legislature authorized $1 million for the commission and Governor Smith signed the bill. Perkins immediately moved to reinforce accounting procedures to ensure that state monies and private contributions would be tracked with great care, sensing that a fresh opportunity was opening up for the commission to significantly advance its agenda because of the attractiveness of the Rockefeller-motivating financial challenge.

Within the park system, the pace of activities was hectic. Residents of nearby towns were complaining about traffic gridlock on the roads to the parks weeks before heavy summer visitation was expected. "I want to call your attention again to the dreadful condition of the County Highway between Fort Montgomery and Highland Falls Village. The Bear Mountain Buses run up and down it every day and have helped to wear it in great ruts and holes so that it is hardly safe for life and limb," wrote Herbert L. Satterlee, partner in the Wall Street law firm of Satterlee, Canfield & Stone. The mayor of Englewood Cliffs was irate about "the worst congestion of automobile traffic that has ever occurred" and demanded more commission police to direct traffic on the approach road to the Dyckman Street Ferry.

On June 6, 1920, the *New York Tribune* ran a major piece by Emma Bugbee headlined, "New York Crosses the Hudson to Get Back to Nature."

> Sunday on the Palisades is an institution far more typical of New York than the Zoo. What other city in the world has a wall of mountains for a playground, within a distance of one mile and a 10-cent ferry trip, and yet as remote as roaring campfires and tumbling waterfalls? On a warm Sunday afternoon, it is almost impossible to find a fallen log or a natural fireplace that has not already been preempted by those mysterious and omnipresent persons who never seem to be going anywhere, but have always already arrived. Their beefsteak is always cooking over an already perfect bed of coals before you get there; their bathing suits are wet and their shoulders already sunburned, their Victrola already singing to itself under the trees, and father is already displaying a string of infant fish which he has caught from the pier. One is frequently disillusioned in finding the cliffs more full of Girl Scouts than columbines, and in learning that aching muscles last three days, while the new moon reflected in the water lasts only to the subway.

To gather information for the article, Emma had placed herself in the knowledgeable care of a friend she described as "the best Palisader I know":

> It was she who taught me that a narrow skirt, no matter how old and shiny, is not efficient in cliff scaling, and that solid alcohol will burn when rain-soaked twigs will not. She knows that a taxi will take you to the end of the Hendrick Hudson Drive, saving your legs for the harder hike over the unfinished portions of the trail further north. This is the most romantic portion of the Palisades, with a gravestone now and then to remind visitors of a once prosperous fishing village clinging to the foot of the cliffs, and with dying honeysuckles and lilac bushes telling of the unconquerable instinct of woman to beautify her dooryard.

Her friend was a fashion expert, too. "No matter how firm a believer in freedom a woman may be and no matter how long she may have lectured on the subject of proper clothing for hiking, the moment of her first appearance without a skirt is somewhat delicate." Emma's friend wore a skirt during the ferry crossing, but with "trouserettes" in hand, "you're sure to meet your most churchgoing friends in their best clothes," she cautioned. The two women "clung to the beach for the first mile," scanning with an anxious eye the picnicking crowd. Suddenly the friend "stopped and gazed delightedly at a particularly hilarious group of bacon toasters. Look, she whispered, that girl's got 'em on. Sure enough she had, in the full sight of the entire Sunday parade."

The cartoonist for the *Tribune* added his own touch to the Bugbee article. One drawing showed a fat man and woman enjoying a generous picnic with the caption "The rotund couple hiked ten miles upstream in eleven hours flat—primary idea involved being 'reduction of avoirdupois'—they then partook of food—some of which may be seen in the picture—only to find upon reaching home—that each had gained four pounds."

With preparations under way in anticipation of another record-breaking visitor season, including management of the group camps that were under contract to sixty-seven diverse and often demanding charitable organizations, the *Clermont* and *Onteora* were delivered.

Perkins had seen to the purchase of these steamboats out of his own pocket, with the understanding that the Laura Spelman Rockefeller Memorial Fund would provide reimbursement. Title to the steamers was in his name, so Perkins "loaned" the steamers to the commission. The steamboats could carry a combined total of 5,000 passengers and joined the privately

Dayliners *Clermont* and *Onteora* at Bear Mountain Dock (PIPC Archives)

operated *Mary Powell, Highlander, Grand Republic, Seagate, Mandalay, Hendrick Hudson, Sirius, Robert Fulton, Monmouth, Albany, Benjamin J. Odell,* and other excursion boats that provided access to the parks.

Writing years later, Capt. William O. Benson described an August 1920 trip as a nine-year-old boy aboard the *Onteora*. The trip, held vividly in his memory fifty-two years later, had made a "tremendous impression" on Benson. He kept a written list of the names of every steamboat he saw on the river that day. "When we approached Bear Mountain," Benson reported, "I could not take my eyes off the steamboats." Several were standing out in the river, waiting for dock space. In crystal-clear weather, Anthony's Nose loomed on the opposite shore. Passengers were strolling up the trail from the docks to the playfields, picnic areas, and inn at Bear Mountain. Barge-towing tugboats and river freighters vied with sailboats in the narrows. Perhaps somewhere an artist was trying to capture the classic Hudson River School scene. From the distance of a half-century, Benson could remember almost every detail of his weekend adventure, reflecting similar impressions gathered and cherished by hosts of other steamboat passengers who claimed their own personal, first-time discoveries of the river.

Welch and his staff were finding that the operation of two large steamboats on the Hudson River was vastly challenging. Firing up the engines and pointing the bows into the river were the least of Welch's problems. Finding dock space in New York City and along the Jersey shore and refitting, maintaining, and crewing leaky wooden vessels of such large dimensions were nautical tasks unlike anything that the staff had encountered before. Among other considerations, the steamboats needed music programs to be competitive. Welch went in search of an electronic version of orchestra music, and commissioners Lindabury and Averell Harriman used their contacts to assist him with the dock problem. Joseph B. Harris & Son offered free dock space in Jersey City, thanks to Lindabury, and the Erie Railroad Company responded to Harriman by providing temporary space at one of the company's docks in the city.

The commission's "navy" was launched. The *Onteora* remained in operation until 1936, when it caught fire and burned while moored at the Bear Mountain dock. The *Clermont*, renamed the *Bear Mountain* after World War II, was sold for scrap in 1950, falling victim like other river steamers to the ease of automobile access across the Hudson River on bridges and improved roads.

The distractions of making ready for the summer caused few on the commission staff to note the chilling fact that George Perkins's health continued to fail. In a routine response to a letter of inquiry about commission activities, Perkins's secretary, Mary Kihm, wrote on May 27, 1920, "Mr. Perkins is out of town at present and is not expected to return for several weeks." Perkins's physician prescribed additional extended rest in an attempt to reverse the threatening physical ailments that were dragging him down. In a June 11 letter to Welch, George F. Kunz, president of the American Scenic and Historic Preservation Society, suggested that Welch drop by at Tiffany & Co., where Kunz held a senior management position, to discuss "a matter concerning the park," adding a postscript: "Am awfully sorry about Mr. Perkins, may he be well soon."

On June 18, correspondence was flowing to and from the commission as usual. Welch declined a proposal to place a "dancing tent" on the Palisades shore where the Dyckman Street Ferry delivered its passengers. By memorandum, camp director E. F. Brown advised Welch of a pending meeting with the staff of the Museum of Natural History to develop a nature guide for the parks. Commissioner Elbert W. King sent a memo to Commissioner Sutro about the need to convince New Jersey governor Edwards to approve a supplemental budget appropriation.

Then, late in the day, King wrote to Mrs. Evelina (Ball) Perkins, "May I express to you, on behalf of every employee of the Palisades Park Commission, the deep sorrow that is felt in the death of Mr. Perkins; a sorrow that carries with it a most heartfelt sympathy for you in your loss." Within the commission, the assumption had been that Perkins would soon regain his strength and step back into his accustomed leadership role, but at age fifty-eight he suffered a heart attack, rallied for a brief time, and then was gone. Welch, in shock, felt that the world "had turned upside down."

The following day, a tribute written by Frank A. Munsey appeared in the *Sun and New York Herald*: "In all my acquaintance with men, I have never known one of more generous soul; have never known a better friend, or one more ready to go far, very far, to serve another. To those of us who knew him best, who knew the true impulses and purposes in his heart, who found delight in his buoyant, cheery, strong nature, the world will be dulled by his passing." The obituary in the *New York Times* ran for four long columns. Funeral services were held in Riverdale-on-Hudson, New York, where Perkins first heard, saw, and felt the explosions on the Palisades. The man who had started his career by managing a small grocery store owned by his father in Cleveland, Ohio, left an estate estimated at $10 million.

With the passing of George Walbridge Perkins, the world of the commission indeed had turned upside down.

9

▝▚▘

Jolliffe

By direction of the vice-president, a special meeting of the Commission of the Palisades Interstate Park, New York, will be held at the office of the Commission, 90 Wall Street, Manhattan, on Friday, June 25th at 2:00 pm.

GEORGE PERKINS HAD been working on the "third gift fund" only days before he died. In preparation for the June 25, 1919, meeting, Commissioner Richard Lindabury wrote to George Perkins Jr., asking for a status report on the fundraising initiative. Perkins Jr. listed the pledges from J. P. Morgan's son, Jack ($25,000), Arthur C. James ($50,000), Colman DuPont ($50,000), Edward S. Harkness ($100,000), Cleveland H. Dodge ($50,000), George F. Baker ($100,000), E. E. Olcott ($15,000), and George Perkins Sr. ($50,000). These pledges, plus the $1 million authorization from New York, placed the commission at about the halfway point in the "third gift fund" effort, not close enough to win the $500,000 Rockefeller challenge grant.

At the special meeting, Lindabury reported the results of his inquiry to Perkins Jr. The commissioners reviewed a list of revenue options, including a recommendation by William Welch that the commission publish 30,000 booklets containing information and a park map, to be sold for 10 cents each. Just prior to the meeting, Welch challenged the vendor of "Sight-Seeing Map of the Hudson River," contending that the map was full of errors and inaccuracies. The vendor, who had printed thousands of the maps on the assumption that they would be sold aboard the river steamers, including the *Clermont* and *Onteora*, tried to brush off the inaccuracies.

He complimented Welch on his knowledge of the river and invited him to suggest changes that might be incorporated in future printings. The vendor did not miss the opportunity to remind Welch, too, that the commission stood to make 7½ cents on each map sold to the thousands of passengers expected aboard the steamboats during the summer.

Welch was in no mood to be diplomatic or flattered and, despite financial needs, reported to the commissioners that he had refused "to allow this map to be placed for sale on our boats. I think it is an imposition on the public to sell them this thing," adding that since the vendor was in the map business for profit, he should employ someone "really familiar with the present conditions" to make the necessary corrections. Welch concluded by saying, "I am quite sure that the Hudson River Day Line will join me in insisting that the map be withdrawn from sale."

At the close of the meeting, and with no evident progress on fundraising or revenue enhancement, a statement was attached to the minutes in honor of George Perkins. In part, it stated,

> The Palisades Interstate Park as it now exists was the conception of Mr. Perkins, and it was he and he alone who raised the money for its acquisition and development. As the Park grew in size and usefulness his ideas and enthusiasm grew with it, until in the latter years of his life he conceived a playground large enough and complete enough to afford rest and recreation to all the people of New York and New Jersey who lived near enough to enjoy it, and where all the facilities for such enjoyment should be furnished at cost without profit to concessionaires or others. To the development of this idea Mr. Perkins gave the best there was in him.

Summer activities were in full swing when Perkins was lost to the commission. In a letter to Horace Albright, Yellowstone National Park's superintendent, Welch predicted that visitation to the commission's parks would increase by more than 90 percent that year. His prediction proved to be conservative. People were pouring off the steamboats onto the narrow beaches at the foot of the Palisades and the docks at Bear Mountain. Others were arriving via the growing network of roads, including an increasing number of paved miles. Still others rode the trains.

An incentive for some visitors might have been Welch's decision to purchase an "automatic orchestra" from the Rudolph Wurlitzer Company. This "orchestra" was placed in the dance pavilion at Bear Mountain, much to the delight of the younger crowd. When the numbers were added up at the end of the year, visitation stood at 2,166,455 (New York, 1,307,089; New

Jersey, 859,366). Nothing of this scale in state or national park use had ever been experienced, anywhere in the world.

With the swarm of visitors came complaints that were deposited on Welch's desk, ranging from botched ferry schedules to food-service problems at the Bear Mountain Inn. One of the commission's own, Commissioner Charles W Baker, found need to write to Welch about his own experience: "I was over on the Palisades Sunday afternoon and had invited a friend to accompany me for a tramp, who was unable to do so. I was frankly glad afterward that he did not, for I must confess I was ashamed of the appearance of things at our front door." Baker described the scene he encountered at the ferry landing on the New Jersey shore. "The most slipshod, ragged, neglected place in the whole park that I know is right there as one leaves the ferry."

Welch responded to Baker with a lament that would be repeated time and again for years to come: "Do you realize that we have barely had sufficient money in the New Jersey end of the park to employ three laborers from Bloomers to the New York State line and six laborers from Bloomers to Fort Lee? These nine men have had to clean up that entire twelve-mile section. These men have barely been able to keep the bathing beaches, bath houses and shelters and the main path clean. The picnic groves are in shocking condition as so is much of the shorefront and the whole reason is that the appropriation which we received from the State was only sufficient to allow us to keep this small force on."

Despite personnel shortages, many of the property transactions set in motion before World War I were being completed. Commissioners Lindabury, Edward L. Partridge, J. DuPratt White, Frederick C. Sutro, and Elbert W. King stepped into the breach left by Perkins, supported by Welch and the seemingly ageless George A. Blauvelt, long retired from his Senate seat in Albany but continually responsive to the commission's mission as a practicing attorney. Twenty land acquisitions and one right-of-way transaction added 32,922 acres to commission holdings in 1920 at an average per-acre cost of $26.

The commission's expansive parklands did not escape the attention of "sportsmen" in New York who began pressing again to have thousands of acres opened for hunting. Welch polled each commissioner by letter about this issue, receiving a unanimous response: the hunting restrictions, set in place when this question was first raised in New Jersey almost two decades previously, should be maintained. Seeking a legal basis for this decision, Welch gained a "game and fish preserve" designation from the New York Department of Conservation. Pressure through the years to open parks to hunting has sometimes been intense. This was a critical moment for the interstate commission. Had the commissioners bowed to the sportsmen, the ripple effect across parks everywhere could have been deadly for the wild, unsuspecting residents of these protected places.

But there were explosive moments, nonetheless. The commission was using TNT in road construction and had received permission from the New York State Industrial Commission to keep 100,000 pounds stored in a powder magazine. When an Industrial Commission representative made an unannounced inspection of the site, he discovered that 200,000 pounds of TNT were stored, including "100,000 pounds now stacked on the ground about 300 feet from your licensed magazine, and covered with canvas."

The commission had received two railroad cars full of TNT, which was duly stored in the magazine. Then, because of an error in paperwork, three more TNT-laden cars arrived. Welch explained this mix-up to the unsympathetic inspector. The TNT mini crisis prompted Welch to inquire about storing the extra explosives at the navy depot on Iona Island. The local commander was unresponsive, so the commissioners turned to a friend in Washington, D.C., with the hope of influencing the commander. Unfortunately, their timing was off. Franklin Delano Roosevelt had just resigned as secretary of the Navy and was unable to lend immediate assistance, even though this son of the Hudson River Valley and his wife, Eleanor, would later become active champions of the commission. The crisis was resolved when Welch pledged to complete a second powder magazine "as rapidly as possible, and to post a double force of armed guards, day and night," to watch over the illicit stack of canvas-covered TNT until it could be properly stored.

While Welch and staff scrambled to stay ahead of the parks' demands, inquiries continued to arrive from interested observers. Writing for the *Wanderer of the Pittsburgh Dispatch,* Mary Ethel McAuley asked Welch for his comments on an article, "How Can We Improve Our National Parks?" Referencing his experiences, Welch offered five simple points: 1. Obtain more appropriations from Congress. 2. Build better roads. 3. Pay concessionaire executives well but retain a large percentage of concession receipts for the government. 4. Operate parks solely for recreational and educational purposes. 5. Place the National Park Service under the control of a nonpartisan commission. Had even some of Welch's advice been heeded, many mistakes in the future management of U.S. national parks, especially the eventual monopolistic grip gained in many parks by concessionaires, might have been avoided entirely.

Welch's friend Enos Mills was keeping an eye on the recently created Rocky Mountain National Park, Colorado, for exactly this reason. "The policy adopted by the National Park Service to allow for monopolistic concessionaires is wrong," he had written to Welch in October 1919. Mills attributed this wrongheadedness to NPS director Stephen Mather, who was pushing hard for construction of hotels and lodges to prove to Congress that national parks were worthy tourism assets.

Wrongheadedness aside, the collaborative association between Mather and Welch was strong, as demonstrated when Mather asked Welch in February 1920 to hurry as soon as possible to the Presque Isle Peninsula, near Erie, Pennsylvania, to independently study the feasibility of making the peninsula into a national park. Mather reminded Welch that "Perry's fleet was sunk in Misery Bay, a little harbor at the edge of the peninsula." With the permission of the commissioners, Welch journeyed to Misery Bay and reported back to Mather, "Presque Isle can be made, with very small expenditure, a wonderfully attractive and useful park, and provide opportunity to make one national park close to great centers of population." Belatedly, Mather wrote back to Welch, explaining that there was "some uncertainty" about the "status" of the Presque Isle Peninsula, apparently because a key portion of peninsula land, thought to be owned by the federal government, had been returned to the state of Pennsylvania. State officials were not enthusiastic about the idea of allowing title to the land to be transferred back to federal stewards. This Mather initiative was dropped.

A few months later, Everett G. Griggs, representing the National Park Association of Washington State, sought Welch's advice about park projects in Washington and Oregon. Welch wrote of his "dream of an interstate park in Washington and Oregon to protect the Columbia River Gorge." Welch had made this same recommendation to political officials in Oregon while he was directing the war effort to produce spruce for the aircraft industry. This Welch dream, too, was unfulfilled.

Those from other states who were seeking counsel from Welch about park management would have been fascinated by the debate sparked within the commission by Anthony H. G. Fokker, inventor of the synchronized airplane machine gun and manufacturer of the planes flown by German pilots in World War I, most notably Manfred von Richthofen, the Red Baron. Several years after the war, Fokker had immigrated to New York. Fokker and a group of financial backers thought that New York City would be the ideal market for a commercial seaplane base using the latest technology. Their choice of sites was the commission-owned dock named for Sanford P. Ross, a former owner, that extended outward from the base of the Palisades into the Hudson River.

In exchange for a twenty-year lease to use Ross Dock, Fokker's Aeroplane and Motor Company, represented by J. C. Mars, proposed to invest $200,000 for a bulkhead and $50,000 for a hangar. Company representatives proposed to pay the commission 5 percent of the company's first-year gross income, to be increased by increments to 15 percent annually based on future company earnings, asserting that hangars and machine shops would be attractive and of interest to park visitors. They professed that the development of aircraft should be encouraged and that devoting Ross Dock to this specialized purpose would not be any more limiting to the park's use by the general public than bridle paths were in Central Park.

The proposed use of Ross Dock as an aeroplane base was refined and strengthened. Mars recommended a refinement to the earlier concept, a "passenger elevator" that would carry visitors from the cliff top to the aeroplane base below for a per passenger price of 5 cents. Mars assured the commissioners that river-steamer passengers would not be able to see the elevator's structure, made of steel beams anchored to the cliff face. Ross Dock would offer a public flying-boat anchorage, passenger air transportation, ferryboat service, and hangar facilities. In return for the requested twenty-year lease, the company guaranteed the commission a minimum payment of $10,000 per year, plus liquidation of an existing $100,000 mortgage that the commission was carrying on the Ross Dock property. Mars suggested, too, that company representatives would be interested in inspecting the Dana mansion as a possible cliff-top assembly point for visitors who wished to take the elevator ride.

Unrelated to the proposal, but related to the commissioners' search for funds, was a contact from another entrepreneur about the prospect of "storing" between ten and twenty oceangoing steamships on moorings just offshore the commission beaches in New Jersey. The entrepreneur offered to pay the commission $1,800 annually per steamship. After brief consideration, the commissioners declined this opportunity, but the matter of the aeroplane base remained under serious consideration. The motivation was money.

The proposed seaplane base did not seem to fit with Welch's philosophy that parks should be used "solely for recreational and educational purposes," but the money and excitement of being involved in the development of commercial air transport in the United States were tempting to the commissioners. In the meantime, Fokker purchased a five-acre plot of land atop the Palisades and announced that he would build his house there. The commissioners went so far as to draft a lease for the seaplane base, and Fokker began constructing the foundation for his house, but neither project moved forward. Fokker and his business associates chose, instead, to leave water landings behind and move their aircraft enterprise to a grassy field in Teterboro, New Jersey (now the busiest airport in the nation for privately owned aircraft), and he decided to live in Nyack, New York.

Revenue from another source caused a significant change in the staff ranks of the commission. Rental fees from the group-camp program netted $12,154 during the 1920 summer season. This income was promptly reinvested in group-camp expansion, but camp manager Edward F. Brown, looking at the profitable numbers and probably thinking of his eight-year pioneering effort to develop the camps, took the opportunity to ask for a pay raise. The commissioners declined his request, prompting Brown to submit his resignation. Brown's departure was adorned by mutual expressions of high regard and appreciation between Brown, Welch, and the commissioners. He left behind a remarkable record of personal achievement, having taken the group-camp program from an uncertain, almost accidental beginning to a level of major service to metropolitan residents. By the time of his departure, the camps were a snapshot of the urban society that surrounded Harriman Park, with special emphasis on the keystone program serving underprivileged children.

With Brown's exit, the commissioners turned to an unlikely successor named Ruby M. Jolliffe, director of camps for the New York City YWCA. She would remain at the helm of the group-camp program for the next twenty-eight years. "I've heard people say she was a curious mixture of dignity and devil-may-care," commented Jack Focht, director of the Trailside Museum, many years after the person simply and fondly known as "Jolliffe" retired. "She was the first one down the toboggan hill in the winter of 1922 when it opened, clocked at seventy miles an hour." Another time, Jolliffe hopped on an idling police motorcycle and ran it into a wall. Staring out from a photograph taken of her in winter attire soon after she was hired by the commission at a pay rate of $3,000 per year is a prim and proper woman, cloche cap pulled tightly down to her eyebrows, large eyes, larger spectacles, a slight, tight smile (the kind that seemingly says I hate to have my picture taken), a crisp white shirt and large striped necktie covered by a heavy wool jacket, gloves, wool knickers, socks up to her knees, and laced, polished boots. Jolliffe is standing as if at attention, the first female executive on the commission's staff.

Camp Director Ruby Jolliffe (PIPC Archives)

Born in Montreal, she earned a Bachelor of Arts degree from the University of Toronto and a master's degree in modern languages from Bryn Mawr College. After studying in Europe and teaching in New Jersey and the state of Washington, she assumed responsibility for the New York City YWCA camping program in 1912. When the commission took over the old rifle range at Blauvelt Park on South Mountain, Jolliffe was there to establish the YWCA camp.

When war caused the U.S. Army to reclaim the rifle range for the duration, W. Averell Harriman welcomed Jolliffe and the YWCA campers to property he owned at Summit Lake on his family's Arden estate. She was already well known to Welch and the commissioners when Brown stepped down from his camp manager post, and the decision to hire her was prompt. Before she retired almost three decades later, Jolliffe would "meet the sons, daughters, and grandchildren of her first campers," according to a report in the *New York Times*. She became a compelling presence in the camps.

A young camper named Mildred Rulison particularly impressed Jolliffe, who wrote to Rulison during the winter of 1920–21, asking that she "take over" one of the nature museums that Jolliffe had established in Harriman Park. Mildred was not inclined to accept the offer. "I drove to Bear Mountain and told her, 'I think Bess McClelland should have the job. She knows more.' Jolliffe said, 'But I asked you.' I gulped and said, 'I'll take it.'"

Florence Ball also remembered Jolliffe: "Saturday was beef stew day and the Director thought that the type of meal was not the right one for so honored a guest as Miss Jolliffe. She told the chef to change the menu to the Leg-of-Lamb scheduled for Sunday. The chef became so incensed that he began slamming articles around, tried opening #10 cans with a butcher knife, kicked them all, then packed up, got his pay, and left. I was called to cook the lamb dinner. The dinner went off without a hitch and Miss Jolliffe took the time to come into the kitchen, say hello, and said she enjoyed the meal."

The Jolliffe presence, imposing enough to create beehive activity simply in anticipation of her arrival as a dinner guest, seemed made up of equal parts disciplinarian, confidant, health officer, storyteller, naturalist, role model, sports advocate, moralist, commander, loyalist, educator, inspector, and stern administrator. The *New York Times* referred to Jolliffe as "the hand that struck the match for more campfires than any other in the country."

The scores of camps, all directed by different personalities, each pursuing its own purposes, had one common bond: they had to be "ship-shape" at all times because the directors never knew when Jolliffe might appear unexpectedly. She patrolled constantly. Like Welch, Jolliffe won allegiance for being fair and quietly charming, listening carefully, and, when necessary, acting tough-as-nails. She made clear that her duty was to ensure the well-being of the camps and campers.

Her domain reached the incredible number of 102 group camps, then eventually settled in at seventy-one. Among her ardent fans were Franklin and Eleanor Roosevelt, who regularly visited and participated in activities with young campers. Children, awed by Jolliffe, would carry vivid memories of her far into their adulthood. Welch would discover that he had a formidable colleague who saw no limitations to the good that could be accomplished in the parks.

Sharing this view was a group of men and women who were discovering that the vast acreage controlled by the commission offered an opportunity to step away, even for a moment, from the sounds and intensity of urban life, to wander along woodland paths, to pause to listen to nature's gentle voice, to stretch body and mind, and then to return refreshed to the workaday world. More and more visitors were choosing hiking as a highly personal form of recreation, far from the crowded park picnic areas, beaches, camps, and playgrounds.

In September 1920, Welch received an invitation from Dr. John H. Finley, Judge Harrington Putnam, Albert Handy, Bayard H. Christy, and Dr. George J. Fisher to meet at the Waldorf Hotel to discuss the formation of a "League of Walkers." At the Abercrombie & Fitch Building on October 5, this discussion was extended to include the Appalachian Mountain Club, the Fresh Air Club, the Green Mountain Club, the Tramp and Trail Club, the Associated Mountaineer Clubs of America, and the New York State College of Forestry. Meade C. Dobson of the Fresh Air Club, had written several articles for the *New York Evening Post* about the commission's park system. Raymond H. Torrey, editor of the Outing Page for the *Post*, attended and assured media clout for the venture. In a follow-up memorandum to the commissioners, Welch explained that the purpose "was to try and take some steps to better our trail system and make our park more useable to pedestrians and tramping organizations."

Welch continued, "I told them I was sure the Commission would be very glad to have any help they could give us to accomplish this and to take up with a Committee from the several organizations the question of the location of pedestrian trails in the Park." But Welch had cautioned the group that the commission "rather discouraged volunteer work on trails in the past because we did not want any amateurs marking out these trails nor attempting to clear them, but that we would be very glad to have the cooperation of experts in this sort of work." Feeling that he may have overstepped, Welch concluded the memo by asking, "Will you please give me your opinion on this question and let me know if you object to my going this far with this matter?"

The reply came from Partridge: "On the contrary, I approve!" Through these gatherings at the Waldorf and Abercrombie & Fitch, the commission tapped into a constituency of hikers who would become volunteer planners, workers, and stewards of a trail network that would expand to more than 1,300 miles. The expertise Welch was hoping for emerged from among a spectrum of professionals, students, spouses, and young people who became devoted to spending their weekends and holidays building and maintaining trails. In the process, the trail volunteers became avid defenders of open space.

The formal organization of the trail groups began with the meetings in 1920. By 1922, a "Trail Conference" had been formed specifically to lend volunteer assistance to the commission and to be watchful of the commissioners' policies and practices. Today, with a membership of about 10,000 and an intimate knowledge of the park system earned with sweat equity, the New York–New Jersey Trail Conference is a force to be reckoned with. Its officers occasionally demand accountability for what the members perceive to be ill-considered commission policies and actions, but far more frequently, the conference stands as a skilled and dedicated defender of basic park purposes.

Welch was key to this generally good chemistry. He found camaraderie and common ground with the hikers and instantly assumed a leadership role, becoming an active proponent for opening up the parks, footstep by footstep. At a November 1920 meeting of the hiking group, Dobson reported in the *New York Evening Post*, "Major Welch expressed heartiest cooperation of the Commission, and of himself, in the trail scheme. He made several sound suggestions, which were adopted, as to routes, blazes, markers, and shelters." Welch particularly emphasized "the great idea of a path from Delaware Water Gap, across New Jersey and New York into New England." Dobson included a comment that Welch made at the end of the meeting concerning a "housing problem of a colony of beavers placed in the park several weeks ago. The building operations undertaken by the beavers evidently have been unhampered by strikes, for their winter quarters are in readiness, together with the necessary dams and runways." At the meeting, committee officers were chosen: Albert Britt, chairman, Frank Place Jr., vice chairman, and Dobson, secretary.

The year 1920, so sorrowfully marked by the loss of Perkins, but so successfully concluded by leaps in activities on all fronts, was complicated by the fact that Perkins had raised only about half the funds needed to trigger the $500,000 Rockefeller challenge grant. His fellow commissioners did not have the kind of personal access to potential major donors enjoyed by Perkins that was needed to fill the fundraising gap. Ahead loomed the challenge of convincing Albany and Trenton to do more to support the parks financially.

One of the financial challenges was the care of the *Half Moon*, the replica of Henry Hudson's ship that had been in the care of the commission since the Hudson/Fulton celebration in 1909. New York State owned the ship, but the commission was paying for the cost of its upkeep. One cost-cutting option under consideration was to ground the *Half Moon* close to shore at Bear Mountain and encase the hull in concrete. This option was not pursued. The replica, moored at Bear Mountain for many years until eventually moved upriver to Albany, met the fate of most other wooden boats whose owners did not have the money for continuing maintenance. It fell apart and disappeared.

The commission's dayliner boats were receiving different attention. Commissioner Baker, who was serving on the "Hook Mountain Committee," contacted comptroller King to suggest "the possibility that one of the Commission's boats may be put entirely in the Hook Mountain service, making two round trips per day at a fare as moderate as to attract the really poor people." Unstated, but implied, was that the "really poor people," mostly African Americans, could be sent to Hook Mountain, physically separating them from the more well-to-do, mostly white Americans who frequented Bear Mountain and other park locations. The commissioners initiated the development of facilities at Hook Mountain for this purpose.

Assistant treasurer Elbert King wrote to J. DuPratt White, now president of the commission, "I hate to trouble you when you are not feeling well, but we really are very short of money." The commissioners had approved several more construction projects and earmarked $25,000 to "carry" the operating departments through the winter months of 1921, "leaving me with practically no funds on hand," King reported.

The "Third Gift Fund" pledges obtained by Perkins were materializing slowly in the form of checks, including $50,000 from Baker, $10,000 from James, and $12,500 from Dodge, but no new pledges were in sight. King even expressed concern about the price the commission was charging for cigars, worrying that the federal government might cap the price. His fear on this point proved unfounded, but a much more unsettling turn in the search for income came in the form of New Jersey Assembly Bill 546, sought by the commissioners and approved by Governor Edward Edwards, "giving the Commission powers to sell, lease, or grant easements over its properties on top of the Palisades."

The commissioners were seeking to establish a $200,000 reserve fund. Reasoning that most of the holdings on top of the Palisades had been purchased with private funds provided by New Yorkers, they decided to "gladly consider these properties as an asset of the New York Commission." Further claiming that the cliff-top properties had been purchased only to provide for "entrance" to the beaches and shoreline below, the commissioners identified "several said properties no longer necessary for park purposes," including the Dana property and mansion, Greenbrook Park, known for its lush natural appeal, and even the summit portion of the Carpenter brothers' quarry site.

The total estimated value of these properties was $322,537, more than enough to create the reserve fund, if sold. The commissioners' action, twenty-one years after the inauguration of the effort to save the Palisades, confirmed that attention in New Jersey remained riveted on the shoreline and cliffs.

But the fact that the Dana property was on the hit list, even though Perkins had worked so diligently to acquire it, suggested that the commission was drifting badly off course. Angus King, in a follow-up letter to Blauvelt, the commission's counsel, worried that sale of the Dana property would have to be handled with great care, reminding Blauvelt that the property had been rented for several seasons to the Brooklyn Children's Aid Society. King raised no objection, however, to the sale itself, seemingly unconcerned about two crucial considerations. One was that the commission was about to set an ill-advised precedent that could open the way for future land sales to cope with budget needs. The second was that the park vision in New Jersey seemed barricaded by the cliffs of the Palisades. Sweeping views of the Hudson River and the cityscape beyond could be found at practically every point along the summit of the cliffs. Terrain extending westward from the brink of the Palisades was gently rolling and gardenlike. The "millionaires' row" houses already constructed on the summit confirmed that only time stood between the existing natural beauty and heavy development in such a spectacular location so close to New York City.

Yet the commissioners were seeing financial assets, not park benefits, on top of the Palisades. Albert S. Riker, writing from the Standish Arms Hotel in Brooklyn, was among the first to express interest in the Dana property. He introduced himself in correspondence to King as a real-estate agent specializing in shorefront properties, including Riker's Island in the East River of New York City. Riker thought that the Dana property could be readily turned into a hotel and motion-picture studio. Perhaps because of continuing concerns about the Brooklyn Children's Aid Society, the commission was not prepared to deal with Riker, and his interest faded.

The commission's search for funds might have been forever resolved by Commander Clellan Davis of Englewood, New Jersey. On May 5, 1921, a headline in the *New York Times* read, "May Drill for Oil Under the Palisades." Davis claimed that an "oil film" was constantly appearing on the Hudson River near Edgewater, adjacent to Fort Lee. He asserted that this film must be coming "from some underground petroleum stream." Commissioner Lindabury, representing the state and the commission, said, "We will do everything in our power to determine whether there is a great lake of oil under the strata of New Jersey soil." The determination was quickly made. No lake of oil was found. The commission's holdings atop the Palisades remained at risk.

While the commissioners grappled with problems of money and policy, the police grappled with the public. James D. Moore was appointed as a "traffic magistrate" to deal with the growing number of motorist problems. "Our troubles from automobiles are not the speeding variety," King explained to Partridge. "Principally, trouble is created by the inconsiderate, selfish types who try to jump ahead of the line waiting to take the ferry. We also have difficulty with those individuals who might be called 'chesty,' who either know a Commissioner or some other influential person." The police made certain that, "chesty" or not, motorists stayed in line at the ferry terminals. Moore's appointment established a commission-operated traffic court for the New Jersey section of the park system. The court became an unplanned but unique organizational fixture. Unlike the erstwhile lake of oil, the court began to produce significant revenue because the "speeding variety" of motorists was becoming more common. Fines collected from these violations began to strengthen the police budget.

Moore became the first of many magistrates to hold sway in the commission's traffic court. He was succeeded by Abram A. Lebson, who at the remarkable age of twenty-one had already been appointed a judge in Bergen County, New Jersey. He later became cofounder of the respected law firm Lebson & Prigoff.

Not all violators of park rules faced the magistrates. Capt. James Conway, chief of the commission police, had an answer for "foolish canoeists" who crossed the river from the city under good weather conditions, stayed too long, and then were tempted to return in waters roiled and chopped by the wind. Conway was not interested in earning "medals for heroic lifesaving." He and his men took the canoes into custody and ordered the paddlers to walk to the nearest ferry for their return trips home.

Welch faced a blend of issues, as usual. To Jolliffe, he responded that a YWCA request for six four-hole latrines was "preposterous," pointing out that this number would be sufficient for 480 campers. The YWCA camp accommodated only seventy-five. At Bear Mountain, construction of a general office was under way in 1921 as recommended by commissioners Sutro and Baker. The new office was intended to improve management efficiency. In New Jersey, the commissioners and Welch continued to move toward the sale of the Dana property. A restaurant owner came forward as a prospective purchaser, suggesting the construction of a large hotel "on top of the Palisades," estimated at a cost of $4 million. The idea was that the hotel would be a "show place, a very magnificent edifice copied after one of the French castles." As with the earlier Riker proposal, the conflicted commissioners put this inquiry, too, on hold.

Back on the New York side, communications were taking place about the possible donations of the 114-acre "George Grant Mason Farm," including implements, livestock, and sheds valued at $25,000. Also in discussion was an 800-acre property owned by Dr. Ernest G. Stillman on Storm King Mountain. Stillman's father, John, was the banker who had provided Perkins with his first entree to Wall Street. Partridge was in contact with Stillman, while Harriman expressed a hope to Welch and his fellow commissioners that the 29.5-acre Storm King Stone Company could be acquired in conjunction with the Stillman transaction to "control all of the lands between the top of the hill and the river." A gift deed was subsequently received for the Mason Farm. The Stone Company property was offered for $2,000 and was purchased for $1,500.

Partridge also alerted Welch to a pending visit by Preston Clark, "who has in mind a future park upon Cape Cod." From Albany, George D. Pratt, commissioner of the New York Conservation Department, asked Welch to help block a bill that, if approved, would gut the department by taking away its "game protectors [game wardens]." "I know that you are influential with Governor Nathan Miller," Pratt reminded Welch. Welch delivered the message, and the game protectors stayed in the Conservation Department.

Frank Mason, a highly decorated ex-Army sergeant "covered with medals, service and wound stripes, and recipient of the Croix De Guerre," presented Welch with another challenge. Welch was advised that Mason was a "Negro" but did not see that as a problem. What did concern Welch was that Mason offered to bring his six-piece jazz band to Bear Mountain to provide a free weekday concert. Welch and King debated the matter, agreeing that they "did not care" for jazz, but decided nonetheless to accommodate the highly decorated war veteran. Mason and his band played at Bear Mountain to an appreciative audience. By contrast, more "traditional" music offered on the ferry was a different matter. One passenger lamented,

> I've listened to music melodic and sad,
> And music harmonious and merry;
> But I never heard stuff I considered so bad,
> As that on the Dyckman Street Ferry.

There was also the persistent matter of the Women's Memorial. Writing to White, Ida W. Dawson reminded him that she had been president of the New Jersey Federation of Women's Clubs in 1908 "when the park agreed to erect a memorial on the plot of land on which the Federation holds a deed. Can you inform me as to what has been done about it in the last thirteen years, and what is the present condition of the matter?" Acknowledging that he well remembered Dawson, White replied, "As I recall it, the Commission understood to set aside the plot in question for a Women's Memorial Park, which was done, and then to erect thereon some suitable monument when the money for the purpose should be furnished. A certain amount of money was furnished and has been carried by the Commission in a separate account ever since. I don't know the exact amount that the account now shows but will be glad to find out and give you the information if you desire."

Dawson wrote back, proposing that the commission use the money on hand to at least place a token marker on the plot, "while all are living who are interested." White, referring to the plot as "that very out of the way place," worried that a marker would be vandalized. "Our experience is that the public is ruthless in its care of such offerings and it is safer and better to erect only massive, substantial structures that cannot invite depredations." White assured Dawson that he and the commissioners would give the matter further thought.

At this time, Welch was recruited for a special task that took him away from his daily activities for six weeks. He was requested to "make an inspection of all national parks in the United States." The federal government paid his expenses.

Welch returned in time for the October 29, 1921, dedication of the long-awaited Henry Hudson Drive, a triumph of engineering and perseverance that allowed the motoring public to tour along the base of the Palisades for a distance of about 5.5 miles. A figure-eight loop at the northern end of the drive reversed the course for motorists. No connection existed between the northern terminus of the drive and the New York State line. Total New Jersey appropriations of $628,747 had been invested in the construction of this recreational drive. An additional $181,063 in private funding was contributed from the commissioners' New York gift fund, marking the third occasion in which substantial funding from New York was used for a project in New Jersey.

"The Drive is in no sense an automobile speedway," reported the commission. "It is more in the nature of a trail, affording ever changing scenes of wonderful beauty." In the spirit of inventive park management, the concept of roads designed especially for leisure driving and the enjoyment of nature, as contrasted with A-to-Z speed corridors, was affirmed by the roadway under the Palisades.

But other project bills still had to be paid. A sign was posted on the Dana property that read:

FOR SALE
This property of about 30 acres, with 1514 feet of Cliff frontage including mansion of 23 rooms
Inquire—Palisades Park Commission, 90 Wall Street

10

⣿

Trail and Bridge

THE FOR-SALE SIGN stood conspicuously on the Dana property, prompting inquiries from real-estate brokers and potential buyers. Because of the property's high asking price of $200,000, progress was expected to be slow in finding a person or organization with the means to seriously pursue the sale.

While the commission struggled with its stewardship options on the summit of the Palisades, another person who would greatly influence the national conservation movement and the commission stepped to the fore. Writing for the *Journal of the American Institute of Architects* in 1921, Benton MacKaye of Shirley Center, Massachusetts, proposed the creation of a hiking trail that would extend from Maine to Georgia, a distance of more than two thousand miles. This trail, according to MacKaye, would be "a sort of backbone, linking wilderness areas to dwellers in urban areas along the Atlantic Seaboard."

He saw the trail as integral to wise regional planning, as a foil to urban pressures, and as a simple, affordable means of enjoying the outdoors. Years later, MacKaye explained in an article for the *Scientific Monthly* that his advocacy for the Appalachian Trail (A.T.) was to encourage hikers and walkers of all ages and abilities to become "acquainted with scenery; to absorb the landscape and its influence as revealed in the earth and primeval life. . . . Primeval influence is the opposite of machine influence," MacKaye contended. "It is the antidote for over-rapid mechanization. It is getting feet on the ground with eyes toward the sky—not eyes on the ground with feet on a lever. It is feeling what you touch and seeing what you look at. It is the thing whence first we came and toward which we ultimately live. It is the source of all our knowledge—the open book of which all others are but copies."

Known for his "capacity for friendship, good conversation, and intellectual exploration," MacKaye found a ready audience in the hikers who had formed the Palisades Interstate Trail Conference. On April 25, 1922, the conference members reorganized to become the New York–New Jersey Trail Conference. The first elected officers of the new organization were Maj. William Welch, chairman; W. W. Bell, vice chairman for New York; C. P. Wilbur, vice chairman for New Jersey; J. Ashton Allis, trails committee; Frank Place, publicity; and Raymond Torrey, secretary.

In little more than a year, Torrey and Place, joined by pen-sketch artist Robert L. Dickinson, produced the first *New York Walk Book*, a masterful guide complete with detailed maps intended to assist anyone who might have interest in exploring trails in the lower Hudson River Valley. MacKaye's vision of a trail that spanned fourteen states, supplemented by a "cobweb" of connecting trails and embellished with recreational and educational facilities, galvanized this group to action.

At the forefront was Torrey, described as a "roundish figure who could maintain a moderate hiking pace all day." Torrey, the newspaperman who had attended the first Abercrombie & Fitch meeting hosted by Welch in 1920, was a "gentle man" who at an early age had discovered the joys of woodland trails near his hometown of Georgetown, Massachusetts. His many articles in the *New York Evening Post* reflected this personal interest. Echoing the values of his wide audience of mostly urban residents, Torrey captured in written words a strong philosophical allegiance with mother nature.

Once, while hiking on a Sunday, Torrey challenged a "beetle-browed" man who was painting graffiti on rocks. Not liking the challenge, the man shouted at Torrey, "You ought to be in church!" Torrey shouted back, "You ought to be in jail!" Torrey dutifully reported this incident to his readers. He would soon experience another confrontation, this time with the immovable object named Robert Moses, who was just entering the park scene in New York.

Welch and Torrey developed a close and informal partnership to build trails in Harriman Park. Welch was laying out trail routes and providing logistical support for Torrey, who organized and coordinated a group of volunteers to begin building the trails. The Welch-Torrey team was quick to respond to MacKaye's vision. Welch, using his drafting skills, designed a

Appalachian Trail Symbol designed by William Welch (PIPC Archives)

distinct marker that would ultimately identify the entire route of the Appalachian Trail.

Welch and Torrey saw shovel strokes taken on the first segment of the Appalachian Trail in the summer of 1922. This first sixteen-mile trail section was completed on October 7, 1923, beginning on the shore of the Hudson River, rising to the summit of Bear Mountain, then dropping down into Harriman Park and extending westward to the Ramapo River. MacKaye's vision had been transformed into reality. These first sixteen miles represented a tiny fraction of the total mileage envisioned for the trail but confirmed that volunteers of all ages and persuasions would step forward to build and maintain it.

Torrey spent years urging on the volunteers and cajoling landowners to join in a loose alliance with the multistate Appalachian Trail Conference (A.T.C.), founded in 1925 in Washington, D.C., as a greatly expanded version of the New York–New Jersey Trail Conference. Welch and Torrey assumed accustomed leadership roles with the A.T.C., Welch as chair, Torrey as treasurer. The New York–New Jersey Trail Conference became one of many branch units of the A.T.C. until it became an independent membership organization in 1931.

Today, the 2,158-mile Appalachian National Scenic Trail, so designated by Congress in 1968 and administered by the National Park Service, is complete. For many, just standing on the Appalachian Trail is to feel temptation and connection. "Through-hikers" who commit to the full length of the trail follow Welch's markers for weeks from Maine to Georgia or vice versa. For Welch, Torrey, and a host of volunteers spanning decades, the

A.T. represents great accomplishment through sweaty deed. For MacKaye, the Appalachian Trail is his legacy.

The lowest point in elevation for the trail is the Hudson River crossing at Bear Mountain. In a sense, hikers walk downhill either from Mount Katahdin, Maine, or from Mount Ogelthorpe, Georgia, to arrive at the tidal river, almost at sea level, and then begin walking uphill. When the first shovelful of earth was turned for the A.T. at Bear Mountain, no bridge crossing of the Hudson River existed for 150 miles between the open sea and Albany, except for a railroad bridge at Poughkeepsie, New York. Once trail sections were added on the east side of the river opposite Bear Mountain, hikers had no options other than to swim, find a boat, or cross by ferry, either upriver at West Point or downriver at Nyack.

The demand for automobile access across the Hudson River was to solve this problem. From May to November 1922, the Dyckman Street Ferry carried 325,000 automobiles from New York City to the Palisades. At the Nyack, New York, ferry crossing, 107,000 autos were hauled across the river. It was obvious that the demand to cross to the parks and communities on the west side of the Hudson River was rapidly reaching beyond the capacity of ferryboats. Articles began appearing in the *New York Times* and other newspapers, speculating about the possibility of bridging the Hudson River at the narrow river crossing between Anthony's Nose and Bear Mountain.

Proposed for the crossing was what would be the world's largest suspension bridge, a structure 2,500 feet in length, with a clear single span of 1,632 feet arching 135 feet above the river's high-water mark. Two steel cables, each woven of 7,252 strands of wire to create cable bundles eighteen inches in diameter, would be slung between twin four-hundred-foot-tall towers attached to concrete anchors on each shore. The concrete roadway plates to be attached to the bridge's superstructure would be thirty-eight feet wide and include walkways for pedestrians and A.T. hikers.

"Both Farmers and City Dwellers Gain by New Bridge Across the Hudson River," announced one of the *Times*'s headlines. Another confirmed, "Mrs. Harriman Backs Hudson Bridge Plan." Part of the justification for the bridge was that large portions of Westchester County, a rural area on the east side of the Hudson River, could be developed into a system of parks, poetically described as "chains of pleasure grounds," that would rival or even exceed in scale the PIPC's holdings on the west side. New York assemblyman Seabury C. Mastick argued, "There are not a few men and women in Westchester County who own estates, and who are really public spirited and of sincerely generous feeling. I am confident that they will be glad to follow as far as expedient the splendid example set by Mrs. E. H. Harriman, who gave thousands upon thousands of acres to the public which now form so valuable a part of Bear Mountain Park."

A few critics grumbled that such a bridge would simply increase the value of the Westchester County estates, but momentum was on the side of the proponents who claimed that the Bear Mountain Bridge would link motorists from as far away as Buffalo, Boston, Albany, New York City, and Philadelphia, allowing them to cross the Hudson "anytime, day or night, regardless of weather conditions."

There was one problem. The estimated cost for the bridge was $5 million, and New York claimed empty financial pockets. This problem was solved by a unique legislative maneuver. The financial force of the Harriman and Perkins families prompted the New York legislature to approve a bill permitting the formation of a private enterprise to construct a toll bridge at Bear Mountain. A charter was granted to the Bear Mountain and Hudson River Bridge Company organized by W. Averell Harriman, his brother, Roland, and George W Perkins Jr., acting on behalf of their families and other investors, to build the bridge.

The PIPC made available the necessary approaches to the bridge on the west bank. New York took similar action on the east bank of the river. The charter provided that the state could acquire the bridge for a sum of $4.5 million at the end of five years or claim outright no-cost possession at the end of thirty years. In the meantime, the investors could charge tolls for bridge crossings ranging from 80 cents from small cars to $1.75 for large trucks.

Ironically, a policy to "wipe out" all toll bridges in the state had been adopted in Albany only two years before the charter was granted to the Bear Mountain and Hudson River Bridge Company. Despite the fact that the state had recently purchased and opened the last three toll bridges in New York to cost-free transit, public support for the Bear Mountain venture, championed by motorists who were routinely experiencing hours of waiting at the crowded ferry terminals, overcame the policy. Construction of the Bear Mountain Bridge began in the spring of 1923.

By November of the following year, the privately financed bridge was in place, ushering in a new era in the Hudson River Valley. To the eye of almost everyone, the bridge was an artistic delight as well as an engineering success, sweeping from shore to shore just north of Iona Island and the Bear Mountain Dock, almost exactly where Gen. George Washington's troops stretched the famous chain barrier across the Hudson River from

Bear Mountain Bridge (Frank Becerra Jr., *The Journal News*)

Forts Clinton and Montgomery to the opposite shore. Somehow, this suspension bridge designed for purely utilitarian purposes seemed to fit aesthetically into the bold terrain of the Hudson River narrows.

On November 26, 1924, the West Point Military Academy Band greeted invited officials, eager for a ceremonial drive across the bridge in the almost three hundred automobiles that had transported them to Bear Mountain. Wilson Fitch, chief engineer in charge of construction, "proclaimed a new record in bridge building." Praise was given to Frederick Tench of the engineering firm Terry & Tench for conceiving the bridge plan and seeing construction through to completion.

The moment, though, really belonged to the Harrimans. Roland Harriman, president of the bridge company, honored his mother and gave her full credit for "accomplishment of the bridge project." The memory of E. H. Harriman was close at hand. The crowd was reminded that the elder Harriman had been "about to start construction of the bridge when he died eleven years ago." Perkins Jr., W. Averell Harriman, and former governor Benjamin B. Odell, vice chairman of the Bridge Celebration Commission, participated in the ceremonies. Mrs. Mary W. Harriman unveiled a plaque dedicated to all who had participated in the project, then cut the ribbon to open the bridge.

Unknown to the dignitaries assembled for the ceremonial motorcade crossing of the Bear Mountain Bridge, newlyweds Lt. E. S. Hopewell and the former Miss Jessie Welch had upstaged them. They had been married at the West Point Chapel about two weeks prior to the dedication ceremony. The concrete roadway was already in place on the bridge, even as final preparations were being made for the official opening. Major Welch and his wife watched proudly as their daughter and new son-in-law unofficially drove across the bridge to celebrate their wedding and claim bragging rights.

In 1922, before the bridge construction project got under way, New York governor Nathan Miller had appointed Perkins Jr. to the PIPC. His appointment and that of George T. Smith, an executive of the New Jersey Title & Guarantee Trust Company, Jersey City, were prompted by the sudden losses of John J. Voorhees and Otis H. Cutler, who died within one week of each other. Voorhees had served from 1915 to 1922 and had been particularly effective in delivering the PIPC's message in Trenton. Cutler, who was chair of the American Brake and Shoe Company and had commanded the 14th Division of the American Red Cross during World War I, had been appointed in 1921 to fill the vacancy left on the PIPC by the death of Perkins Sr.

PIPC assistant treasurer Elbert King had written to Cutler at the Royal Palm Hotel in Miami in February 1922, reminding him that the crucial second installment of the $1,000,000 Rockefeller pledge to the PIPC was due to expire on March 1. King had reported that New Jersey governor Miller "did not appear to be very much impressed with my statement that we might lose the balance of the Rockefeller pledge if the State did not do something at this time" and had asked Cutler to intervene. Time ran out for Cutler before he could respond.

On March 15, 1922, Perkins Jr. was appointed to take Cutler's place on the PIPC and, in so doing, assumed the mantle of his father. For the next thirty-eight years, until 1960, Perkins Jr., who, as an infant at the family's Glyndor estate had awakened crying because of the detonations on the Palisades, affirmed a tireless family commitment to conservation and park stewardship.

A graduate of Princeton University, Class of 1917, and, like his father, dedicated to the high ideals of the "dollar-a-year" businesspersons who heeded the call to public service when asked, Perkins Jr. assumed many important responsibilities for the U.S. government. His first government assignment was not at the request of an elected official, however. He voluntarily enlisted as a private with the U.S. Army American Expeditionary Force, 1st Division, immediately after graduation from Princeton, returning from Europe at the conclusion of World War I with the rank of second lieutenant.

After receiving a master's degree from Columbia University, Perkins Jr. accepted his first civilian assignment in Washington, D.C., as executive secretary to the postmaster general. He was executive vice president of Merck & Company and director of its Canadian subsidiary when World War II intervened. Back in uniform, this time with the rank of colonel, Perkins Jr. saw action in Europe and the Pacific and was awarded the Legion of Merit. After the war, he returned briefly to Merck but then agreed in 1948 to become chief of the Industries Division of the Economic Cooperation Administration (Marshall Plan) in Paris.

In 1949 Democratic president Harry S. Truman appointed Republican Perkins Jr. as assistant secretary of state for European Affairs. His principal assignment was to represent the United States in the formation of the North Atlantic Treaty Organization (NATO). President Eisenhower followed Truman's example by appointing Perkins Jr. as permanent representative to NATO in 1955 with the rank of ambassador. Throughout these years, he remained devoted to the always struggling park organization that had been so mightily influenced by the style and commitment of his father.

The arrival of Perkins Jr. and Smith came when both praise and criticism were being directed toward the PIPC. The praise came from more than 150 attendees at the Second National Conference on State Parks, held at the Bear Mountain Inn in the spring of 1922. Speaking for the attendees, Albert M. Turner wrote that he had experienced "pioneer work of the highest order, produced by that extraordinary combination of artist, economist, and engineer now known as Major William A. Welch." Turner reminded his colleagues that there was "no manual or guidebook for park managers" but that the PIPC had provided "a working model on the scale of twelve inches to the foot." He said that Major Welch "had apologized wickedly for the joyous noises of hundreds of visitors coming from outside" while the conferees were trying to hear one another inside the inn. Turner was also impressed by the one regulatory sign he saw in the park. It read, "Please do not pick the wildflowers, others want to see them." Otherwise, Turner said, there were no posted regulations, no ponderous listing of living commissioners or "graven images of the mighty dead."

The conferees were conducting earnest business. They heard from Stephen T. Mather and park representatives from practically all the states. Resolutions were passed urging the designation of a "National Conservation Day," the preservation of the Redwoods in California by "saving" them, the purchase of more National Forest land in the Appalachians, and the retention of park revenues for development and maintenance within the parks. Welch advised the conferees of another resolution he had sponsored at a meeting of the American Road Builders Association during a meeting in Chicago. The Road Builders had unanimously resolved to "prohibit advertising signs on public highways."

Welch ensured that the conferees toured the park system. Turner was impressed: "Back in the wooded hills are twenty-seven lakes, with scores of camps, the campers all fed three times a day with hot food from the Bear Mountain Inn, delivered by motor trucks for distances of up to twenty miles at a cost of $0.25 a meal. The craftsmanship displayed on the round timbers of the camp buildings was especially noteworthy, for the neatness and perfection was executed by woodsmen and impossible to the ordinary carpenter. Down below in the placid cove lies at anchor the replica of the *Half Moon*, that fragile shell of a child's toy in which Henry Hudson discovered these waters. Over the fireplace in the Inn are two shrunken links of the great chain which the optimists of '76 stretched across the Hudson to mark their dead line for the Britishers, and hanging above them skins of a real Teddy bear, shot and presented by himself, the great Theodore. A cordial welcome was extended by the civic authorities at Washington's Headquarters in Newburgh, an old stone house, on a tract of six or seven acres on the riverbank. This house was set aside as a State Park in 1849, the first State Park in these States."

En route to Washington's Headquarters, the conferees visited the newly constructed chapel at West Point and were amazed by the daring route of the PIPC-constructed Storm King Highway. Turner and the conferees were unaware that New York's official Winter Carnival ground was being moved from Saranac Lake in the Adirondacks to Bear Mountain. Bobsleds, toboggans, ice skates, and snowshoes were already available for visitors who were beginning to flock to the park during the once-quiet winter months. A ski jump to be constructed near the Bear Mountain Inn was on the drawing boards. The longest toboggan run, measuring 1,200 feet, ended with a rocket-like hurtle across the icy surface of Hessian Lake. For the first time, winter sports at Bear Mountain were confirming that parks could serve the public in all seasons.

Nor were the conferees advised of Elbert King's concern that "the Palisades section appears to be becoming a graveyard for stolen automobiles— two were pushed over the cliffs last Friday." Faced again with an unexpected byproduct of opening the Palisades to public access, Welch and the police took the only step available to them, since they did not have recourse to the thieves. They demanded that the owners of the stolen cars remove from the base of the cliffs what was left of their automobiles.

At the concluding conference banquet, held at the Hotel Pennsylvania in New York City, Dr. Edward Partridge, speaking on behalf of his fellow commissioners, announced that Dr. Ernest G. Stillman had donated 800 acres on Storm King Mountain to the PIPC in memory of his father, John Stillman. None of the attendees at this highly successful gathering of the nation's earliest class of park managers could have imagined that, years later, Storm King Mountain would become the focus of an epic environmental battle that would dramatically and permanently change the manner by which the United States safeguards its natural resources.

But the budding strength of the national environmental movement was already evident in the banquet room. The principal host for the evening was the American Scenic and Historic Preservation Society. The PIPC was prominently listed with other organizations that had cooperated with that society in making the dinner arrangements, including the Adirondack Mountain Club, the American Alpine Club, the American Automobile Association, the American Game Protective Association, the American Museum of Natural History, the Appalachian Mountain Club, the Associated Mountaineering Clubs, the Association for the Protection of the Adirondacks, the Bird and Tree Club of New York, the Boone and Crockett Club, the Bronx Parkway Commission, the City History Club of New York, the Explorers Club, the Fresh Air Club, the Green Mountain Club, the Sierra Club, and the Zoological Society of New York. At the conclusion of the banquet, the conferees gave the PIPC and Welch a standing ovation, pleased with what they had seen and eager to take away the lessons they had learned.

While Welch presided at the conference, King was dealing with a controversial problem that was prevalent throughout the country in the society of the 1920s: the use of parks by African Americans. A challenge to the commission on race relations came from John E. Robinson, editor and manager of the *New York Amsterdam News*. Robinson contended that he had received "many complaints" that the commission "discriminated against colored people."

In response, King wrote, "The Palisades Interstate Park is distinctly a Park for all the people irrespective of nationality, color, or religion." To reinforce his point, King added, "We have had a great number of excursions of colored people to the various sections of the Park. Particularly, I recall one group from a church of colored people in Harlem, who, for a number of years, held excursions at our Forest View Grove. This group stood out above all others in the manner in which they used the Park property, leaving it in even better shape than they found it, picking up and destroying all the papers and rubbish from the excursion, and in every way cooperating with the Commission's employees. These fine actions were brought to the attention of the President of the commission, and he personally complimented this organization on its treatment of the Park. We have had a great many excursions of colored people to Bear Mountain, and we have found them very desirable."

During the first quarter-century of the commission's life, its record of nondiscrimination was apparent. Group camps were available for African American children, and general use of the parks was available to all ethnic groups. Still, the commission, like other public organizations, struggled with the race issue. The "really poor people" King had in mind for the special, low-cost ferry service to Hook Mountain would, on arrival, find themselves physically separated by miles from any other group of park users. African Americans were confronted in the parks with the same attitudes that prevailed in the cities. They were welcome so long as they kept to themselves. In his criticism of the commission, Robinson struck a raw nerve and hoisted a warning flag that racial practices would remain under scrutiny in a nation obviously tainted and infested by blatant racial discrimination.

Racial segregation was not the only separate-but-equal practice that challenged the commission. Writing to former senator George Blauvelt, the commission's counsel, King said, "Two years ago the Commission determined that the mixing of the sexes in camp on any one lake in the Park was unwise and in fact a dangerous plan. In view of the dangerous situation, the Commissioners decided upon a distinct segregation of the sexes between the several lakes in the Park, reserving only one lake, Tiorati, where camps for the two sexes might be maintained, only because this lake was so well adapted to family camps."

The Brooklyn Bureau of Charities did not see it this way. At its camp on Lake Stahahe, considered by the commission to be strictly a "boys" lake, the bureau was alternating use of its camp between groups of girls and boys. King explained to Blauvelt that the commissioners had made a temporary exception for the bureau in recognition of the difficulty and cost of moving a camp but that "the exception could be made no longer; the danger was too great." With legal encouragement from Blauvelt, the bureau moved its camp to Lake Tiorati. Despite the boy/girl policy, the group-camp program kept booming. In 1923 alone, fourteen new camps were established, including Hebrew Orphan Asylum, the Boys Club of New York

Tabernacle Church, and the Association for Improving the Condition of the Poor.

The commission's chronic funding problem continued. Writing to Stillman, J. DuPratt White worried that "with the exception of the amount still due from the Laura Spelman Rockefeller Memorial, all of the contributions have been exhausted by investment in construction of the great camps." There was no guarantee that the Rockefeller money would be forthcoming. Theoretically, the commission could claim an additional $380,000 from the Memorial Fund, but to do so, it had to match the amount. Fortunately, Governor Al Smith, newly returned to office in the election of November 1922, was an advocate. In 1923, the New York legislature appropriated $500,000 to the commission for land acquisition. Among properties on the commissioner's list was a 127-acre holding on Dunderberg Mountain just south of Bear Mountain that was owned by Charles Edison, whose father, Thomas A. Edison, had purchased it years earlier with the intent of opening an iron mine. The commissioners agreed to purchase the Edison property for $3,000.

The costs of developing, operating, and staffing the park holdings took the commission in some odd directions. In New Jersey, FOR SALE signs were posted on four more plots of land. King had written to Lindabury that "to meet the Ross mortgage, we can use the balance of one of our old gift fund accounts, $30,000, and raise an additional $70,000 by selling properties on top of the cliffs." FOR SALE signs subsequently appeared on the E. I. DuPont de Nemours Powder Company, Carpenter brothers, W. W. Phelps, and Henry Torrence properties: total appraised value, $143,500.

In New York, the commission's need for development funds took a truly bizarre twist. At the recommendation of Welch, and with encouragement from Governor Smith, the commission decided to go into the quarry business. The target was none other than Hook Mountain, still battered and scarred by the discontinued quarry operations that had been acquired in 1917 through the best persuasive and financial efforts of George Perkins Sr.

Under the *New York Times* headline "To Move a Mountain for a Great Park" were the subheadings, "Palisades Commission Proposes to Transfer Hook Mountain from River Bank," followed by "To Sell Rock to Pay Cost" and "Governor Smith Approves $5,000,000 Quarrying Job Near Haverstraw." The lead paragraph in the accompanying article stated, "A proposal of the Palisades Interstate Park Commission to remove the greater part of Hook Mountain and establish in its place a beautiful playground, like Bear Mountain Park, was disclosed today to Governors Smith of New York and

Silzer of New Jersey, who were guests of the commission on an all-day inspection tour."

Accompanying the governors aboard the *Onteora* were state senators, assemblymen, and invited guests. The rationale for dismantling a mountain was explained in a subsequent paragraph: "Hook Mountain is one of the highest peaks on the west shore of the Hudson River, just below Haverstraw. Its four and a half miles of trees and foliage are bitten into by deep quarries, defacing much of the mountain's beauty. The Commission plans to remove the remaining portion of the mountain as far inland as the deepest excavation, rounding it out to restore it to its former natural state."

Welch was quoted in the article as saying, "More damage has been done to the scenery along the Hudson above Nyack by the excavations than anyone could think of. We have to remove the rest of the rock, virtually removing the mountain, and letting the rock pay for the work." Governor Smith assured Welch that he would do everything he could to make the commission plans real, saying, "I don't see how there can be any opposition. Go to it and go to it quickly." Accompanying the large group on the inspection tour were commissioners White, William P. Porter, Partridge, Perkins Jr., Lindabury, Franklin W. Hopkins, and Frederick C. Sutro.

A letter to the editor in the *New York Times* seemed to capture the public's reaction to this proposal. Under the caption, "It Seems to Be Alright," the writer suggested, "One's first thought on hearing the proposal to cut away Hook Mountain and sell its material to pay the cost of removal is apt to be antagonistic to the scheme. The case for Hook Mountain, however, is special, in that the quarrymen already have cut great gashes in its side. What the PIPC wants to do is cut the mountain back, giving it a more equal and natural surface to create a new pleasure ground, with a riverside roadway that would greatly facilitate the visits of automobiles to West Point and points north. There is nothing obviously reprehensible in this, and much to commend, especially if it can be done at no expense to taxpayers."

Flattening a hard rock mountain to provide for playing fields and picnic areas certainly falls into the category of a novel idea, but penciling out the cost/benefit caused the commission to put the project on hold. Not so other less daunting projects. The waters of many "lakes" in Harriman Park rippled against well-constructed dams.

In Albany, Governor Smith, soon after his return to office in 1922, delivered a special message to the legislature that seemed innocuous, but would result in raising public-works projects to new heights under the guise of park development. The governor advised the legislature, "It is clear that the time has come when we must have some central control and planning for our parks and places of scenic, scientific, and historic interest. I am sending you a bill which provides for the establishment in the Conservation Commission a State Park Council. I suggest that this new State Park Council prepare the budget for all of the park properties in the State."

According to the governor's plan, member organizations of the council would be the conservation commissioner (as ex-officio chair), the Palisades Park Commission, Allegheny Park, the State Reservation at Niagara, the American Scenic and Historic Preservation Society, the State Museum, Roosevelt Memorial Park, and the Finger Lakes State Park Commission (soon to be formed).

No mention of public-works projects was included in the message, but poised to take over the Council was a young man named Robert Moses. According to biographer Robert Caro, "Four years earlier, at age 30, Moses was standing in line outside City Hall in Cleveland, applying, in vain, for a minor municipal job."

Born in New Haven, Connecticut, in 1888, Moses grew up in New York City and held undergraduate and graduate degrees from Yale, Oxford, and Columbia. He had been invited into the Smith administration during the governor's first term, 1919–20, to aid in the reorganization of the state's administration. In a newspaper article entitled "The Master Builder," George DeWan reported that, during the governor's first term, Moses developed a lifelong friendship with Smith, whom DeWan described as a "poorly educated, gruff-voiced Irishman from the Lower East Side." The governor apparently found an alter ego in the highly educated Moses. Smith's return to office permanently solved Moses's employment problem.

The concept of the State Park Council was classic Moses: establish central control and grab the purse strings. By force of personality and imposing physical presence, Moses would oversee the expenditure of billions of dollars in public funds during the next four-plus decades on "bridges, tunnels, parkways, expressways, power projects, public housing, sandy beaches, concert halls, and tens of thousands of acres of parklands," mostly on Long Island.

His association with the commission would extend over many years, not always to the comfort of the commissioners. The commission's reaction to the newly created State Park Council was predictable, being not at all inclined to be restricted in the exercise of its budget-making and administrative powers. With a wait-and-see attitude, the commissioners were willing to cooperate with the council, so long as they were "left free" to pursue policies "peculiar and inseparable in the interstate nature of the enterprise."

One of those peculiar enterprises surfaced unexpectedly in New Jersey when the Police Benevolent Association managed to move a bill through the legislature that would require the commission to employ police officers on a full-time, all-year basis. No additional funding was attached to the legislation. The commissioners agreed with Welch and King that police officers employed on a seasonal basis were enough to maintain order in the New Jersey section of the park system. But the bill, carrying an imposing price tag, had gotten through the legislature and was on its way for signature by Governor Silzer.

Sutro, Welch, and King rushed to Trenton to head off the pending financial disaster. Aided by Col. Myron W. Robinson, who had been appointed to the commission in 1919, King was able to report to White that "on last Monday night, when we were just ready to retire after a very hard evening, Col. Robinson brought the Governor and his secretary to our rooms at the hotel. Conceive, if you can, the three of us in our pajamas welcoming the Governor and his secretary." The pajama discussion went well; Silzer vetoed the bill.

As a result of the police caper, the commissioners moved to consolidate all police activities under one administrative line of authority, anticipating the appointment of a superintendent of police for the Palisades Interstate Park. To find just the right individual for this important position, the commission turned for guidance to Col. Herbert Norman Schwarzkopf, chief of the New Jersey State Police. Schwarzkopf offered to assist "wherever needed" and recommended hand-picked candidates for the position. The salary of $2,400 per year for the position was too low. The two top candidates declined the commissioners' offer. Despite Schwarzkopf's effort and his continuing interest in the commission, the strategy to consolidate New York and New Jersey police operations proved unsuccessful.

On the matters of Hook Mountain, the proposed State Park Council, and the police, money was the core of the problem. The absence of Perkins Sr. was reflected in the status of the "Third Gift Fund." Of the $500,000 offered as a matching grant by the Laura Spelman Rockefeller Memorial Foundation, the commissioners were able to raise only $136,573 by the end of 1923, including a $25,000 contribution from Frederic A. Juilliard.

And still no women's memorial.

11

▗▗▗

Uncle Bennie

IN A LETTER sent to William Welch from Washington, D.C., on July 15, 1924, Robert Sterling Yard, president of the National Parks Association, said, "The longer I think it over, the more I am convinced that I ought to write the real story of Bear Mountain and circulate it at this particular stage of recreational development. The appeal to me is two-fold. It is so wonderful a tale that my pen fingers fairly itch to get at it. That's the personal side. And I can't escape the influence it will have on recreational thinking at this formative time. Today, lunching with Butler of the American Forestry Association, I told him a little of what you were doing, and he went crazy over it; demanded from me (and got the promise) an article for his forestry magazine. The point I'm making is that notwithstanding the great fame of your undertaking, no one knows the size, proportions and human greatness of what you've got. It is time that the world knew."

The world may not have known of the existence of the commission, but Welch's reputation as a leader in the park movement was well established among his peers. More and more, he was being invited away from day-to-day management matters to attend to policy at the national level and to share his experiences with like-minded conservationists in other states.

At the request of Stephen Mather, director of the National Park Service, Welch received an important assignment from Secretary of the Interior Dr. Hubert Work, who asked that Welch provide his expertise in the search for suitable national park sites in the southern Appalachian region of the United States. Welch and his fellow committee members visited many potential locations, hearing from anxious proponents and opponents of the plans in various states. To reduce the number of park proposals to a workable list, the committee determined not to consider sites of less than 500 square miles. From a list of more than sixty potential national park sites, Welch's committee recommended two to Secretary Work and Director Mather. During the process, Welch exchanged correspondence with many federal and state officials, attended public meetings, and gave numerous statements to news reporters about the benefits of park conservation based on his experiences in New York and New Jersey.

In Welch's opinion, the Great Smoky Mountains in North Carolina and Tennessee "easily" ranked first on the committee's list because of the "height of mountains, depth of valleys, ruggedness of area, and the unexampled variety of trees, shrubs, and plants." The Blue Ridge Mountains of Virginia ranked second. Writing to Col. W. B. Greeley, chief of the U.S. Forest Service, Welch explained, "I do consider that we should have a National Park or two in the southern Appalachians that we may preserve for all time typical sections of these very wonderful mountains." Welch, Mather, and others were effective advocates in their testimony before various congressional committees. Congress responded by authorizing the establishment of Shenandoah and Great Smoky Mountains National Parks in 1926. The Appalachian Trail, anchored at Bear Mountain, would ultimately wind its way through both parks.

Welch's inspired committee work was widely applauded. The Northern Virginia Park Association, whose motto was "A National Park Near the Nation's Capital," hosted a celebratory dinner in Washington, D.C., in 1926. The letter of invitation to Welch explained that "there will be no prearranged speeches, the occasion being one of general congratulation and jollity, but it is hoped that you will not fail to give us a few words of encouragement and advice on the subject of the celebration. Your presence will be both a pleasure and a cause for satisfaction to others present; indeed, it is not too much to say that the happy occasion positively demands it."

While Welch was enjoying well-earned recognition on the national stage, back in New York, a dispute occurred between the commission and Robert Moses when Moses interposed himself in the line of communication between the commission and Senator Nathan Straus, a longtime supporter of commission activities. Moses claimed that Straus was obliged to turn to him about a camp matter "after he made numerous efforts to deal directly with the Commission through Major Welch and others." J. DuPratt White was startled by the implication that Moses had to intercede on behalf of the senator to get action from the commission. "I am sorry that you should feel irritated about having such a matter brought to your attention," Moses wrote to White. "If this irritation is due in any way to your feeling that the State Council of Parks is attempting to interfere with the administration of the Palisades Interstate Park, you are entirely mistaken."

The young Moses, so new to the park scene, closed the letter by preachily reminding White, who, by then, had been involved with the park movement for almost thirty years, that "all the parks in the state owe a good deal to Senator Straus." In response, White took the obvious route by contacting his friend, the senator. Without further assistance from Moses, the two met for lunch, and the matter of the senator's inquiry about a camp was resolved.

Much more to the commission's benefit, the fledgling State Council of Parks decided to launch a bold initiative to place before the voters of New York a bond-issue referendum that, if approved, would provide $15 million for expansion and improvement of parks statewide. The park-bond referendum was an early hint of what would become Moses's approach to public-works projects in general: going aggressively for the big money without wasting time on nickels and dimes.

The bond referendum won strong endorsement from Governor Al Smith and a bipartisan majority of the legislature. Prompted by the bond issue idea, W. B. van Ingen wrote "A Wilderness Transformed" for the *New York Times Magazine*. He took the reader on a chronicled journey looping through Westchester County, across the Bear Mountain Bridge, through the commission's *forty square miles of recreation*, as he termed the interstate park system, and back across the river to the city. "These accomplishments are so great that one can list them only in a bewildered sort of way, but it is clear that the principles of combination of effort which have made for our wealth can be applied to the service of our health: for this is exactly what has been done. We cannot live by bread alone."

Another strong voice of support, though one not usually involved with park matters, was announced by a headline in the *New York City Evening World*: "Cardinal Hayes Endorses Park as Aid to Children." His eminence, the cardinal archbishop, was aware of a Harriman Park camp operated for boys by the Catholic Charities of the Archdiocese of New York and threw his considerable political weight behind the bond initiative. Statistics, too, favored this proposal. In addition to the Catholic Charities camp, eighty-four other camps were in operation in the park, serving more than 30,000 children. In the *New York Times* article "City Migrates to Camps," a reporter captured the mood of the camps and park activities on just one summer day:

> In distant groves Boy and Girl Scouts may be seen wig-wagging. Bugle calls rise and fall in echo-like cadences. Elderly matrons sitting on running boards of cars strip off their stockings and presently are wading with toddling babies. Men in their shirt sleeves lie at ease smoking foreign looking pipes. Youngsters run and shout and skitter stones over the waters. The side paths present a human stream of campers coming and going in either direction. They come out of the water in dripping one-piece bathing suits, youths and misses, and sprint for hot-dog stands. There are scores of signs bearing Indian names which designate the various camps, sometimes completely encircling a lake, and with more hidden in the wooded recesses behind it. At 4 o'clock in the afternoon an entire city seems to have migrated afoot or awheel to these lakes. A polyglot Babel fills the air. The camps teem with visitors, the waters with bathers, the woods with hikers. Their language, their manners, their clothes proclaim the majority not of the aristocracy and the limousine, not the income surtax payer, but the commoner and the flivver, the wage earner income taxpayer. To them a journey of fifty miles, with a day in the open ahead of it, is no more than a ride to Central Park. They, with thousands of others of their kind, are those who find and use new park lands as fast as they are thrown open.

In 1924 the PIPC was counting more than 5 million total visitors, had planted 4 million trees, and had ninety miles of marked trails and seventy-five miles of roads. The commission was in a position to demonstrate to a statewide electorate that parks were proven assets for communities. An indirect benefit was that the assessed valuation of properties bordering the parks was rising, confirming that preserved open space was a measurable financial asset to adjoining communities.

Commissioner Frederick C. Sutro and comptroller Elbert W. King attended a State Council of Parks meeting in September 1924, only two months before the scheduled election, and asked the logical question, "How will the Bond Issue campaign be publicized?" No one seemed to know. The startled Sutro proposed that publicity committees be organized for western, central, and metropolitan New York and that each committee develop informational booklets, posters, and newspaper materials for the publicity campaign.

The Metropolitan Committee was comprised of Moses, representing Long Island, Jay Downer, representing the Westchester County park system, and King as chairman. Sutro thought that the Metropolitan Committee should produce at least 150,000 booklets, estimated at a cost of $3,000. Moses said that he had "already drawn too much extent on private contributors for previous financial needs" and could not raise a dollar for the campaign. Downer responded that as a representative of a county organization, he could promise no public funds but would attempt to accomplish some private fundraising. Sutro and King, feeling boxed in, took their plea for bond-act publicity funding to the commissioners, who agreed to fund almost the entire publicity campaign at a cost of $3,800. Downer sent in a check for $150. Moses sent nothing.

Proposition 1, the bond initiative, passed handily in November 1924, the favorable vote in the metropolitan region outweighing the predictably negative vote in upstate New York. But winning voter approval of the referendum was one thing; dividing up the bond money was another. Under the headline "Battle Over Parks at Albany Hearing," the *New York Times* reported, "Two Long Island factions opened verbal batteries on each other today at a hearing before the combined Senate Finance and Assembly Ways and Means Committees on bills appropriating $6,000,000 for State parks. At times half a dozen speakers were shouting recriminations at one another simultaneously, and after the disorder at last caused adjournment, groups continued to wrangle for half an hour."

Writing to a friend who had left the meeting early, Sutro reported, "So far as practical accomplishment was concerned, you missed nothing. The afternoon session flouted the glories of the October sunshine by dragging along its dreary course through a maze of figures, portraying the needs of innumerable small parks and historic buildings all over the State. After four-thirty, the Council finally reached the real business of the meeting and made up the budget of the bond issue money to be appropriated by the legislature next year. There seemed to be some hesitation on the part of Moses to agree to a substantial amount, but I think I convinced him and the rest that it was to the interest of the State to meet pressing needs, especially if money could be saved thereby, as in many cases of land purchases. My view of the meeting is that there was set up an immense and unnecessarily expensive machinery to accomplish the very simple result of collating the budget requirements of the several park systems."

When Moses called a subsequent meeting to create park regions within the state, Sutro gave his proxy to Welch and King, saying that he "had gone beyond the limit of time I can afford to spend on the State Council."

Fallout from this dispute was minimal for the commission, but the hot-button question of State Council of Parks authority to forcibly appropriate land, and, by extension, the authority of the commission to do a similar thing, would remain a point of caustic debate for many years. Even so, for the "extension and improvement" of state parks, bond-referendum funds were allocated according to a predetermined formula. The commission's share was $3.5 million, standing to receive a first-year appropriation of $1.5 million, the rest to be appropriated in succeeding years. Half the money was earmarked for land purchases, the rest for maintenance and development projects; none, this time, for New Jersey. The commissioners hoped that New Jersey would take note of the bond referendum north of the border and loosen its purse strings for the commission in the next legislative session.

Group camps in Harriman Park were not on the list of projects that would benefit from the bond funds. From the distance of Albany, the camps were viewed primarily as revenue sources for the commission rather than as a unique service for children. Welch did not want to argue the point. He had many other maintenance and development projects on his plate that needed money, but he was concerned, too, that the private funds that had carried the camp program to such a high level were all but exhausted.

Welch turned again to the trustees of the Laura Spelman Rockefeller Memorial Foundation with a request for $120,000 for prompt expenditure in the camps, followed by $100,000 per year for four years. He added $30,000 for a dormitory at Bear Mountain, bringing the total request to an even $550,000. The request landed on the desk of Kenneth Chorley of the foundation staff, who understandably thought that such a large request deserved careful scrutiny.

He had not looked far before reporting to Col. Arthur Woods, his superior, that the Boy Scout Foundation of Great New York was pulling out of its site in the park. Members of the Boy Scout Board gave several reasons for this decision: their location in Harriman Park was not isolated enough; the rent was too high; the buildings were not suitable for their program; park authorities were not cooperative and did not understand Boy Scout needs; and there was too much red tape involved in management of the camps.

Chorley queried several other camp directors and found mixed complaints, mostly about red tape, but he noted, too, in a follow-up memo to

Woods, that very little turnover occurred, remarking that only "two or three" of the eighty-seven camps had changed hands in several years. Woods added, "Miss Jolliffe impressed me as being a very efficient woman, but I still believe that she is not the sort of person who ought to be Superintendent of the Camp Department. It seems to me that the department would be much more efficiently run if there was a man at the head of it." He also stated that Welch had told him that "since the death of Mr. Perkins, Sr., there is nobody to raise funds but myself, and I do not have the time." If the Rockefeller trustees were to be enticed to provide another major contribution, the commissioners and Welch would obviously have to present a better, more compelling case. Rockefeller support was not going to be automatic.

In addition to being thrown financially off balance by the less-than-eager response from such stalwart friends as the Rockefellers, the commission suffered the loss of former senator George Blauvelt on October 16, 1924. Blauvelt had worked with the founders of the park system from the very beginning, providing political and legal guidance since before the turn of the century.

Soon after Blauvelt's death, more bad news reached the commissioners. A *New York Times* article announced, "Trail Typhoid Peril to Palisades Brook." Sixty-eight cases of typhoid fever, four of which were fatal, were traced to a brook that flowed through the hamlet of Englewood Cliffs, New Jersey, tumbled over the Palisades, and flowed into the Hudson. Three cesspools on land outside the park were identified as the culprits. Most of the victims were children, including eighteen Boy Scouts from one troop.

The *Times*'s reporter confirmed that signs were posted within the park warning visitors not to drink from the streams, but he added that the signs were in English, while "many hundreds" of the daily visitors either could not read English or could not read at all. The commission was not helped by the fact that the article placed the crisis in a boy's camp in Harriman Park, New York, rather than in the New Jersey Palisades. Welch had enough to worry about in the camps in Harriman Park. Four confirmed and eighteen suspected cases of scarlet fever had been reported at a YWCA camp at Summit Lake.

The commissioners and Welch were learning the hard way that streams, ponds, lakes, and rivers, however pristine and inviting in appearance, were potentially severe health hazards. A program to pipe treated water to public-use areas, a practice now taken for granted, was vigorously accelerated by Welch, his engineers, and maintenance workers in the mid-1920s and was at the time considered cutting-edge in the management of parks.

In a moment of buoyancy amidst these struggles with health calamities and bad press, King wrote to Sutro to let him know that the new hit song, sung aboard the *Clermont* by members of the orchestra, was "Does Spearmint Lose Its Flavor on the Bedpost Overnight?"

Finding an answer to the mystery of bedpost spearmint was not high on the educational agenda, but nature education was becoming an increasingly important aspect. Nowhere was this more evident than in a small, third-floor space tucked in the far corner of a large room at the Museum of Natural History in New York City dominated by a huge stuffed whale suspended from the ceiling as if swimming in midair. The space was occupied by Dr. Benjamin Talbot Babbitt Hyde, a patron of the Department of Anthropology and educational director of the Boy Scouts Foundation of Greater New York. Hyde's corner space was partially hidden behind "exhibits of live snakes, birds, skunks, and the like," according to Eleanor Adolph, a visitor who first encountered Hyde on a Harriman Park trail.

Adolph was so amazed by this happenstance meeting that she captured her impression on paper: "A heavy, measured tramp, tramp, as of many feet, beat upon the murmurous silence of the woodland road. At first faint and indeterminate, the regular clop-clop grew louder and ever neared, until, suddenly, around a steep bend, strode the army of the Twentieth Century Crusaders. Straight little fellows of ten and twelve years, this band of Boy Scouts, perspiringly, but with uncomplaining courage, followed the lead of a tall, tireless figure at their head. A leader he was. One felt it, this gift of leadership, in every movement of the big, lithe, khaki-clad frame, in the tanned face, with its firm, kind mouth, most of all, in the observant, slate-grey eyes, humorous and sympathetic, yet keen, behind their bowed glasses."

And, she should have added, a bald dome.

"He looks like Teddy Roosevelt," Adolph remembered whispering to a companion. "He is Teddy Roosevelt in his way," was the companion's reply. "Ask any Boy Scout. There isn't one who doesn't know and swear by 'Uncle Bennie.'" Adolph was impressed by what she termed "a peculiar quality of forcefulness for he is, above all things, restful. He never seems hurried." Hearing weeks later that Uncle Bennie's plans for future summer programs were to include women and girls, Adolph trekked to the museum the following winter to seek out and interview Hyde.

Uncle Bennie's theory of outdoor education was that children should learn to appreciate woodland inhabitants, not fear them. Snakes were his specialty. At any given moment, one might crawl out of his shirt pocket or slither from under a sleeve. He inaugurated a series of rustic wildlife "museums" in Harriman Park in 1920 in association with the Boy Scout camps. "Boys are, by nature, very active and curious, and they are, for the most part, much interested in natural history," he explained to Adolph. "That they really prefer the study of the natural objects about them to idling or playing town games, is proven by the fact that, last year, ten thousand boys, in camp, voluntarily devoted their time to this work." As if confirming the point, Adolph's interview was interrupted by a young Boy Scout who wandered into Hyde's office. Uncle Bennie, "stooping, unfailingly kind and interested," gave the boy his full attention. "That's the way they do, they come all day, at any time. Very often, their people don't understand or care. Someone ought to. They need it."

Expanding on the subject, Hyde told Adolph that part of his purpose as a naturalist and educator was "development of the character of boys by throwing them upon their own resources, to make them self-reliant. They are never punished for shortcomings, but a constructive, straight-from-the-shoulder talk is given. There are no rewards, aside from merit badges. The value of the job well done is kept always before them and, as they develop a sense of responsibility, they are assigned to positions of importance. With work comes pleasure, with trust, responsibility, and that their duty is to give service to their homes, their community, and their country. We added forestry to the subjects in 1922 and 1923 with the interest of Mr. Franklin Roosevelt."

Adolph then asked the questions foremost in her mind: "Do the girls in the camps take to the nature work as readily as the boys? Do they like it? Are they as quick?" Hyde gave a firm, one-word answer. "Fully," he said.

This man who so fascinated Adolph had been for ten years the president of one of America's largest soap-manufacturing companies. Heir to "Babbitt's Best Soap," which was founded by his grandfather, Hyde joined the soap company after graduating from Harvard in 1901 and rose to the top management position before stepping away a decade later. His decision to leave behind the daily routine of overseeing the manufacture of soap was not surprising. While attending Harvard, he and his brother, Frederick, were greatly influenced by anthropologist Frederick Ward Putnam. Encouraged by Putnam and backed by their own inherited wealth, the two brothers mounted the "Hyde Expedition" to the remote western region of New Mexico known as Chaco Canyon in 1896 while still undergraduates at Harvard.

This was not the first trip to the Southwest for the two native sons of New York City. In 1893 the Hyde brothers had associated themselves with Richard Wetherill, a member of the famous ranching family credited with discovering the renowned Mesa Verde cliff-dweller ruins in southern Colorado. Working with Wetherill and others in the Grand Gulch area of Colorado, Hyde coined the phrase "basket makers" to identify an ancient Native American culture. The phrase is permanently embraced in the lexicon and literature of today's professional archaeologists. Artifacts shipped east from the Grand Gulch venture established the first formal contact between the Hyde brothers and the American Museum of Natural History.

But it was the "Hyde Expedition" in 1896 to Chaco Canyon that "opened a new chapter in anthropology at the Museum," according to J. E. Snead, who wrote extensively about the archaeological adventures of the Hyde brothers. Excavating Pueblo Bonito in the canyon, the expedition members shipped an entire railcar load of artifacts to the museum, including many exquisite examples of Native American pottery and turquoise art. Their task was not easy. Reaching Chaco Canyon required an overland wagon trip of nine days.

Once in the canyon, working conditions were severe, and water and foliage were scarce. "In subsequent years," Snead observed, "the Chaco Canyon collection has had significant effect on the relationship between the American public and the cultural heritage of Native American peoples." Hyde was not just collecting artifacts. Through his association with the Museum of Natural History, he was opening to scientists and the public an important view of Native American culture. Chaco Canyon today is a National Historical Park and World Heritage site.

Mounted in his trusty Model "T" Ford, anointed the "flying squirrel," Hyde took his passion for exploration and shared discovery into the Harriman Park camps. Boys and girls were so excited when they saw him coming that they would spontaneously cheer. He remained director of nature education for the Boy Scouts and was a mainstay in the camps until 1927, when he moved permanently to Santa Fe to establish the Children's Nature Foundation.

Hyde reflected the same skills and ability to communicate with young people that Dr. Frank E. Lutz had earlier demonstrated at Harriman Park's Station for the Study of Insects. Lutz and Hyde could take a beetle or a boulder and open up a whole new world of delight for young, inquisitive minds. Their mastery was in helping people simply see and appreciate the intricate and intertwined natural environment that encompassed them.

With a flair for the snake-assisted dramatic, Hyde helped to firmly plant the anchor of nature education in the commission's park system. His influence had real staying power, as affirmed by a young Boy Scout named

William Carr who encountered "Uncle Bennie" while waiting on a train platform for the journey back to New York City. Inspired by the man and his message, Carr would add immeasurably to the substance of nature education in the parks when his turn came.

At the grassroots level, blossoming educational initiatives were probably not much on Welch's mind. In the summer of 1925, twenty-eight people were rescued from drowning in the Hudson River under the Palisades. First aid was administered to thousands of visitors. Thirteen forest fires were reported in the New Jersey section alone, caused primarily by campfires left unattended.

In the constant search for improved sources of revenue, the commissioners decided to charge five cents per person for each dance in the pavilion at Bear Mountain. Twenty burros were purchased for rent to aspiring urban cowboys. While continuing to count every penny to keep operations going in New York and New Jersey, Arthur King had to confess by telegram to the New York comptroller that state treasurer's check number 36144, in the amount of $62,941.84, somehow had been lost, and he needed to ask for stop-payment and reissuance.

At the same time, King warned Welch that failure of the New Jersey state legislature to approve an anticipated supplemental appropriation meant that the New Jersey park operations would have to be put on a "starvation" basis. Efforts to sell the Dana property continued, even though the FOR-SALE sign was becoming weather-worn and drooping.

King took exception to another sign posted in the New Jersey park, this one reading, "This Park Don't Extend Beyond This Point." He directed Scott R. Knowles, the New Jersey Superintendent, to correct the grammar on this sign and to take down the FOR-SALE signs on the Carpenter and DuPont properties.

By the end of 1925, King wrote to William Shephard Dana "that the Commissioners have sold the so-called Dana property subject to the mortgage which you hold. The purchaser is Mr. Achille Ermeti of Englewood Cliffs, who proposes to develop the property for high class residential purposes with possibly a few stores." With this action, a crucial piece of land atop the Palisades seemed to slip from the commissioners' hands, contradicting almost everything the commission was attempting to accomplish.

While funds for land acquisition and daily operations were almost absent from the commission's purse, funds for construction projects remained available in a different account. The *Nyack Evening Journal* reported in the article "Big Engineering Feat Now Going on in the County" that Welch and his staff were completing a six-hundred-foot-long dam at Sandyfields in

Harriman Park that would create a 375-acre lake. The project was mundanely labeled "Dam #10."

Prospects for other construction projects were reflected in Welch's letter to P. H. Elwood Jr. at Iowa State College, in which he stated, "Last Saturday afternoon, a solid stream of motor cars from New York City, northern New Jersey, and the ferries, was not able to move more than five miles an hour, and didn't entirely clear out until 1:00 am Monday morning." Welch speculated that "parkways" were needed to relieve the congestion. He may have had in mind that the Port Authority of New York had gained permission from the commissioners to erect triangulation towers and concrete baseline monuments on the Palisades as part of a mapping project for the "Hudson River Bridge," proposed to span the river from 180th Street to the New Jersey shoreline.

Private funding for the commission's activities was increasingly difficult to garner. Only two other grants were imminent: $25,000 from Edwin Gould and $100,000 offered by Commissioner George F. Baker for the construction of a camp for his bank employees in Harriman Park. The commissioners agreed that "Baker Camp" would be a suitable name.

Continuing support by John D. Rockefeller Jr. still was on hold because of complaints received about commission operations, but Rockefeller's staff chief, Colonel Arthur Woods, in an internal memorandum to his colleagues, said, "Mr. Rockefeller has had it in mind that it might be well for him to make a substantial contribution in the form of new land. In taking these matters up with the Park people, we found some unsatisfactory conditions, and in the course of conversation with Mr. Harriman he said it would be of enormous value to the Park, and he felt that it would clear our ideas, if we could have a study made of the whole situation, showing up the weak points and making suggestions to strengthen them. In view of our interest in the Park, of their need and request for our help, and of the great service it can be to the public, if well handled, I cannot help feeling that this survey would be an excellent thing to make."

The study was initiated in 1925, and the result would have significant impact on the commission.

With park matters ricocheting in various directions, the reach of the commission was captured by a headline in the *Livingston Enterprise*, a Montana newspaper: "Mather and Welch en Route to [Yellowstone] Park Are Fresh from Dynamiting Big Saw Mill in Glacier National Park: Inspect Tetons." The article described the blunt way a sawmill owned by a hotel operator in Glacier National Park was removed to improve the scenery. The purpose of the nationwide inspection trip was to further refine and adjust the boundaries between national parks and the surrounding federally owned forest lands. Mather and Welch visited Glacier, Yellowstone, the Grand Canyon, Sequoia, Crater Lake, Mesa Verde, and the Grand Tetons national parks. In the meantime, King held the fort at the commission as best he could until Welch returned at the end of August 1925, just in time to host the fourth meeting of the New York State Council of Parks at Bear Mountain.

With private funds depleted, increasing demands for visitor services, the honor and burden on Welch to aid with park policy at the national level, and the initiation of the Rockefeller study, the vulnerability and resiliency of the commission were even more severely tested when a riderless saddle horse came trotting home to a barn on the Meadowbrook Estate of Richard V. Lindabury near Bernardsville, New Jersey. Searchers found Lindabury dead beside a country lane, where he had been thrown from his horse. A commissioner for fourteen years, Lindabury could open any door in New Jersey, carrying with him a well-earned reputation as an esteemed corporate lawyer, civic leader, and fair man. The commission had greatly benefited from his experience and access on many occasions. His fellow commissioners keenly felt the loss.

12

⚏

Black Thursday

BRAINSTORMING AMONG THE commissioners to find a replacement for Richard Lindabury led to an impressive list of possible candidates. Governor Moore signaled that he would appoint whomever the commissioners recommended. After considerable investigation and dialogue, the name of William Childs was advanced to the governor.

Childs grew up on a small farm owned by his father in Basking Ridge, New Jersey. After he and his brother, Samuel, roamed west in an unsuccessful search for better farming opportunities, they returned to New Jersey, pooled their collective financial resources of $1,600, and in 1889 opened a restaurant near the family farm. Forty years later, a chain of Childs Restaurants with an estimated value of $36 million extended across the nation and into Canada, delivering an estimated 50 million meals per year to satisfied customers.

Customer satisfaction was somewhat curious because Childs was a confirmed vegetarian. Reflecting his personal dietary inclinations, the meatless and partly meatless menus in the Childs Restaurants were increasingly suspected of hurting revenues, leading eventually to a stockholder revolt that deposed Childs from his own company in 1929. By then, Childs was a self-made, independently wealthy, and admired man and a substantial contributor to the Democratic Party. He was promptly appointed to the commission to fill the void left by Lindabury.

A few months after Childs's appointment, the commission suffered a second loss when William H. Porter, a banker who had joined J. P. Morgan &

Company just as George W. Perkins Sr. was leaving, died suddenly of a heart attack. Porter had served as a commissioner for twenty-one years, beginning in 1905. In the spring of 1927, Frederick Henry Osborn, a Hudson River Valley resident and an investor in railroads and banking, filled Porter's chair. Standing six feet eight inches tall, Osborn was a presence wherever he went. By the age of forty he had made a fortune and retired from business, but he would remain a commissioner for forty years, pausing only during World War II when, with the rank of general, he was placed in charge of Army morale.

Childs and Osborn were joining the commission just as a four-hundred-page critique of activities compiled by attorney Mark M. Jones for the Trustees of the Laura Spelman Rockefeller Memorial Foundation arrived on the desks of Welch and King. Alerting J. DuPratt White to the report, King cautioned that "on a number of points, the report is highly critical," adding that he and Welch had read the document without pause from 9:00 A.M. to 6:00 P.M. Although the criticisms were hard to take, a hint of the overall message contained in the report had come weeks earlier in an internal memorandum from Jones to the Rockefeller staff in which he had offered the observation that the "Commissioners have made remarkable progress . . . considering that they have had to work through an organizational structure which in itself is an obstacle to good management." Jones reaffirmed this theme in the text of the full report: "The Palisades Interstate Park is worthy of generous financial support. Genuine progress has been made toward realization of its main purpose of making outdoor recreation available to an urban population of over ten million at a low cost. Use of the Park is increasing steadily, and the Commissioners are laboring under considerable difficulty in their endeavors to keep up with demands—."

Noting that the commission still was dealing separately with park matters in New Jersey and New York divided by the state line, Jones said, "It has frequently been proposed that the adoption of an interstate compact (treaty) be brought about as a means of providing greater assurance of continuity and further protection against political influence. The adoption of such a compact (treaty) requires action by the Legislatures of both States as well as by the Congress of the United States. At the time of this writing, steps are being taken to formulate an instrument that might serve this purpose, with the view of presenting it to the Legislatures as soon as possible."

Turning to a thorny issue, Jones reported, "The State Council is an unfavorable factor in the situation. It is a comparatively new agency of New York

State and was organized in 1924, chiefly at the instance of Robert Moses. Its purposes are stated to be advisory in nature, yet Mr. Moses appears to aim toward building up a supervisory organization. He has strong backing from Governor Smith. He seems to think that the Palisades Park is endeavoring to escape supervision, and for that reason he may resist the interstate compact. It is my understanding that he is thoroughly disliked by the New York State Legislature and that this attitude toward him may result in favorable action on the compact by the Legislature, although such action would otherwise be difficult."

Then, in two sentences that must have chilled Welch and King, Jones said, "The interstate feature is so definitely of advantage to the Park that it should be preserved at any reasonable cost. Until the corporate structure is simplified, however, we would not consider that there is a sufficiently solid legal foundation for the Park's activities to warrant large contributions being made to it directly by private organizations and individuals."

Referring to land acquisition, Jones's advocacy was apparent: "The Park proposes that seventeen tracts, consisting of about 30,000 acres, all of which are situated in New York State, be acquired at an estimated total cost of $2,670,000. If this aim is realized, it will increase the size of the Park over seventy per cent to a total of almost 70,000 acres."

Under the heading "What Mr. Rockefeller Might Do," Jones may have considerably reduced the Welch/King chill by stating, "We should like to see the entire cost of the real estate made available to the Commissioners from Rockefeller sources." However, Jones continued, "Whatever sum is given should be conditioned upon the Commissioners' raising a substantial sum from other private organizations and individuals to cover a large part of the rest of the development program. The decision as to the time when action should be taken presents a difficult problem."

Returning to his earlier statement, Jones reinforced his view that "it is important to secure an interstate compact at this time, not only to assure the unity and the continuity of the Park as an interstate enterprise, but also in order that there may be assurance on continuing the interstate feature as a means of resisting political interference."

Jones felt that the public/private characteristics and successes of the commission set it apart from other government entities: "Of the capital investment represented by the Palisades Interstate Park about fifty-one per cent has come from private organizations and individuals either in land or cash. . . . We believe that this fact should place this Park in a different position from that of other State parks, so far as present and future State supervision is concerned." Nonetheless, the days of commission support by the

Laura Spelman Rockefeller Memorial Foundation staff soon would end. Further financial assistance, if any, would come personally from John D. Rockefeller Jr.

Prompted by the Jones report, J. DuPratt White and William Welch were eager to convince New York and New Jersey state legislators and Congress of the wisdom of legally affirming by interstate compact a unique park organization that had existed for a quarter century in substance, not in fact. The commissioners retained the services of Judge Charles Evans Hughes to pursue the compact goal, but its achievement would prove to be anything but easy. Resistance would come from at least one predictable source: Robert Moses.

In the typically animated style of the commission, while major policy matters were being pondered in boardrooms and in the quiet sanctuaries of estate libraries, the park experiment was plunging forward. With a transfusion of New York State bond money, the much-debated development of Hook Mountain was initiated but, fortunately, not as originally envisioned. The mountain would remain in place. Instead, removal of waste-rock and development was concentrated along the shoreline at the bases of the old quarry sites, resulting in a very popular public use area to be improved even more years later by the Civilian Conservation Corps (CCC).

Throughout Harriman Park, bond funds allowed for construction of even more dams and roads, supplemented by the privately funded construction of the "Baker" and "Harding" camps. The latter camp was named at the request of donor Frederic A. Juilliard in honor of former president Warren G. Harding. Privately funded camps were not welcomed by camp director Ruby Jolliffe. She strongly preferred commission-sponsored camps open for use by charitable organizations. Baker Camp would exclusively serve employees and their families of four city-based banks: First National, New York Trust, Bankers Trust, and United States Trust.

Large crowds were beginning to gather at Bear Mountain in the winter to observe and participate in an increasingly delightful schedule of winter activities. King, writing to B. C. Wallin of the International Newsreel Corporation, described one of the many events: "One race—that on snowshoes with about twenty women competitors, no one of whom had ever been on snowshoes before—was a scream."

By the following winter, a ski jump was in place, based on plans provided by the Scandinavian Ski Association. But getting the plans and building the jump were two different things, Welch discovered. He wrote to P. H. Elwood Jr., "To my surprise, I find that the proper design of a real ski hill is quite a complicated matter and one on which none of the Swedish and Norwegian engineers entirely agree. About the only thing they do agree upon is the fact that the hill below the take-off should be a thirty degree angle and that the steepness of this hill should increase with its distance from the take-off to care for the increased velocity of the jumpers in proportion to the length of their jumps."

Whatever the engineering challenges, the hill proved to be a great success. The New York State and Mid-Atlantic Ski Jumping Championships were held for the first time at Bear Mountain in 1927, attracting scores of Scandinavian jumpers and hundreds of spectators. Outdoor speed-skating races were added to the venue, ranging in distance from 440 yards to three miles. In the first series of races, one hundred men and twenty-one women were in competition. Thrill seekers continued to hurtle down the ever-popular toboggan runs.

While Welch tried to keep up with winter and summer demands, he continued to field inquiries and requests from all over the United States to speak at hearings and attend meetings on park-management matters. In rare moments, he could focus on more basic concerns. Taking drawing pen in hand, Welch designed a rustic trail shelter large enough to accommodate a dozen or more hikers. Using boulders, logs, and flat slabs of granite gathered at the site, a park maintenance crew hand-built the prototype shelter on Tom Jones Mountain in Harriman Park. One boulder used in the construction weighed six tons. Great care was taken to position the shelter behind shrubbery slightly away from the main trail and to limit construction scars.

There also was the problem of three lovesick female deer at distant Allegheny State Park in western New York State. DeHart Ames, writing to Welch on behalf of the Allegheny State Park Commission, described the problem: "As you will recall, we lost one of the deer which you so kindly furnished for the Park and then we procured a buck to place in the pen with the three does and he met with an accident and broke his neck and therefore we are without a buck to run with the does . . . is there any way that we might be able to secure a buck from you?" Welch said yes, but not until summer. "Last year's fawns are pretty husky now and would be hard to catch," he wisely counseled.

Other wildlife matters arriving on Welch's desk were not so easily placed within the context of professional park management. One man wrote, "Gentlemen, I am in the market for some Beaver musk, and I saw your ad in the paper where you had just marooned 50 Beavers in their homes for the winter. Now, I will explain myself to you. I want the musk from a healthy Beaver during mating season, in the midst of Beaver heat, before she is pregnant with young. I would like to know your price—and if you have got musk taken from under the jaws or throat or in the thigh of the animal, let me hear from you at an early date."

The inquiry was filed with no record of response by Welch. So, too, was a letter containing a business proposition: "I have heard through a friend that you are about to open a park of some kind in Bear Mountain. I have relations in Tampa, Florida, who would like to come north with about 250 Alligators, ranging from 10 inches to 12 feet in length and perhaps they could have space with you . . . let him know direct just what the rent will be, the license that must be paid, and what percentage he must pay you."

A more telling indicator of prevailing wildlife-management attitudes was highlighted by a *New York Times* article that announced, "Park Invites Fox Hunters to Shoot 500 Which Kill Game." In the article, Welch was quoted as saying, "Any reputable citizen with a fox hound and a trigger finger . . . can take all the foxes he can hunt," adding that foxes were considered "vermin" and could be hunted any time of the year.

Responding to the article, E. Childs of the Bon Ami Company and vice president of the Wyandanch Club, Smithtown, Long Island, wrote to Welch, suggesting the club might be interested in buying live foxes for sporting purposes. Childs counseled, too, that sending hunters and hounds into the commission's parks was no way to get rid of foxes and offered detailed advice on sure techniques to catch and kill foxes using steel leg-busting traps. "I use a wire cage with a chicken or pigeon in it," Childs explained. "All you need is a stick for a perch, if you use a pigeon, and a couple of tin cans and some water and some cracked corn. Set it in the middle of a field, or along an old hedge row, or fox run, and put four steel traps, one on each side, double spring. If you are after hawks as well, stick up a pole about twelve feet high, within ten or fifteen feet of the cage, with a small steel trap on the top."

Childs "guaranteed" that the menacing traps would do the job. In the late 1920s, the concept of balance-of-nature had not been proposed. Welch was acting on the common scientific logic of the day. There is no confirmation that he took Childs's advice, probably preferring, instead, to stick with the local fox hunters.

In the forestry arena, friends from California shipped eastward to Welch a fine supply of redwood and sequoia seedlings for transplant in Bear Mountain and Harriman parks. The idea was to establish redwood and sequoia forests in the Hudson River Valley, a prospect that Welch thought might prove unsuccessful because of alien soil and weather conditions. He was right.

Beaver musk, alligators, fox hunters, and redwood and sequoia trees aside, Welch reported with pleasure to the commissioners that the New York Fine Arts Commission had approved the design for an "entomological and historical museum" to be constructed on the site of old Fort Clinton at Bear Mountain. The American Association of Museums made a tentative offer to pay the estimated cost of $7,800 for the project.

Part of the vision for the site was that wildlife, insect, and plant specimens could be put on public display for educational purposes at Bear Mountain as a small-scale version of Uncle Bennie Hyde's collection at the Museum of Natural History. A factor not considered in the enthusiastic rush toward this project was the possible impact of constructing a building on the very ground where the remains of old Fort Clinton could still be seen. As a parting gesture of support for the commission, the trustees of the Laura Spelman Rockefeller Memorial Foundation agreed to provide $7,500 to construct the museum. The small, masterfully constructed stone building that resulted was among three funded in 1927 by the foundation, the first of their kind in the nation. The other two museum buildings were constructed at Yosemite National Park, California, and Grand Canyon National Park, Arizona. William Carr, the young Boy Scout who had been so impressed by Hyde during their chance meeting on the Bear Mountain train platform, enough to follow in Hyde's footsteps, was now on the staff at the American Museum of Natural History and was appointed acting director of the newly anointed "Trailside Museum" at Bear Mountain.

The commission parks eventually would add immensely to knowledge of the natural world. But as with fox hunts, an understanding of ecological values still was in its infancy in the 1920s. This was confirmed in an exchange of correspondence between Welch and Harold M. Lewis, the author of *The Regional Plan of New York and Its Environs*. Lewis alerted Welch to the search around New York City for suitable airplane-landing fields, suggesting that a field was needed at Bear Mountain. Welch replied that the commissioners were "considering seriously filling in the Iona Island marsh" for this purpose. Using his engineer's eye, Welch felt that making the necessary arrangements to fill in the marsh would take "many years to work out."

Fortunately, the "work out" never happened. The Iona Island tidal marsh is an exquisite natural area, left behind in a horseshoe bend when glaciers straightened the course of the Hudson River just below Bear Mountain. The marsh nourishes the river and is home or way station for a rich variety of upland birds, waterfowl, fish, and scores of aquatic species. Today it is one of four areas that comprise the federally designated Hudson

Iona Island and Marsh (American Cruise Lines)

River National Estuarine Research Reserve and has been designated by the National Park Service as a National Natural Landmark.

Airports were not the only concern highlighted in Lewis's report. For seven years, this planning group, chaired by Frederic A. Delano and funded by the Russell Sage Foundation, had been studying city-growth patterns and needs within a fifty-mile radius of City Hall. The words of Ralph Waldo Emerson, "Build therefore your own world," were a rallying cry for the planners, who saw around them a haphazard pattern of development and waste of aesthetic opportunity. Using only the "power of recommendation," the planners issued their findings in May 1928 in an attempt to guide and encourage better distribution of population, a more effective balance between residential and industrial needs, and reduction of "frictions of space" by providing for parks and parkways.

Looking across the river to the Palisades and anticipating the impact of the Hudson River Bridge project (the George Washington Bridge), scheduled for completion in 1932, the planners urged construction of a parkway "for about twelve miles along the crest of the towering cliffs." "Unless immediate action is taken to place the land along the crest of the Palisades under public control, the New Jersey skyline along the entire sweep of the cliffs will be marred by apartment houses and other structures built close to their edges," the planners cautioned.

Realtors and developers, already speculating in land in the vicinity of the bridge, responded even before the *Regional Plan* was published and claimed that apartment houses could be built along the Palisades without detriment to its beauty, to which Loula D. Lasker of *Survey Graphic* replied that the Palisades would turn into "a gap-toothed horizon of skyscrapers . . . billboards, water tanks, and Coney Island shows."

Anticipating "tremendous industrial growth" in northern New Jersey and a projected population of 6.5 million, the planners also urged extension of the commission's holdings to include a large part of the Ramapo Mountains in Bergen, Passaic, and Morris counties, New Jersey. Welch estimated the cost of acquiring a five-hundred-foot-wide parkway right-of-way along the crest of the cliffs at $25 to $40 million. One person who took careful note of the plan in general and the Palisades Parkway recommendation in particular was the commission's staunch supporter John D. Rockefeller Jr.

In an entirely different sphere, Welch was grappling with the perplexing task of setting in motion the machinery necessary to raise millions of dollars for land acquisition at the proposed Great Smoky Mountains National Park, Tennessee/North Carolina, and the Shenandoah National Park, Virginia. To energize the national campaign for the two national parks, Welch visited Adolph S. Ochs, publisher of the *New York Times*, to convince him to chair the campaign. Reporting back to Senator Mark Squires of New Jersey, Welch remarked that Ochs "feels very positive it would be a great mistake for him to assume the Chairmanship for this work as it would prevent all other publications, particularly in this section of the country, from taking any interest in the matter." Welch added, "I am going to see Franklin D. Roosevelt and see if we cannot induce him to take it." Roosevelt, too, declined the fundraising task.

Even so, Welch stood at a privileged point of communication when Rockefeller Jr. wrote to him in September 1927 confirming a pledge of $1.5 million toward the $4.5 million fundraising goal for acquisition of lands at the "Big Smoky Park." This was astonishingly good news. Prompted by the Rockefeller Jr. pledge, Edsel B. Ford pledged $50,000 in support of the initiative.

In January 1928, Rockefeller Jr. wrote in confidence to Arno B. Cammerer of the National Park Service to inform him that the two pledges to Welch were being canceled in favor of an increased pledge from the Laura Spelman Rockefeller Memorial Foundation of $4.5 to $5.0 million that was intended to match an equal sum jointly provided by the states of North Carolina and Tennessee. This extraordinary Rockefeller gift, first urged by Welch, eventually allowed for the purchase of about one-half of the eight-hundred-square-mile Great Smoky Mountains National Park.

Then, abruptly, Welch dropped out of further efforts to create the Great Smoky and Shenandoah National parks. Writing to Daniel P. Wine of the Shenandoah National Park Association, Welch said, "It is impossible for me to take any further hand in the Shenandoah and Great Smoky Park projects. I cannot explain the details that make this necessary, but I simply must drop it." He officially resigned from his various southern Appalachian duties on January 24, 1928. A hint of Welch's motivation may rest in a letter he sent to a friend a month later in which he lamented the "long illness" of his daughter and the "tremendous hospital bills." Dr. Howard J. Benchoff, headmaster of the Massanutten Academy, near the Great Smokies, spoke for many when he wrote to Welch, "You have done a piece of work that will stand as long as the hills of Tennessee. You deserve the title, 'Prince of Diplomats.'"

For a different reason, Raymond Torrey, who had traded in his newspaper reporter's hat and was now handling publicity for the commission, also contacted Franklin Roosevelt. Torrey, whose voice for development of the Appalachian Trail remained strong and persuasive, was delighted when Roosevelt, serving as chairman of the Taconic State Park Commission, steward of an evolving park system just across the Hudson River from the commission's holdings, had contacted landscape architect Benton MacKaye, proposing that work be started on a section of the Appalachian Trail extending from the Bear Mountain Bridge to the Connecticut border.

The trail was unfolding, isolated section by isolated section. But three years after a trail crew began work at Bear Mountain, the only significant, marked, and newly constructed segment of the A.T. ran through Harriman Park. The rest of the segments consisted of existing trails that happened to be in reasonably close proximity to MacKaye's Maine-to-Georgia corridor. Through the power of his journalist's pen, Torrey made certain that the public remained informed of MacKaye's grand plan and of the progress being made. Roosevelt's credible name was a wonderful addition to the message.

With plans for major new land acquisitions to expand the New York parks under debate, the commission continued to display its split personality regarding the Dana property on the summit of the Palisades in New Jersey. Title to the property finally passed to Achille Ermeti after last-minute snags were resolved; sale price, $252,000. The commissioners also moved to clean up debts on some of their other properties and to refresh the moribund gift fund. At almost the same time, Elbert King wrote to attorney L. O. Rothschild, thanking him for his recommendation that the commission "purchase more land atop the Palisades to protect park resources below the cliffs," and promised to place Rothschild's views before the commissioners at their next meeting.

The commission also broke the mold by leasing the steamers *Clermont* and *Onteora* to the McAllister Steamship Company. Driven by the vessels' need for major refurbishment and faced with the complexities of operating the two large boats and the difficulties in maintaining schedules and dealing with passengers' complaints, the commissioners decided to hand over the operation of the steamers to a private contractor.

This was a tricky bit of business. The commissioners were keenly aware that Perkins Sr. had won the first major contribution of $500,000 from Rockefeller sources specifically to purchase the steamers. At the time, he made the case that the commission must own and operate the steamers to ensure city dwellers low-cost access to the parks. He had firmly maintained that all commercial services within the parks, except for minor vending, be operated by the commission as a guard against profiteering. Leasing of the *Clermont* and *Onteora* ran counter to Perkins's philosophy and opened the door to other commercial concession contracts.

After expressing skepticism at first about the wisdom of this significant change in policy, the Rockefeller trustees acquiesced and agreed to the lease arrangement, guided by the report prepared by Jones recommending just such a shift in policy. Despite this shift, King, responding to a business inquiry early in 1927, stated, "There are no concessions for let anywhere within the Palisades Interstate Park. The Commission operates every Park facility."

On June 15, 1927, ground was broken at a ceremony on top of the Palisades to begin construction of the Hudson River Bridge (the George Washington Bridge). The commission cooperated with the Port of New York Authority by accommodating survey parties, arranging for the ceremony, and, above all, agreeing to sell the twenty-acre DuPont property to the Authority, including riparian rights, so that the western segment of the massive structure could be anchored to the western shore, just north of historic Fort Lee. The only concern for the river steamer operators was that the bridge be high enough to allow for their unencumbered passage underneath.

At about the same time, the deal with McAllister to take over the commission steamers was placed in jeopardy by a court ruling. The commissioners had tried to reserve access to the Bear Mountain docks to the McAllister and Hudson River Day Line Steamship companies, contending that the two companies had contracts with the commission that provided exclusive use of the docks. In a shouting match at the docks, the crew of a boat sent by the Delaware-Hudson Steamship Company attempted to debark 1,500 passengers, only to be chased away by a combined force of commission and McAllister employees.

In the lawsuit that followed, attorneys representing the Delaware-Hudson company argued that any member of the public should have reasonable access to the park. In a precedent-setting ruling, Justice Joseph Morschauser of the New York Supreme Court agreed. He ruled that in conformance with logical regulations, "every reasonable opportunity should be given the public, either on the land or water, to have access to this park."

The judge was careful not to randomly throw open the park to any commercial operator who happened to appear on the scene, but he concluded that exclusive commercial rights, such as the use of the Bear Mountain docks, must be based on competitive contract bids. He found that the commissioners had bypassed competition by granting docking rights only to McAllister and the Hudson River Day Line.

In a follow-up letter to commission attorneys, King stated, "Our recourse now is to the Appellate Division. This matter strikes at the very root of our authority to regulate the public use of the Park, and it will be fought through to the end." The commissioners worried that Judge Morschauser's ruling would turn the parks into a commercial grab bag. The subsequent court fight did not take long. Within a few months, the Appellate Division of the New York State Court ruled in favor of the commission by overturning the lower court finding. Tight-fisted access to the Bear Mountain docks was back under the control of the commissioners.

The commission's next legal challenge would not be so easy. The quarry wars, dormant for so long, burst open again at Tallman Mountain, just north of the New Jersey/New York border near the village of Piermont. The Standard Trap Rock Corporation arranged with the owner of a 171-acre property to open the largest quarry operation ever envisioned along the banks of the Hudson River.

On a trip aboard the *Stingrist*, Averell Harriman's eighty-seven-foot yacht, the commissioners viewed Tallman Mountain from the water. Already, the quarry company was constructing a giant crushing plant and forcing a channel through the beautiful Piermont Marsh to provide barge access. White, taking a page from the style of Perkins Sr., sought urgent financial help from Rockefeller Jr. The commissioners hoped to initiate a friendly negotiation with the traprock company through which the property would be acquired for about $500,000, the appraised value. Rockefeller advisor Col. Arthur Woods supported White's plea. Rockefeller Jr. was not inclined to pledge the entire amount but signaled that he would contribute $300,000 if the remaining amount could be raised from other private sources. Thomas W. Lamont, a wealthy landowner who had recently acquired property on the Palisades summit near the state border, offered to contribute an additional $100,000. Averell Harriman then matched the Lamont pledge.

By September 1928, the PIPC was in a position to recommend the Tallman Mountain purchase at a State Council of Parks meeting in Ithaca, New York, but there was a problem. The Standard Trap Rock Corporation claimed that the rock it intended to quarry was worth $2 million. The property owner who was leasing the site to the quarry company added an additional demand for $950,000 in compensation. Assuming that the commission would not accept these prices and knowing that it held the power of "appropriation" and could take the property if it chose to do so, the traprock operators proposed an alternative. They would "carefully" quarry the property and give it to the commission after their work was done. The commissioners upped their offer to $600,000, and the quarry operators countered at $4 million.

In 1928 Welch, King, and staff member Ellis Horwell traveled to the State Council of Parks meeting in Ithaca, New York, to present the case for the appropriation of Tallman Mountain. On their return to Bear Mountain after the trip, King, the ever-careful accountant, listed the expenses: bridge tolls, $1.80; ten gallons of gas, $2.05; suppers in Ithaca for three people, $4.35; Ithaca Hotel, four nights, three people, $30.50; breakfasts for three people, $2.55. The grand total was $41.25. Apparently, the group did not eat lunch. In 1928, a dollar was a dollar.

A horse was also a horse. Writing to Welch that summer, Anne Tracy Eristoff inquired, "How is the horse, Harry, getting along . . . when he comes useless, I shall take him off your hands with the greatest pleasure." Sheepishly, Welch wrote back to Eristoff, "Horse Harry died of colic while I was away . . . regret that you were not notified."

In October 1928, the commission served papers on the Standard Trap Rock Corporation and the Sparkill Realty Corporation and "appropriated" the Tallman Mountain property, expecting to pay no more than the appraised value. The following March, the *New York Times* announced, "Park Commission Sued." Standard Trap Rock challenged the constitutionality of the commissioners' authority to appropriate property for conservation purposes. No one at the time realized that the U.S. Supreme Court would resolve the legal odyssey on which the contending parties were embarking.

In June 1929 the PIPC lost in the New York appellate court and was enjoined from keeping control of the Tallman Mountain traprock property it had taken. But in February 1930 the *Times* confirmed, "High Court Reverses Palisades Case." Except for deciding the level of compensation, the commission prevailed in the Tallman Mountain case on appeal to the Supreme Court.

Writing later to a friend, White put the case in perspective. The commission "blazed the way for development of the idea that the sovereign may assert a right to preemption for the preservation of scenic beauty quite disconnected from recreational purposes," he said. An important legal precedent was established in the *Standard Trap Rock/Sparkill Realty vs. Palisades Interstate Commission* litigation. Preservation of scenic beauty by appropriation was deemed by the U.S. Supreme Court to be a legitimate responsibility of public park agencies. This finding would benefit park preservation nationwide.

The question of deciding the level of compensation for the plaintiffs would drag on for years and finally settle near the original appraised value, but Welch ordered that roads into the Tallman Mountain property be blocked off the moment the high court ruled. As far as he was concerned, the commission owned the property, and he intended to defend it.

Another responsibility went much better for Welch. He had been in contact for several months with Mrs. William H. Osborne (no relation to Commissioner Frederick H. Osborn) about the long-delayed New Jersey Federation of Women's Clubs Memorial project. Using his own engineering talents, Welch presented a "tower design" for the memorial that was accepted by the federation. On April 20, 1929, the newly constructed Watch Tower, tucked in the woods just steps from the precipice of the Palisades, was dedicated. Frederick C. Sutro represented the New Jersey commissioners. J. DuPratt White represented his New York colleagues. Mrs. John A. Holland (Cecilia Gaines), who had been president of the federation in 1897, was in attendance, joined by Elizabeth Demarest and a large crowd of federation members and commission staffers. In a crypt within the Watch Tower, they helped place a "casket of records" attesting to the early and stubborn struggle by women of the federation to protect the Palisades.

Elizabeth Vermilye was remembered for her leadership. A bronze plaque affixed to the tower reads, "This Federation Memorial Park is dedicated to the successful efforts of the New Jersey State Federation of Women's Clubs and of those men and women who aided in the opening years of the twentieth century in preserving these Palisades Cliffs from destruction for the glory of God who created them and the ennobling of the generations which may henceforth enjoy them." After a mere thirty years of asking and waiting for recognition, the women had their memorial.

Federation of Women's Clubs
Memorial, Palisades (PIPC
Archives)

Writing afterward to Welch, Osborne said, "Words entirely fail me in trying to express to you my feelings about our Memorial—It is dignified and suitable, standing guard over the cliffs and that noble stream, providing one a quiet, remote spot to rest and meditate right in the midst of the city's teeming millions."

Unknown to the commissioners, Welch, Osborne, and others in attendance at the memorial dedication, memos were circulating internally within the Rockefeller organization. Perhaps largely prompted by *The Regional Plan of New York City and Its Environs* report, staff member Charles O. Heydt advised his boss, Rockefeller Jr., that purchase of a large portion of land on the crest of the Palisades would probably cost about $10 million. The memo hinted that Rockefeller Jr. might wish to pursue these purchases on his own, rather than through the commission.

Almost at the same time, the so-called "Mandigo Property," an 836-acre mountainous piece of land west of the village of Fort Montgomery near

Bear Mountain, came on the market at an asking price of $40/acre. The "New York Gift Fund" of private contributions, assembled so long ago by Perkins Sr., was all but exhausted. Commissioners Harriman and Osborn decided to respond to the opportunity by buying this property, to be held until such time as the commission could reimburse them.

A management routine was appearing within the commission. To a large extent, the park experiment had matured into park-business-as-usual, with some exceptions. One of those exceptions was an attempt by William Welch to schedule the 1932 Winter Olympics for Bear Mountain. In a letter to the International Olympic Committee in Lausanne, Switzerland, Welch affirmed that daily crowds of up to 160,000 were coming to Bear Mountain to enjoy winter sports activities, including 15,000 spectators on average to watch the ski jumping. Congressman Hamilton Fish Jr. of New York, himself an avid sportsman, strongly supported Welch's initiative. Fish cautioned that Welch should send a telegram to the Olympic Committee in addition to the letter. Feeling that he had ample time, Welch decided against the telegram, only to learn that the Olympic Committee had selected Lake Placid the day before the commission's letter arrived.

In the summer of 1929, governors Franklin D. Roosevelt of New York and Morgan F. Larson of New Jersey inspected the Palisades parks. Roosevelt was very familiar with the parks and made sure that he and Larson spent time with some of the children in the 112 group camps now operating in Harriman Park. An article in the *New York Times* summarized the comments Roosevelt made during the park tour at the Bear Mountain Inn: "The successful cooperation between New York and New Jersey in the Palisades Interstate Park was cited yesterday by Governor Roosevelt as a striking example of the way cooperation might be effected in many other fields, particularly the economic."

Roosevelt was also quoted on the benefits of outdoor recreation. "It will relieve us of the dangers of overcrowding in the cities, the strain on our nervous systems. In the long run, it can be seen, the park idea is essential to American civilization. The two States of New York and New Jersey have set an example for the whole country." Unmentioned was that the car provided by the commission to Welch was so dilapidated that he had to borrow a car from an acquaintance to transport the governors on the tour.

The pace of administrative affairs was reasonably stable. A large administration building, including boardroom, court, police desk, and fifteen offices, had been completed in the New Jersey section of the park system in 1928. Inquiries on everything from fundraising to transplanting mountain laurel continued to flow to Welch's desk from park advocates throughout the United States. Welch learned, though, to his regret that his good friend, sixty-two-year-old Stephen T. Mather, was stepping down as director of the National Park Service because of ill health. Horace M. Albright, who sent the elk to Bear Mountain from Yellowstone, would succeed Mather.

Dialogue continued regarding the "treaty" between the two states that would reconfigure the identical-twin interstate commissions into one governing body. The commission minutes, dated March 1929, confirmed that "the Chairman of the State Council of Parks (Robert Moses) still is in opposition to the treaty." Then, on October 24, 1929, "Black Thursday," the stock market fell by 500 points. A slight recovery occurred on Friday. On the following Monday, full-blown panic set in. The onset of the Great Depression was at hand, and the commission would feel immense impact.

13

The Compact

"ON YOUR AUTHORIZATION to expend not more than $10 million for property on the Palisades," Charles O. Heydt wrote to John D. Rockefeller Jr., "we have expended up to the present time approximately $8,050,000 . . . we have come to the point where we must deal with holders of the larger tracts . . . Mr. Osborne and I have discussed the matter at length and we recommend that authority be given to bid up to $6,000,000 (in addition) for the Paterno holdings."

In a follow-up communiqué four days later Heydt added, "There is an immense amount of curiosity as to who is buying the Palisades property and we have done everything possible to conceal the operations underway by operating through perhaps a dozen different corporations to confuse the issue. Representatives of the Palisades Park Commission have been to see Mr. Osborne, but he has, of course, been very evasive and non-committal."

Heydt was a senior real-estate specialist on Rockefeller Jr.'s staff. John A. Osborne was a real-estate broker in Bergen County, New Jersey. Through these two men, Rockefeller Jr. had set out to purchase the crest of the Palisades from Fort Lee, New Jersey, to the New York border, a distance of about thirteen miles. Prompted by his familiarity with the commission and strengthened in resolve by *The Regional Plan of New York and Its Environs* report, Rockefeller Jr. was secretly buying Palisades land, including the Dana property that the commissioners had sold only a few years earlier.

His respect for the commissioners and William Welch did not extend to including them in the strict circle of confidence surrounding his land-buying activities. At the direction of Governor Morgan Larson of New Jersey, Welch and Elbert King were busy gathering landowner information along the crest in anticipation of legislation that might be sought to allow the commission to extend its condemnation authority inland from the edge of the cliffs. The commission was running on a track parallel to that of the secret land buyer.

Welch and King, along with scores of local residents, land speculators, builders, and elected officials, were wondering just what was going on. Rumors circulated as the scale and pace of the land-buying activity began to sink in. Rockefeller Jr. was prominently mentioned, but other rumors had it that a "group of public-spirited citizens" was buying the land to preserve it, or that Columbia University was amassing property for a new campus. Osborne, the front man, was a human vault; the secret could not be pried from him, not by reporters, other realtors, Frederick Sutro, or the other commissioners.

If the commissioners had been privy to a letter written by Rockefeller Jr. to Heydt as 1930 ended, they would have been even more astonished. Working the numbers down to the penny, Rockefeller Jr. confirmed in writing a discussion of the previous day with Heydt at the University Club at which Heydt verified that $7,508,400.23 had been expended for outright purchases. Heydt added that contracts, mortgages, interest, and taxes "coming due year by year" until 1935 amounted to another $13,348,096.79. Rockefeller Jr., having committed a grand total of $20,856,497.02 to the clandestine acquisition project, directed that the "buying program of the Palisades" be "terminated."

He added this intriguing thought: "We were of the opinion that a broad motor road should be laid out along the property, safely back from the edge of the cliffs, so as not to destroy the beauty thereof—." These words were the first hint that Rockefeller Jr. would be a champion of the Palisades Interstate Parkway. For the time being, though, he planned to stand back and await further events, being particularly encouraged by the thought that the commissioners might, in fact, succeed in gaining broad legal authority to "take" those properties that could not be obtained in willing seller–willing buyer deals. The commissioners remained in the dark about Rockefeller Jr.'s activities, strategies, and hopes.

A saga in scenic preservation of more human and heartrending form was unfolding in Harriman Park. Governor Franklin D. Roosevelt received a petition from fifty-three residents and allies of Sandyfields, a mountain hamlet tucked in a remote section of Harriman Park, who opposed the construction of a dam and the creation of a 250-acre lake. Claiming that the project would "flood several of our homes and completely wipe out our entire village, which has been in existence for over 100 years," the residents pleaded for help from Albany to stop construction of the dam.

In response to the petition, Welch confirmed to the governor's staff that most of the Sandyfields residents had voluntarily sold their properties to the commission. Only a handful of owners remained, and of these, most had expressed interest in selling. But the existence of the "mountain folk" was a poignant connection with the past. Writing for the *New York Times Magazine*, Diana Rice said, "Up Sandyfields way the mountaineers of the Hudson River Highlands, near Bear Mountain, are still singing their hymns."

Acknowledging that many of the residents, feeling the pressure of park development, were selling out, Rice posed the question, "But what does cash mean to a mountaineer bereft of his mountains, his shotgun and his hound-dog? Five miles or so west of Stony Point, which lies along the Hudson River not far from Haverstraw, the settlement boasts but a handful of houses. A general store, a little schoolhouse and a Methodist church are its gathering points. There is no post office. Mail comes by rural free delivery from Stony Point; packages by parcel post. The mountaineer is an ardent reader of catalogues, and his perusal of these thick volumes often result in heavy work for the mailman. Two hundred dollars in one day have been known to leave the mountain en route to mail order houses in the Middle West.

"Womenfolk of the Highlands do not waste time shopping. Few of them ever leave their homes. They order their gingham, cotton stockings, and heavy shoes direct from the great emporiums whose advertisements read like fairytales. If the old folks cannot read, the young ones can. For even in the mountains, life is growing more complex as sons and daughters journey out to the strange world beyond the rim of hills and bring stories of different standards of living."

In addition to Sandyfields, the mountain settlements included Baileytown, Bulsontown, Johnsontown, Pine Meadows, Pittsboro, Queensboro, Woodtown, and Doodletown, the hamlet through which British troops marched on the attack against forts Clinton and Montgomery. The hamlets were settled by pre–Revolutionary War woodcutters, hardscrabble farmers, and hunters. Hessian soldiers, deserters during the Revolutionary War, found their way into the Highlands to join the earlier settlers, as did runaway slaves and the occasional convict looking for a remote hiding place. As described in one news article, "They set up their lean-tos and cabins in the Ramapo forests, fished in the streams for trout and pickerel, hunted cottontails and snow-shoe rabbits and deer and completely ignored the world outside. They . . . were sufficient unto themselves."

Because so many of the Sandyfields residents had already voluntarily sold their properties, the petition to stop the dam project failed. The last holdouts sold their land and moved on. In 1935, Ramsey Conklin, sixty-two years old, closed the door to the isolated cabin that his grandfather had constructed in 1779. He had never been more than a few miles from the cabin, but when waters from another of the artificial lakes began lapping at the pasture where he kept his cow and heifer, he and his family also agreed to be resettled in an abandoned schoolhouse outside the Harriman Park boundary. Referring to his new home, Conklin said, "This here is a right busy spot, right off a road with them motor cars goin' by, as many as seven or eight in a single day. It won't be good for the stock."

In a way, the hamlet of Doodletown was more fortunate. It was not immediately in the path of park development and continued to thrive, even gaining population as parklands were acquired around it. For understandable reasons, the hamlet's population sharply declined during the Depression, but Doodletown persisted and adapted, providing solitude for its residents and some employees of the commission while other hamlets in more remote forest vales disappeared one by one.

Still, the days of the last remaining hamlet were numbered. Park development plans in the 1960s convinced the last of the Doodletowners to sell out. The hamlet ceased to exist in 1965. It continues to live only in the pages of the book *Doodletown*, tenderly written by a former resident of the hamlet, Elizabeth "Perk" Stalter, and published by the commission.

In addition to the impact on the pioneer lifestyle of the Hudson Highlands, the Depression created great uncertainty for the commission, coupled with a search for new answers to keep the parks functioning. Welch, writing to Robert Moses, suggested a special push for funding from Albany to hire additional park workers. "The unemployment situation in the vicinity of our park . . . is most deplorable. A number of large factories have closed up . . . there is practically no State work . . . no county work and very little municipal work . . . private construction work is at the lowest ebb I have seen."

Red ink began flowing across the ledger of the Bear Mountain Inn, and the commissioners voted to sell the dayliners *Clermont* and *Onteora*. They succeeded when the McAllister Navigation Company made a bid of $300,000 for the two vessels. McAllister's timing was bad. A combination of the Depression and the 1931 opening of the George Washington Bridge was too much for the company, which had begun providing river excursions to Bear Mountain in 1913. In the late 1930s the company vanished when its assets fell under the hammer of the deputy U.S. marshal at a bankruptcy sale.

"Retrenchment," as Elbert King referred to it, was under way in all maintenance and operating departments. Wages were cut across the board, and a hard squeeze was placed on part-time employees. Raymond Torrey, the newspaperman-turned-publicist for the commission, found his salary cut in half, from $3,000 to $1,500 per year.

Torrey was fortunate, nonetheless, to have any salary at all, because a year earlier, Moses had tried to strangle him. Torrey wore several hats, including service as secretary to the State Council of Parks. His strongly held environmental views clashed with Moses's vigorous push to build the Northern State Parkway on Long Island. At a meeting in Albany, the confrontation came to a head when Moses cursed at the rotund, usually gentle Torrey. Torrey lost his temper and called Moses a "big noisy kike." Moses leaped forward and grabbed Torrey by the neck in a vice-like grip. Restrained by Jay Downer, executive director of the Westchester County Park Commission, who was a long-time friend of Welch, Moses released his grip only to grab a heavy smoking stand and throw it at Torrey as his antagonist beat a retreat out the door.

Torrey sent a letter of apology to Governor Roosevelt and resigned from his position with the State Council of Parks. Fortunately for the commission, the Moses-Torrey incident was kept at arm's length. In several instances, King referred to "our good friend, Bob Moses," in correspondence on various park and State Council matters, and Moses, his wife, and their two children were regular users of a guest house at Bear Mountain, including one extended stay while Moses recuperated from a serious illness.

As the commission grappled with the challenges of the Depression, King alerted Sutro in May 1930 that "our dear friend, Dr. Partridge, died last Friday." Stricken while attending a dinner at the New York Academy of Medicine, Edward Partridge had served as a commissioner for seventeen years. Only vaguely remembered by those who eulogized him was his turn-of-the-century advocacy for the preservation of the Hudson Highlands as a national park.

No one seemed to remember that Partridge had influenced and reinforced the thinking of E. H. Harriman, leading to the decisive donation of land and money that placed the commission firmly on track toward building an expansive park and historic-site system. Despite the fuzzy recollection, Partridge would have been pleased by the selection of his successor.

Acting promptly, governors Larson of New Jersey and Roosevelt of New York appointed former New York governor Alfred E. Smith to the post. Smith would serve for the next thirteen years. The commissioners, Welch, and King were delighted with the prestige of the Smith appointment.

King was attending to the various rules that guided commission activities, including the continuing restriction against single women who wanted to camp. He had consistently turned down women's camping requests, making exceptions only for married women accompanied by their husbands. This restriction was challenged when a young man, aged nineteen, who had been crippled by infantile paralysis at age two, asked that his twenty-two-year-old sister be allowed to attend him on a camping trip to Harriman Park.

King turned to Sutro for advice. King and Sutro were leery of single women, who seemed to expect the same freedoms in the park as men, but their practical judgment prevailed. Permission was granted. As a result, a slight dent appeared in the armor of the men-only camping policy. King and Sutro may have been influenced by their personal knowledge of the physical challenges faced by polio-stricken Governor Roosevelt.

King had clearly become the gatekeeper for the commission. He was at his desk in all seasons, tightly controlling the purse strings, responding to thousands of letters, staying in constant contact with the widely traveling commissioners, working with Welch to stamp out controversy and short-circuit criticism at the first sign of trouble, rushing to Albany and Trenton to deal with the latest budget crises, seldom away from his post for personal reasons. Persnickety, precise, loyal, King had become as invaluable and irreplaceable as anyone can be in an organization.

But in response to an inquiry in May 1931, King's assistant, Jessie A. Marvin, responded, "Mr. King is at home, ill, and may be there for some days." Elbert W. King, who had joined the commission as a clerk in 1913, died on June 7, 1931. The commission minutes memorialized the man who was so dedicated to his responsibilities: "The years of his service were the years of the growth of the Park. As the Park grew, so grew his versatile ability in a wide field of unique and exacting duties, grew his enthusiasm for the park and zealous devotion to its development, grew the wide circle of his friends and the high regard in which he was held by them and especially by his Commissioners."

The vacuum left by King was filled a few months later in an unlikely manner. The Depression caught up with Frederick C. Sutro. His company, Sutro Brothers Braid, Inc., went under. After eighteen years as a commissioner and successor to Richard Lindabury as president of the New Jersey

arm of the commission, Sutro resigned. His former commissioner colleagues immediately hired him as executive director, assistant secretary,
and assistant treasurer, the same titles that had been held by King. Sutro
was making a difficult transition by stepping down from a board of directors' position to become a staff member within the same organization. He
would now work as second-in-command to Welch and answer to the commissioners with whom he had been a copartner for so long.

With barely a hitch, Sutro made the transition. He began immediately
to tighten business practices, demanding strict compliance with budgets
that he critiqued in detail, department by department. He cut Torrey's
activities even more and discharged the commission's Albany lobbyist.
Wages paid to the employees of the Bear Mountain Inn were reduced.
Welch, who had assumed routine approval of funds to replace worn-out
ice-hockey equipment, found himself closely questioned by Sutro and
Commissioner Charles Whiting Baker regarding this need. This sent a signal to the staff not to assume anything and to be prepared to defend every
expenditure. The businesslike Sutro greatly admired Welch and the legacy
left by King, but he intended to firmly apply his own standards of corporate
management. Over the next three decades, his influence on park conservation grew to the point that his voice was always heard and his advice never
ignored.

Adding unlikely event to unlikely event, Sutro's replacement on the
commission was Abram De Ronde, who had already served from 1900 to
1912 but had fallen out of favor with his own Democratic Party associates
and was not reappointed by then-governor Woodrow Wilson. After an
absence of nineteen years, De Ronde was reappointed by republican Governor Larson.

These changes were taking place just as Welch received the good news
that New York, through New York governor Franklin Roosevelt's newly
created Work Relief Program, was making available $750,000 to the com-
mission to provide employment for able-bodied men. To the press, Welch
stated, "This will be work, not charity." The first members of this new work
force soon arrived by train from New York City. "They were as uneasy as
strangers arrived in a new land," Welch remarked. And well they should be.
Most of the men knew nothing of vast woodlands not dissected by pavement, free of concrete walls where the sound of wind in the trees replaced
the grind of traffic noise on city streets.

Within weeks, 2,500 men were employed in the park system, 1,200 of
them traveling daily from New York City. The urban contingent crossed the
river by ferry at 6:15 A.M. to board a northbound train, arriving at Bear

Mountain at 7:30 A.M. to begin walking or being trucked to the work sites. The workday ended at 4:00 P.M., followed by a contest of shoving and pushing to determine who got the seats on the train or who had to stand for the long trip back to the city.

Many of this vanguard of laborers had held "white collar" jobs that evaporated with the Stock Market crash. Their former status was confirmed by the suits, ties, and street shoes many of them wore to work at Bear Mountain during the first days of their assignment. These clothes were soon replaced by sturdy pants and boots commandeered by Welch. The men were grateful for the work clothes and, even more, for the $4 per day they were paid, minus 50 cents for transportation. For the first time since his days in the spruce forests of Oregon during World War I, chief engineer William Welch found himself in command of a small army of men whose eagerness for any kind of work superseded their questionable qualifications.

Welch was ready with a long list of projects, ranging from the clearing of fallen and dead trees from thousands of acres of forest and the planting of seedlings to sophisticated and complex construction projects, including buildings, dams, roads, trails, utility systems, and a variety of public-use areas. He even hoped to construct a "through highway" from the recently opened George Washington Bridge to Bear Mountain, reminding anyone who would listen that "now, on Sundays, it often takes eight hours to travel from Bear Mountain to New York City." At the top of Welch's list was the construction of the long-envisioned "George W. Perkins Memorial Highway" to the summit of Bear Mountain. Now Welch had the men to accomplish the task. No heavy machinery would be used to assault the route up the mountain. Instead, Welch directed that the road be built by hand so that it could be placed as gently as possible on the steep slopes of the mountain.

The first Depression-era laborers to arrive at Bear Mountain were soon joined by a much larger work force. The election of Franklin Roosevelt to the presidency in 1932 brought great emphasis to publicly funded work programs. The first two Civilian Conservation Corps (CCC) camps in the eastern United States were promptly established in Harriman Park early in 1933, one at "Beechy Bottom" on an old road over which "Mad" Anthony Wayne and his troops had marched in 1779 to attack the British garrison at Stony Point. The camp occupied land once owned by another Revolutionary War hero, Isaac Van Wort, who, while serving in the American militia at Tarrytown, had helped capture the British spy Major John André in 1780.

The second camp was farther west, located in a "wilderness" location known as "Pine Meadow Swamp." Each camp housed 185 men, but before Welch and his staff could catch their breath, nine more CCC camps were under construction, increasing the complement of men to almost 6,000. By 1934, twelve camps were in place, accommodating a work force of 10,000 men.

A flurry of projects was inaugurated, extending from the New Jersey Palisades to Storm King Mountain north of West Point. Welch marched a troop of men into the Tallman Mountain site to remove quarry equipment and begin park development, even though the Tallman case remained in court. Hook Mountain came alive through landscaping and development of the old quarry sites into picnic areas, playfields, and hiking areas. Construction of a golf course was begun at Rockland Lake adjacent to Hook Mountain. A new administration building made of stone and logs was rising at Bear Mountain. Intricate stonework was a hallmark of projects everywhere, ranging from huge walls built to stabilize the road cuts on the face of the Palisades cliffs to small and large buildings scattered throughout the park system.

Civilian Conservation Corps Project (PIPC Archives)

At Blauvelt Park, Torrey, still working diligently despite his cut in salary, reported, "An experiment in the rehabilitation of single, homeless, unemployed men selected from thousands seeking shelter at the New York Municipal Lodging House, or in the flophouses on the Bowery . . . has attained results within two months . . . leading those responsible to believe that it may point a way to solve some of the problems of human wreckage caused by the Depression."

The 200 men at Blauvelt had been itinerant wanderers who "were driven to seek scanty meals in the breadlines." Placed in charge at Blauvelt was a thirty-seven-year-old Social Sciences graduate from Notre Dame, himself from the ranks of the jobless. The head cook, also jobless until the Blauvelt opportunity came his way, had been an instructor in the army's Cook School. Only seven of the Blauvelt recruits dropped out or were kicked out of the program. The rest, held together by a place to sleep, square meals, and the challenge of real accomplishment, formed themselves into an effective work unit. Part of the incentive was that they could save as much as $30 over three months if they were frugal.

In addition to the resident camps, the Works Progress Administration (WPA) continued to provide a commuter work force from the city. A reporter for the *Ossining Citizen Register* spent a day with these men: "The hardships are seared in the soul of every man who goes to the WPA project at Bear Mountain. It is a story of men, unused to physical labor, moving huge rocks with bare hands in bitter cold; dinners of frozen sandwiches; crowded transportation; rugged pioneer conditions; generations removed from soft, Twentieth Century drawing-room life. . . . Bear Mountain is considered a "Siberia" among the WPA workers." Despite the trauma of the Depression, highest-quality results were expected. Time was not an urgent factor, so care could be taken in attendance to details, training, and project design. Talent and maturity were the common denominators among the work force. The organizational structure was military in design and expectation. The men had to measure up to a stern standard of discipline.

An *Ossining Citizen* reporter thought that the whole endeavor was like fighting a war. White collars merged side-by-side with blue collars. The results were artistic, appropriate, solid, and lasting. The stone-and-log theme of the Bear Mountain Inn, completed in 1915, was reflected in an architectural style throughout the park system. Probably because of the hardships, many of the men who found themselves engaged in the public-works programs of the economically devastated 1930s carried the experience with them as a badge of honor for the rest of their lives, having learned much in a hard school of personal necessity and by accomplishing truly remarkable projects that have lasting public value.

William Carr, guiding the fortunes of the Trailside Museum, took advantage of the public-works programs by increasing his corps of trained naturalists to twenty-two, including geologists, botanists, archaeologists, and zoologists. He sought and won Welch's approval for the construction of new buildings, trails, and exhibits at the museum. In league with Ruby Jolliffe, museum activities were expanded to include four "regional" museums, strategically placed in Harriman Park to more effectively serve the group camps. Carr and his naturalists staffed the regional museums, allowing for direct contact with hundreds of wide-eyed children. Influenced by the style of "Uncle Bennie" Hyde, the education programs specialized in snakes. Mimicking Hyde, a popular technique used by Carr and his staff was to allow snakes to casually crawl out of shirt sleeves or collars while the seemingly oblivious naturalists talked about something else.

Out in Harriman Park, snakes and museums were only part of Jolliffe's constant effort to expand the thinking and joy of her campers. Her most compelling educator existed in the person of Princess Te Ata, a member of the Chickasaw Nation who had once performed on Broadway. As described in *Liberty Magazine*, "Sometimes she wears blue butterflies in her jet-black hair. She is tall, very straight, very slender; voice of clear contralto, tones

Princess Te Ata (PIPC Archives)

quietly modulated. Her name—Te Ata—means 'Bearer of the Morning.' She's an American-Indian girl; on stage she portrays dramatically the ancient arts and legends of her people."

The magazine *Camp Life* echoed the allure: "Princess Te Ata modestly claims to interpret the fine spirit of the North American Indian in her artistry of dance and song and old legends. Yet she goes far beyond the limits of any tribe or any nation. It is not only the soul of the Indian which takes form in her figure, her dress, her voice and poetry of her cadenced movements. It is the spirit of man and womankind expressing in its fullness the beauty of the wild and primal world from which it sprang. This is what makes her art significant for those of us who are privileged to hear it."

This privilege was extended time and again to the young campers on Te Ata's regular tours through Harriman Park. The girls and boys would sit in rapt attention, captivated and carried along by the voice and movements of the princess. Jolliffe felt that Te Ata's importance to her young campers was such that one of the newly created CCC lakes should be named in her honor.

Jolliffe was joined by her friend and compatriot Eleanor Roosevelt in this request. Welch and the commissioners readily agreed. On the appointed day in July 1932, Mrs. Roosevelt, surrounded by hundreds of the campers she cherished, watched the ceremonial arrival of Te Ata, carried by canoe across the lake that would bear her name. Mrs. Roosevelt performed the christening by pouring a mixture of waters, collected from all the other park lakes, into "Lake Te Ata." The First Lady later invited the princess to perform at the White House and, again, for King George VI and Queen Elizabeth on their 1939 visit to Hyde Park.

Through his press releases, many of which were picked up verbatim by newspapers, Torrey attempted to extend the educational services provided within the parks by Carr, Jolliffe, and their staffs. Writing about the wildflower bandits, he declared, "Some are just plain dumb; they pluck laurel at the side of a sign with letters a foot high—PLEASE DO NOT PICK WILD-FLOWERS: OTHERS WANT TO SEE THEM. Some are just thieves, and a few are innocent and honest, supposing that the flowers are for everybody, and walk back past the doors of the police station. Visitors of foreign origin forget English. Sweethearts are funny; they try to take the blame for each other. The judge's desk is heaped with wilted flowers." The police desk sergeant was quoted as saying, "If we didn't pick 'em up this way, they'd carry off the park."

But on other matters of natural history, Torrey was more upbeat: "Those who recall the condition of the forest in the Highlands a quarter century ago will have in their minds' eye a memory of sad-looking sprout lands, often burned over and set back for years, which were the result of a century of repeated cutting of saw logs, charcoal, and cord wood. The forest in the park, aided and protected by man in the past twenty years, now shows what nature can do if left alone in our deciduous broad-leaf vegetation in the northeastern United States. The recovery of the larger forest trees is bringing with it the return of shrubs, herbaceous flowering plants, ferns, mosses and lichens, so that the Harriman, Bear Mountain, and Storm King sections, in particular, now present one of the richest and most interesting regions of the East for the study and enjoyment of native flora and fauna."

Reporting on the consequences of flooding caused by the many artificial lakes created in Harriman Park, Torrey observed, "Many species of bog and shore plants did survive in the curious floating island, which was formerly part of the quaking bog around Little Cedar Pond. When the water rose, the bog tore loose and floated around the lake until it finally came to a halt near the southwest end of the present lake. The old bog plants, the insectivorous sundews and pitcher plants, the marsh fern, water loosestrife, cranberries, swamp honeysuckle, poison sumac, red maple, a few small tamaracks, a terrestrial bright yellow flowered species of bladderwort, water arums and cattails, persisted and now afford a very interesting association on the floating island."

As part-time tour guide, in addition to his public-relations specialty, Torrey hosted a visit by scientists to the Palisades cliffs, explaining his rationale to Sutro: "The fossil Phytosaur was found in this part of the Park, a primeval dinosaur, with a duck bill, about as big as an alligator."

While Carr, Jolliffe, Torrey, and most of the staff were in the trenches, attempting to cope with and squeeze advantages from the contradictory forces of the Depression, Sutro was coping with the surreal nature of funding for the commission. Above the line, in the public-works arena, the commission had access to funding and a work force beyond its wildest imagination. Below the line, in the realm of the traditional state budget process, the news was grim. Reacting to an episode in Trenton in which funding for the commission almost was eliminated, Sutro wrote to Commissioner Edmund Wakelee, who had succeeded Sutro as president of the New Jersey branch of the commission, "This incident has impressed on my mind the extreme hold which we, as an interstate body, have on the two States. Either Legislature could practically without notice destroy the work of almost 35 years by simply refusing to appropriate any money to the Park. All this again brings up the question of the vital importance of pushing to a conclusion a thought that has been uppermost in Mr. White's mind for years, vis: the enactment of the interstate compact."

J. DuPratt White picked up this theme in a letter to Rockefeller Jr. in which he reiterated the need "for a treaty between the two states." White continued, "I have been met with opposition in this effort by what I think is jealousy. . . . The formation of the treaty has not received the approval of the State Council of Parks"—meaning Moses.

Even with belts tightened, the commission was able to host the Thirteenth National Conference on State Parks at the Bear Mountain Inn in May 1933. The highlight of the event was a motorcade up the George W. Perkins Memorial Highway, still under construction, that allowed the attendees to be "the first to enjoy by automobile" the magnificent scene from atop Bear Mountain, including views to the towers of the New York City skyline, forty miles away, and a vista that included four states: New York, New Jersey, Connecticut, and Massachusetts.

The next day, White presided over an event at the Trailside Museum at which Eleanor Roosevelt unveiled a memorial plaque in honor of Stephen T. Mather. The principal speaker was secretary of the interior Harold Ickes. Attending were Horace M. Albright, Mather's successor as director of the National Park Service, Mrs. Stephen (Jane T.) Mather of Darien, Connecticut, Mather's daughter, Mrs. Edward R. McPherson Jr., William Welch, and the conference participants.

Below the cliffs of the Palisades, another dedication took place on June 8,1933, when the old "Cornwallis Headquarters" was opened to the public, thanks to yet another public-spirited initiative taken by the New Jersey Federation of Women's Clubs. "Early American treasures of pine, oak, maple, pewter, china and glass, were given by 149 clubs, and 22 individuals," according to the Women's Clubs historical record. "The most precious relic was donated by the Commissioners, themselves, the sea chest of Henry Hudson."

Mrs. John R. (Cecilia Gaines) Holland had earlier written to Sutro suggesting that the Cornwallis Headquarters be "put in order permanently as a museum." She could look back at decades of personal advocacy for the Palisades, and she was pleased by the latest, women-sparked accomplishment.

Mrs. Holland would have been even more pleased had she known of a letter being prepared for the signature of Rockefeller Jr. that arrived in the hands of the commissioners within a matter of days of the Cornwallis Headquarters dedication: "Through various real estate corporations I am now the owner of certain parcels of real estate along the top of the

Palisades on the west side of the Hudson River, commencing at a point about 2,500 feet south of the George Washington Bridge and extending north to the New York–New Jersey State Line, a distance of approximately 13 miles. My primary purpose in acquiring this property was to preserve the land lying along the top of the Palisades from any use inconsistent with your ownership and protection of the Palisades themselves. It has also been my hope that a strip of this land of adequate width might ultimately be developed as a parkway, along the general lines recommended by the Regional Plan Association, Inc."

Describing his understanding of the commission's interest in obtaining funds for parkway construction, Rockefeller Jr. used simple words to affirm his enormous charitable intention:

I am therefore now offering to give or cause to be given to you the titles to all of these parcels of property which I have thus acquired. . . . The gifts of land to be made hereunder are to be made to your Commission from time to time for parkway purposes under appropriate conveyances as the various parcels are required by your Commission for construction of the parkway. This offer is conditioned upon your being able to obtain within a reasonable time sufficient funds to commence the construction of the proposed parkway and thereafter to continue and complete its construction. It is my understanding that the purpose will be eventually to continue the proposed parkway to the north across the State line into the State of New York and up to the Bear Mountain section of the Palisades Interstate Park. In this connection, it would be my hope that it might ultimately be possible, if the lands hereby offered to your Commission are accepted, for a treaty to be arranged between the States of New York and New Jersey for a joint commission, with appropriate powers for policing and maintenance, which would insure for all time a continuity and permanence of interstate administration for both the proposed parkway and the park areas adjacent and connected to it.

Headlines in the *New York Times* trumpeted the proposed gift. With acceptance of the offer a foregone conclusion, White referred to the "princely offer" and confirmed the obvious: that the suggestion of a "treaty has my hearty approval. In my opinion, the matter should receive prompt and favorable consideration of the Legislatures of the two States and the Congress of the United States." Walter Kidde, who had provided a report on the Palisades to the New Jersey State Chamber of Commerce two years previously, was credited with convincing Rockefeller Jr. to begin his land purchases. Anticipating public curiosity about exactly what a parkway was, one of the *Times*'s articles explained,

> A parkway, in fact and by legal definition, differs widely from a highway or boulevard. Specifically, it is distinctive in that the right of access to it of the owners of adjacent land is restricted. This means that its park-like character can be maintained; that crossroads may be admitted only at intervals which fit into the general plan, and that disfiguring structures cannot be erected along the borders of the right of way.

Confidence by White and his fellow commissioners that they could readily find in federal public-works monies the estimated $3.5 million needed to guarantee construction of the twelve-mile New Jersey section of the Palisades Parkway was soon dashed. Despite support, even from President Franklin Roosevelt, White encountered frustration after frustration in his attempt to convince authorities in Washington that funds were needed from the Public Works Section of the National Recovery Administration to help the commissioners keep their end of the bargain with Rockefeller Jr.

In the meantime, President Roosevelt and the First Lady were back at Bear Mountain in early September 1934 for a private inspection of the new road to the summit of Bear Mountain. On October 31, 1934, Roosevelt officially dedicated the "George W. Perkins Memorial Highway," and he applauded Mrs. Linn Perkins for her generous gift of funds that allowed for WPA construction of a memorial tower, made of large blocks of granite, on the mountain's summit.

The sixty-five-foot tower, designed to serve the practical needs of meteorological, aircraft, and forest-fire surveillance, offers a sweeping, 360-degree view of the surrounding landscape, including the ridge lines of Sterling Forest to the west, an area of expansive woodlands and historic iron mines and furnaces just across the Ramapo River from Harriman Park that occasionally prompted entreaties to the commission for its preservation. In one exchange of correspondence, Welch affirmed that the Sterling Iron and Railway Company, a Harriman enterprise, owned Sterling Forest, but he gave no encouragement that the land would come into the ownership of the commission.

White's continuing frustration about lack of funding for parkway construction was finally matched by that of Rockefeller Jr. himself, who, at one point, referred to the "slow moving" commissioners in a letter to a member of his staff. After months of effort, but no results, Rockefeller Jr. decided to remove the original gift conditions and give the land to the commission anyway.

In June 1935, the *New York Times* confirmed the decision by reminding readers that the commissioners were to "obtain additional land and to construct a parkway from the bridge to the State line. Funds for this have not been obtained. No such stipulation is contained in the present gift, although the parkway idea has not been abandoned."

After remaining legal details were resolved, the Rockefeller Jr. deeds were transferred to the commission in late November 1935, including the Dana land, minus the old mansion that was torn down. The Dana property had come full circle. Rockefeller Jr., deliberately concealing his role until the success of his land-buying venture on the Palisades was assured, added splendidly to the development of the commission's park and historic-site system. The threat of "gap-toothed" buildings parading along the crest of the New Jersey Palisades was averted, or so it seemed.

George Gyra, a public-works employee from New York City, had things on his mind other than preservation of the Palisades. Wandering away from his fellow laborers during a lunch break in Harriman Park, "and quite by accident," as later reported, Gyra and one of his coworkers discovered a crevice on the slope of Letter Rock Mountain. "It was barely large enough to admit a man, but he crawled inside. He got in six or eight feet and it was all dark. He lit matches but they made hardly a splash in that great dome of darkness. He picked up stones, when he was able to stand erect, and pitched them as far as he could, trying to hit the opposite wall, but they didn't hit. Then he crawled out."

Park employees and police officers were perplexed when they started hearing what sounded like thunder in the park on clear days with not a cloud in sight. Investigation led to the arrest of four men from the city who were dynamiting the Letter Rock crevice in search of buried treasure. Gyra was not among the culprits, and he claimed no knowledge of how they found out about his discovery. Welch affirmed in news coverage of the arrests that "the whole territory around Letter Rock is rich in legends of buried treasure. One tells of a lode called Spanish Silver Mine, on Black Mountain, a mile east of Letter Rock, where Spaniards who came up from the West Indies were supposed to have mined great fortunes in silver." Offering his usual pragmatic opinion, Welch reckoned, "Don't think there's a dime's worth there, myself."

The persistent Depression maintained a grip on the commission. Sutro had long since issued instructions that no checks would be accepted as payment for services and programs. Only "scrip" or cash would do. The commission issued checks to its employees but would not cash them.

The Dyckman Street Ferry, so important for access by city-dwellers to the Palisades, struggled to reopen in 1935. Use of the ferry by the thousands who annually enjoyed the picnic grounds, beaches, and hiking trails under the Palisades was in jeopardy. Low-income people, especially, might lose the only access they had to the parks. The commissioners appealed to Mayor Fiorello La Guardia for help in developing a revised city contract, minus a $25,000 security deposit, so that the financially threatened ferry operators could continue to offer low-cost services at the Dyckman Street landing. La Guardia agreed, and ferry service was reestablished in May.

Despite the Depression or because of it, group outings to Bear Mountain via the river dayliners or buses by more affluent visitors continued to hold firm. More than one hundred groups reached Bear Mountain during one July-to-September period. Among the groups were the Metropolitan Leather and Finding Club, the Yan Ye Recreation Association, the G. H. P. Cigar Company, the Mother A. M. E. Zion Church, the Sons of Italy, the Swanky Yacht Club, the Lu Lu Temple Patrol Party, the Go-As-You-Please Bowling Club, the New York Stock Exchange Glee Club, the Lithuanian Pleasure Club, the Hartford Steam Boiler Inspection Company, and the Downtown Athletic Club.

But as usual, not all groups were encouraged to voyage to Bear Mountain. Writing to Dan F. McAllister, Sutro confirmed, "All are agreed that they prefer to have especially colored parties go to the North Dock at Hook Mountain rather than to Bear Mountain or the South Dock at Hook Mountain . . . to relieve embarrassment . . . of having colored parties mixed with daily visitors from your regular boat [which] also relieves difficulties experienced at Bear Mountain from having large groups of colored people mingle with the regular visitors there."

Sutro was reflecting the racist attitudes of the times, even though the commission routinely fought off demands that African Americans be excluded entirely from the parks. Segregation of African Americans to avoid "embarrassment" of "regular visitors" obviously was wrong in a nation built on a belief in equality for all citizens, and it seems especially ludicrous given the behavior of some of the "regular visitors."

Reporting on a corporate picnic at Bear Mountain, J. J. Tamsen, superintendent of operations, declared, "The outing arrived yesterday. [They] drank 38 barrels of beer in all. The beer was distributed in pitchers, gallon mayonnaise jars, paper containers, and glasses. These were left strewn over the Park. The drunks lay around on the playground, men staggered, women were falling down, girls and fellows mushing it up. Ladies as well as men did not use the comfort stations. A great number of people got sick. Took until 9:00 PM to clean up."

No such report came from the North Dock area at Hook Mountain to which African Americans were directed.

Fortunately for the commission, most people of all ethnic backgrounds were well behaved. Good behavior was an absolute necessity in the mid-1930s in a park system attracting 14 million annual visitors representing one-ninth of the nation's population—more visitors than all the national parks combined. "We are engaged in the precarious work of attempting to please the public," Welch observed.

Commissioner Victor H. Berman, a 1935 New Jersey appointee to the commission and president of Onyx Oil, saw an opportunity to connect with the public by donating a fifteen-foot-tall Tiffany-sculpted bronze elk head. Berman felt that this spectacular piece of art was emblematic of the wild animals and native plants that the commission sought to preserve. The sculpture, unveiled at a ceremony by Berman's daughters, Joan and Audrey, was affixed to a rock ledge at the Trailside Museum overlooking the Hudson River boat docks. There it remains, silently graceful, now hidden from view by vegetation and almost forgotten.

Not so a statue of Walt Whitman, donated by Averell Harriman in memory of his father and mother. It stands prominently at an intersection of footpaths on the museum grounds, constantly greeting strollers and Appalachian Trail hikers with words from Whitman's poem "The Long Brown Path" chiseled into the statue's rock base.

The Depression slowed but did not stop the commission's acquisition of land. One important acquisition won a Torrey-inspired headline: "Storm King Mountain Now Preserved in Interstate Park." With the addition of a 203-acre purchase, Torrey reported that 900 acres of Storm King, guardian of the northern entrance to the Hudson River narrows, was in the commission's hands "and now stands in the way forever of any industrial development on the river front of the mountain which might mar its grandeur." Torrey's words were prophetic: Storm King was destined to become an industrial-development battleground of enormous importance that would tear at the very core of the commission and realign the national environmental agenda.

But a much more urgent battle was looming in 1936. Under the headline "Joint Board Urged on Palisades Park," New Jersey governor Harold G. Hoffman confirmed his request of the legislature to approve an "interstate compact" for the commission. On the New York side, the solicitor general and Governor Herbert Lehman's counsel recommended approval of the compact, leading to introduction of legislation that would authorize commissioners White, Harriman, George Perkins Jr., and Frederick Osborn, or any three members of this group, to negotiate the final details with New Jersey representatives.

Prominently absent from the list of negotiators was the fifth commissioner, former New York governor Al Smith, who, reflecting the influence of Robert Moses, opposed the compact. Moses's opposition and persuasive strength were made clear in a ten-to-one vote against the proposed compact by the State Council of Parks.

Moses charged that the compact "would lift the commission out of State control and supervision and would interfere with a unified state park policy." Despite Moses's position, the New York Senate voted to approve the interstate compact by a 28-to-22 margin, prompting Moses to complain that New York's action, in concert with New Jersey's, would "take control of the PIP from the State Council of Parks," thus contradicting his oft-stated position that the State Council was an advisory body made up of representatives of the various park commissions, including the commission.

These first exchanges of political fire were muffled and out of the limelight. The noise level dramatically increased when, to the surprise of commission allies, the New York Assembly rejected the Senate's action. "Mr. Moses, in his own slimy way, is determined to defeat these bills but he doesn't come out honestly and openly against them," said Assemblyman Mailler, a sponsor of the legislation. "Instead, Mr. Moses sends his representatives to the Capitol to urge their defeat." Responding, Assemblyman Moffat said, "No man has done more for the people of the State through park development than Bob Moses and it ill becomes a member of this House to refer to him as slimy." Looking for a compromise, the assembly and Senate created a joint committee to further investigate the compact idea.

Moses continued to seek ammunition against the compact. Using State Council of Parks stationery that ironically listed J. DuPratt White on the letterhead, Moses contacted Jay Downer, formerly chief engineer of the Westchester County Park Commission and now serving on the Rockefeller Jr. staff, to inquire about Rockefeller's position on the matter. "The PIP has been agitating for some time for acts of the legislatures of New York and New Jersey, ratified by the Congress, creating the PIPC (Palisades Interstate Park Commission) . . . by treaty and lifting it out of the two states." Downer responded that Rockefeller Jr. indeed had expressed support for the initiative but that no financial pledges or commitments were in any manner awaiting formal approval of an interstate compact.

In November 1936, commissioners White, Perkins Jr., and Baker traveled to Albany to testify before the joint legislative committee. White was serving in his thirty-sixth year of volunteer work for the commission, and he presented a lengthy, eloquent, and well-reasoned statement about the commission's purposes, accomplishments, and the benefits of formalizing a structure between New York and New Jersey that had existed on a handshake for many years. He spoke of the private donations that had been won over the life of the commission, amounting to almost $18 million on the New York side of the border. Stressing the advantages and efficiency of interstate cooperation, White carefully explained the details of the proposed compact, constructed to guard state control and hold the commission accountable to the two states. He reminded the legislators that the commission legally was required to remain a member of the State Council of Parks and would meet a similar requirement in New Jersey, should that state form a similar state council.

Moses, accompanied by Commissioner Smith, took his turn. If the compact is approved, Moses argued, the commission would not be subject to state supervision. The compact would destroy the park system in the state of New York. If the commission became self-perpetuating behind an interstate compact shield, eventually it "would die of dry rot," Moses said, perhaps alluding to the elderly White. Smith, who rarely attended commission meetings, made the remarkable statement that "as soon as one park sees another getting away from the State Council of Parks, all will be looking for some way to get out."

This was not exactly a ringing endorsement of Moses's leadership. White, responding, said that some of Moses's testimony was not based on fact, propelling Moses to leap to his feet to accuse White of calling him a "liar." Mark M. Jones, who had conducted the 1925 survey of commission activities for the Laura Spelman Rockefeller Memorial Foundation and who subsequently recommended an interstate compact, testified that "PIP investment hangs by a thread; in effect, the Park has a corporate existence that extends but from one year to the next."

Following the hearing, Commissioner Smith, discarding his opposition to the compact in favor of working out a compromise, met with White and the other commissioners to work on modified language. The principal change involved how the commission would report its financial activities to the two states and account for state appropriations.

With these changes, White wrote to Thomas M. Debevoise, senior member of the Rockefeller Jr. staff, affirming that "the compact provides that all such land and property shall hereafter be owned by the new Commission and continue under its jurisdiction and be used only for public park purposes, and that none of said lands or any part thereof shall be sold, exchanged, or conveyed except with consent of both States by specific enactments. The mutual pledge of the States to hold in high trust for the benefit of the public the blessings and advantages of the Park . . . would imply a pledge to the various benefactors, of whom there are many and some of whom are not living, of good faith for the future."

Rockefeller Jr. wrote back to White, referring to a "notable step" and saying that the compact would provide "assurance of permanency of the Park." The revised compact was approved by New York on April 5, 1937, by New Jersey on June 2, 1937, and by the U.S. Congress on August 19, 1937. The cross-border handshake that had lasted for almost four decades was replaced by the solid legal foundation of an interstate agency. All New York and New Jersey commissioners, including Smith, were appointed to the new Palisades Interstate Park Commission (PIPC). White was elected president.

14

◼◼◼

The Palisades Parkway

WORLD WAR II understandably brought great changes to the PIPC. The public-works programs gave way to the nation's call to arms in response to Pearl Harbor. Commissioners and PIPC staff alike joined or were recruited into the army, navy, coast guard, merchant marine, and civilian war industry. Gas rationing made travel difficult. For the duration, the *Clermont* was taken downriver to be used as a floating cafeteria at the Brooklyn Navy Yard.

The PIPC's staff, unable to provide continuing care for the forty-two elk that traced their roots to the Yellowstone herd, was directed by the commissioners either to find homes for the animals in zoos or arrange for their slaughter. Fortunately, a farmer from upstate New York decided to provide a retirement home for the herd.

Wartime travel restrictions reduced general visitation to the parks, but new clientele arrived at the Bear Mountain Inn specifically because of the travel constraints. In 1941 the commissioners decided, through contract, to place the inn under the management of a concessionaire, changing for the first time the long-standing policy of George Perkins Sr. and William Welch to keep the inn directly in the hands of the PIPC. The new concessionaire was Jack Martin, described as a "big amiable bear of a man; a human dynamic firework." Martin was a dedicated sports fan. Seeing the need for a place not too far from New York City where travel-restricted professional sports teams could stay and practice, Martin became host to the Brooklyn

Dodgers, New York Giants, Green Bay Packers, and Cleveland Rams (later the Browns).

Martin made certain, too, that accommodations were available for amateur teams. The Notre Dame and Cornell football teams stayed at the inn when they came to play the Army team at West Point. Martin also was a regular host to the Eastern Golden Gloves Boxing Team. Former heavyweight champion Jack Dempsey was a Martin guest while recovering from a serious illness.

On the great lawn adjacent to the inn, Jackie Robinson took batting practice under the appreciative eye of Branch Rickey, aiming his hits at the spires of the Bear Mountain Bridge. Robinson, who had broken the color barrier in major league baseball, was given his own table in the inn's dining room, where he ate alone.

During one season, after the Giants had checked in for three days, Martin strolled up to their coach, Steve Owen, and asked, "Know that fellow sitting at the next table there?" Owen responded, "No, but then you don't expect me to know all my rookie players by name yet, do you? Give me a little time." Martin, grinning, said, "Okay, Steve, I can give you all the time you want but it's costing the Giants money. That guy isn't even on the squad." Bear Mountain became a thriving sports training camp buzzing with activity, winning the informal stamp "happy gymnasium."

Athletes were not the only wartime visitors to the inn. During world conflict, with China at great risk, Madame Chiang Kai-shek, wife of the Chinese premier, arrived at the inn to convalesce for several weeks. There she felt safe, her privacy defended by Martin. On the wall just outside his office, Martin proudly displayed a thank-you letter from his most honored guest.

Athletes and special guests intermingled with soldiers and sailors sent to the inn for rest, including a "shell-shocked" GI who walked up to the front desk and demanded a room, with bath. To make the point, the soldier aimed a loaded machine gun at the desk clerk. Martin helped the young man gain treatment at a veterans' hospital and later reported that the GI was "completely cured."

Bands played, and food kept pace with the appetites of football and baseball players. Sports writers, among others, made sure the bar was kept open. One visitor, reminiscing years later, said, "The Bear Mountain Inn! What a place that used to be. All the big bands played there. Tommy Dorsey. Harry James. Oh, we used to dance! Kate Smith sat at a table at the Inn, composing her signature song, 'When the Moon Comes Over the Mountain.'"

Other changes came to the PIPC before and during World War II. Commissioner Averell Harriman was far away, serving as ambassador to Russia. By the time Colonel George Perkins Jr. and Lieutenant Commander Laurance S. Rockefeller returned to civilian life from their wartime duties, White, Welch, Frederick Sutro, Raymond Torrey, and William Carr were gone. Commissioners Edmund Wakelee, Al Smith, Charles Baker, William Childs, and Abram De Ronde were gone. So, too, was Torger Tokle, the dominant, almost legendary, prewar champion of the Bear Mountain ski jump. Tokle was killed in action while serving in Europe with the famous 10th Mountain Division.

During the war, President Franklin Delano Roosevelt once again visited the summit of Bear Mountain where, on cue, he was treated to numerous spirals of smoke rising on command above the vast forest canopy stretching all the way to the horizon. The smoke spirals were a salute to the polio-stricken president from Ruby Jolliffe and the group-camp managers in Harriman Park. The president could no longer stride into the camps as he once did, so the smoke spirals reached for the sky to mark the locations of scores of camps, signaling a "thank you" from the camp managers to a national leader who had befriended and supported them almost from the beginning.

J. DuPratt White was not witness to the changes brought by the war. He attended his last commission meeting in May 1939, then, because of dete-riorating health, resigned from the PIPC two days later, concluding almost four decades of volunteer service and leadership for the PIPC. He passed away on July 14, 1939, less than a year after the Interstate Compact was approved. Commissioner and former New Jersey senator Edmund W. Wakelee succeeded White as president. Wakelee, a key political sponsor of the original 1900 legislation that initiated the PIPC experiment and an equally effective advocate for New Jersey's approval of the 1937 compact, carried the PIPC through the war years until his death in 1945.

Constantly gaining power and influence, Robert Moses wrote to Laurance Rockefeller in August 1939 urging him to accept an appointment to the PIPC to fill the vacancy left by White. Moses added, "What this Commission needs is a complete reorganization of its administrative staff. The present Chief Engineer (Welch) should be pensioned or made the Consulting Engineer. He has a long and honorable record, and deserves a great deal of consideration, but his usefulness as active head of the organi-zation is over. The other executive (Sutro) never had the slightest fitness for his position, has rendered no service in it, and in my opinion is not entitled to any consideration whatever. He was a Commissioner and suffered some financial reverses, and [then] jockeyed himself into a paid job. A competent new executive could not possibly work for him."

Moses recommended that Kenneth Morgan, the former superintendent at Jones Beach State Park on Long Island, a Massachusetts Institute of Technology graduate in general engineering, and a Moses protégé, be hired as the general superintendent for the PIPC.

Rockefeller accepted an appointment to the PIPC within a matter of days after receiving Moses's letter. He would prove to be an exceedingly worthy successor to White, ironically serving for thirty-nine years, the same length of tenure as White. Rockefeller probably appreciated Moses's encouragement, but his real motivation was a deep and growing personal commitment to conservation, combined with recognition of the compelling interest his father held in preserving the Palisades.

Nonetheless, the caustic message Moses sent to Rockefeller about Welch and Sutro had its intended impact. In November 1939 Perkins Jr. wrote to Morgan on behalf of the PIPC, confirming that Morgan would join the staff on February 1, 1940, with the titles of chief engineer and general manager. "At the same time, Major Welch will retire as Chief Engineer and General Manager and will take the position of consultant to the Commission. Mr. Sutro will retire as Executive Director and Assistant Treasurer and will probably retain a nominal title of Assistant Secretary. His duties, however, will be only those directed by you."

Morgan was offered a salary substantially higher than that received by Welch. The PIPC staff was alerted in a brief memorandum that stated, "It is with regret that the Commission announces that Major Welch and Mr. Sutro have both asked to be relieved of the major part of their responsibilities . . . they will both be available to Mr. Morgan to assist him whenever he requests such assistance."

Welch, who had suffered bouts of ill health for several years, was politely but firmly pushed from his job. There is no record that Morgan subsequently consulted with him. William Welch died on May 4, 1941, leaving behind, by example, a trailblazing legacy of park stewardship that has worldwide benefit.

Sutro, not so politely removed, went on to other work, serving for twenty more years as president of the New Jersey Parks and Recreation Association until age eighty-five. Sutro was consulted once by Morgan, who asked him to write a memorial statement in honor of Welch. Sutro wrote, "Drawing upon long training and wide experience in many fields, he wrought a miracle of transformation. By his magic touch, forests grew in waste places, lovely sheets of water appeared in valleys long since gone dry, roads and trails threaded the woodlands, the deer, the beaver and the elk returned to their ancient haunts in the Highlands, and camps on the banks of lakes echoed the laughter of innumerable children. He loved Nature and used her treasures to make humanity happier."

In the PIPC's minutes memorializing Welch, Sutro's words were not used.

Torrey, the ever-alert publicist for the PIPC and champion of the trails, is commemorated on Long Mountain in Harriman Park, where his ashes were scattered in 1938. Carr, director of the Trailside Museum, was also dealing with health and marital problems. He left Bear Mountain in 1945 for an extended rest in a warmer climate. While he hoped eventually to return to the PIPC and was encouraged to do so by Morgan, events took Carr in a different direction. Always the educator, Carr settled in Tucson, Arizona, to become one of the founders of the Arizona-Sonora Desert Museum, patterned after the Trailside Museum. Today, the Arizona-Sonora Desert Museum is host to many thousands of visitors annually and enjoys international stature as a superb outdoor educational facility, reminding all who visit of the intricacies, beauty, and fragility of the desert environment.

Just after the war, Jolliffe made certain that many of her campers were present when the lake at the old Sandyfields settlement was dedicated to Welch. After almost thirty years on the PIPC's staff and even more years actively engaged in group-camp programs, Jolliffe submitted her resignation in January 1948, leaving behind standards of expectation and excellence that had so positively influenced the lives of thousands of girls and boys.

During this substantial staff transition, the PIPC continued to acquire land. Perkins Jr., Laurance Rockefeller, and Averell Harriman combined financial resources to cause the purchase of a Standard Oil of New York property that greatly expanded Tallman Mountain Park. The craters left behind when oil storage tanks next to a highway were removed eventually filled with rain and snow water to provide a home for hundreds of frogs, turtles, and aquatic plants.

To the north, the threatened development of High Tor, a high ridgeline of diabase rock in Rockland County, New York, prompted local residents to join the PIPC in a mutual effort to protect the jeopardized property. Maxwell Anderson, the playwright, had written the much-acclaimed *High Tor* while living at the base of the ridge. Artist Henry Varnum Poor painted *Gray Dawn*, a scene of High Tor that hangs in the Metropolitan Museum of Art. The poet Amy Murray wrote "Looking East at Sunrise," her ode to the ridge.

High Tor was known historically as well as scenically. From its summit, beacon fires provided warning of advancing British ships and troops during the Revolutionary War. The historic beacon site and surrounding land had been safe in the hands of owner Elmer Van Orden, who, more than once, refused to sell to speculators or quarry operators. When the land and farmhouse passed to Van Orden's estate in 1942, his heirs promptly listed the property for sale for $12,000.

This turn of events sparked a grassroots effort within the Rockland County Conservation Association, the New York/New Jersey Trail Conference, and the Hudson River Conservation Society to attempt to purchase the twenty-three-acre property. In a classic example of galvanized community action, pennies, dollars, and more dollars were successfully collected to buy the Van Orden property for the asking price and to transfer title to the PIPC.

The Van Orden fundraising quest led Mrs. Leonard Morgan of the Rockland County Conservation Association to visit Archer M. Huntington, a wealthy art collector living in Boston. Morgan asked Huntington for a $1,000 donation for the High Tor initiative. After briefly considering the request, Huntington agreed to the donation, but on condition that the PIPC would also accept donation of his 474-acre property, including a seventy-two-room English Tudor-style mansion on the High Tor ridge line next to the Van Orden property.

The mansion had been built by Samuel Katz, son-in-law of Adolph Zukor, and included formal gardens, a swimming pool, and a pipe organ. Zukor, Katz, and Huntington won wealth and fame in the world of entertainment and show business. The mansion at High Tor was rumored to have been built by Katz for his girlfriend. Whatever the genesis of the mansion, Huntington obviously no longer had use for it. Morgan still holds the honor of having made the most amazing fundraising visit in the history of the Rockland County Conservation Association.

The end of the war brought renewed interest in construction of the Palisades Parkway. New commissioners were appointed, including attorney Albert R. Jube, personal counsel to Governor Charles Edison of New Jersey and a graduate of Amherst College and New York Law School. Joining Jube and the other commissioners was Horace M. Albright, former superintendent of Yellowstone National Park, director of the National Park Service, and vice president of the United States Potash Company, producer of Twenty-Mule-Team Borax. Laurance Rockefeller recommended Albright, who won quick and enthusiastic approval from Rockefeller Jr. and New York governor Thomas E. Dewey.

The commissioners were keenly aware of Rockefeller Jr.'s gift of land on the crest of the Palisades and viewed the parkway project as their highest priority. They were optimistic that the parkway, put on hold during the war, would be reactivated without difficulty. The commissioners had not reckoned with Thomas Lamont and his son, Corliss. In a letter to Perkins Jr., Robert Moses said:

> I was somewhat disturbed about our talk yesterday regarding the Palisades Parkway because, as you know, I have gone way out in front of this project, especially as it affects the program in New York. . . . If all this flurry about Tom Lamont amounts to nothing more than opposition on his part, that of Mrs. Lamont, and that of their son, Corliss, based upon their desire to have no traffic and no travelers near their estate, it is of no importance. . . . If, on the other hand, you and Laurance are seriously thinking of deferring to Tom Lamont on the entire Parkway issue, and are prepared to discuss with him whether a parkway is needed or not on the top of the Palisades, then I think it is time to take the matter seriously, and for those of us who after all have taken great responsibilities as to this project, to reconsider the whole matter. There is nothing easier than to take a project or a group of related projects from a position at the head of the list and drop them down to a point where they won't be reached for years. . . . The day has passed when the owner of a big estate can block an important public improvement by wire-pulling, influence and similar shenanigans—that is, unless the public officials who are immediately responsible are impressed with this sort of thing. This is the problem which the members of the Palisades Interstate Park Commission will have to face if they regard Mr. Lamont's opposition as serious. . . .

Perkins Jr. shot back:

> If I were in the habit of getting mad about letters, I probably would about this one, because of the implications in it concerning Laurance Rockefeller and me are anything but complimentary. You ought to know both of us by this time better than that. Of course, we have no intention of deferring to anybody on the Parkway matter on a personal or any other basis. My interest in seeing the Parkway built is of long standing, and I suspect I was interested in it long before you were. Just because we don't agree a hundred percent with you on the way of handling the situation doesn't mean that we are weakening. My own personal policy is to avoid raising opposition whenever possible without compromising the issue. When opposition can't be avoided, then it is time enough to roll up your sleeves.

Perkins Jr. was indeed ready to guide the PIPC through controversy, preferably by winning consensus, but with sleeves rolled up if necessary. At the June 1945 meeting of the commissioners, Perkins Jr. was elected president, succeeding Wakelee. His fellow officers were Jube, vice president, Laurance Rockefeller, secretary, and Victor H. Berman, treasurer. Events affecting or threatening the PIPC would require more and more of Perkins's time, to the point where he would step away from his senior management responsibilities at Merck & Company to devote himself to the PIPC's needs, just as his father had stepped away from Morgan & Company for the same reason.

The first test for the new president was fast to arrive. Opposition to the parkway from Thomas Lamont was known, but it was Lamont's son, Corliss, who fired a journalistic shot across the PIPC's bow. A month after Perkins Jr. assumed policy leadership for the commissioners, a lengthy article written by Corliss Lamont appeared in the prestigious magazine *Survey Graphic*. After describing in intricate detail the beauties and delights of the Palisades, Lamont pointed directly at the PIPC as the misguided proponent of constructing "a new concrete super-highway, running along almost the entire thirteen miles of the Palisades. They would cut down a wide swath of woodland, slaughtering right and left the natural growth of trees, shrubbery and flowers. This would sacrifice much of the wildness of the area; and bring the sights and sounds—not to mention the fumes— of speeding automobile traffic close to the edge of the precipice."

Lamont advocated that a state highway route that paralleled the Palisades slightly west of the proposed parkway alignment be widened to accommodate motorists wishing to drive north toward Bear Mountain. He recommended turnouts and parking facilities along the state route to allow for easy access by hikers and bicyclists who wanted to enjoy the natural beauties of the Palisades.

Lamont gave no credence to the difference between a parkway and a "superhighway," or to the fact that the parkway would be restricted to passenger cars only, whereas the parallel state route was already choked with truck traffic. He found eager political allies in the person of Dr. J. C. Burnett, owner of a sixty-seven-acre estate on the Palisades near the Lamont property, and among elected officials in the New Jersey towns of Fort Lee, Englewood Cliffs, Teaneck, and Alpine. Burnett was motivated by a frightful vision that the parkway would run right through his front yard. The elected officials were worried about the loss of tax revenues because of the possible reduction in property values near a noisy parkway. Lamont, Burnett, and the elected officials offered stiff opposition to the PIPC.

Moses, writing to Morgan, was not impressed with Lamont: "He is a bad actor trained in the communistic tradition and is absolutely unscrupulous in the way he handles facts. The Palisades need no saving at the instance of a phony like Lamont who never did anything for anybody except verbally, and whose mother told me twice that our Parkway should not be built because it brought people and traffic into the region where Corliss did his thinking."

Rockefeller Jr., who, like the commissioners, assumed that a decision to build the parkway had been made before World War II and simply had been delayed, not scrapped, entered the public debate by stating to news reporters, "All my life, I have known, loved and frequented the Palisades." He reminded the press of the $21 million donation in land he made to the PIPC to remove the threat of development, contending that construction of a carefully designed parkway, set back from the cliffs, would help ensure the permanent protection of the Palisades, not destroy them.

By personal invitation, New Jersey governor Walter E. Edge joined Rockefeller Jr. in a tour of the Palisades. Writing to Kenneth Morgan, Rockefeller Jr. counseled, "I am sorry that Dr. Burnett has developed into such an aggressive antagonist of the Parkway project. Perhaps it is best, however, that he has come out into the open and unburdened his mind of the hostile thoughts that have been festering there."

Perkins Jr. appeared at various forums, including a raucous meeting attended by 350 residents of Englewood, New Jersey. He reminded his audiences that the parkway had first been proposed in 1926 and was endorsed by the New Jersey legislature in 1933, and then by the New Jersey Highway Commission in 1935. New Jersey and New York had appropriated funds for planning in 1941. At that time, engineering sketches of parkway landscaping, bridges, and overlooks were made available for public review. Commissioners Laurance Rockefeller and Horace Albright joined Perkins Jr. in describing the benefits of scenic drives, referencing the George Washington Parkway between Washington, D.C., and Mt. Vernon, the Colonial Parkway between Yorktown and Williamsburg, Virginia, and the Blue Ridge Parkway, constructed by the CCC to connect Shenandoah and Great Smoky Mountains National Parks.

The commissioners were making headway, gathering support from many organizations including the New Jersey Federation of Women's Clubs, but not enough to turn the political tide in Trenton. The Lamont-Burnett forces were severe in their criticism of the PIPC and had the support of several weekly newspapers, various town officials, and the ear of New Jersey state senator David Van Alstyne Jr. Van Alstyne was the sole state senator from Bergen County, New Jersey, and he sat on the powerful Appropriations Committee. Governor Edge, mindful of the "acrimonious division of opinion" about the parkway, advised Rockefeller Jr. that "under these circumstances I cannot conscientiously recommend an appropriation for the Parkway at this time."

Donald G. Borg, owner of northeastern New Jersey's most influential daily newspaper, the *Bergen Evening Record*, was watching the debate unfold. His newspaper had published numerous articles about the parkway battle, focusing on the history, construction, and engineering details, as well as the effects of parkway development on neighboring communities. These articles were journalistically neutral, presenting factual information, but then, on August 8, 1946, Borg's newspaper published the editorial "Good Faith and Bad on Parkway." Commenting on the tactics of the Lamont-Burnett group, the editorial stated, "Weapons commonly used by professional propagandists, persons who for money unhesitatingly further an unjust cause, are to hit and run, to distort a fact, to glamorize untruths, and to use pressure-group power in an attempt to further their own selfish interests. All of these devices and more are being used in the current campaign to convince Bergen County residents that construction of a parkway along the edge of the Palisades would be undesirable."

In a booklet being widely circulated by the parkway's opponents, a prominently displayed photograph was labeled, "This Area Closed to the Public by Order of the Palisades Interstate Park Commission" as proof that the PIPC was heavy-handed in its treatment of visitors to the Palisades. Borg's newspaper confirmed that "the important fact is that this particular photograph was taken in New York State, not the New Jersey section of the Palisades Interstate Park. It was taken in an area owned by the Standard Oil Company, so naturally the public was excluded." The new article pointed out that the commissioners, using private donations, were buying the Standard Oil property so that it could be "turned over" to the people for park purposes: "These are the same men who now by the deceptive photo are accused of trying to block natural development of the Park for public purposes." Using other examples of distorted information, the article concluded, "Persons who have been misled, misinformed, or self-deluded can and should rectify their error; their deceivers cannot and will not, but by that very fact are automatically discredited."

With Borg's editorial advocacy, the advantage swung to the PIPC. In April 1947 the *Bergen Evening Record* reported that recently elected New Jersey governor Alfred E. Driscoll had signed a bill approving the "much disputed" parkway route. With the approval came an appropriation to the

PIPC of $500,000 from New Jersey to begin the project, sponsored by the same Senator Van Alstyne Jr. who earlier had been sympathetic to the Lamont camp. With additional funds gathered from various other federal and state sources over the next several months, the commissioners accepted a bid of $834,000 for construction of the first 2.1 miles of the Palisades Interstate Parkway.

At about the same time, Josephine-the-snake returned to Bear Mountain. As chronicled in the *New York Times*, "A changed Josephine has returned here after an absence of ten years. When she made her first appearance for the campers at the Regional Museum, Twin Lakes, ten years ago, she fought her director. Today she put on a sedate show, earning her right to the title, 'the Sarah Barnhardt of the snakes.' For Josephine is a 6-foot-8-inch pilot black snake that has made reptilian history."

When the museum staff first captured Josephine, she was only three feet long but so "scrappy" that the snake was named after heavyweight boxing champion Joe Louis before her sex was determined. Turned over to the American Museum of Natural History, Josephine went on tour, entertaining audiences as far away as Maine and, during the war, making the circuit of numerous convalescent centers for veterans. The seasoned performer "now is handled by the children who visit Bear Mountain. Not only do they pick her up and fondle her and have their pictures taken with her, but they go out into the fields and trap mice for her."

Josephine-the-snake would have approved of a cooperative agreement between the PIPC and the newly formed Palisades Nature Association, a group of volunteers formed to begin managing the Greenbrook area of the Palisades in New Jersey as a wildlife preserve. The idea of preserving natural sanctuaries in a manner that would encourage humans to visit as guests, rather than as manipulators or predators of the wild plants, birds, and animals, was still in its infancy.

Even in the national parks, very little was known of the dynamic relationship and interdependence among naturally occurring life-forms. Bears, vexed by a baseless reputation as dangerous miscreants, were often shot by park rangers for the slightest transgression, such as invading smelly garbage cans. Mountain lions, bobcats, foxes, coyotes, and wolves fared even worse. No transgression was necessary. They stood convicted of capital crimes even before they were born. Bullets, traps, and poisons were common tools for ridding the parks of predators.

Rachel Carson's definitive book *Silent Spring* would not be published for another twenty years, and the word *"environment"* was not commonly in use in the 1940s. But in 1946, sensing the need for better environmental ethics, volunteers of the Palisades Nature Association persuaded the PIPC to draw a line around 165 acres at Greenbrook (a property once offered for sale by the commissioners) on the crest of the Palisades to create a restricted and tightly protected nature sanctuary. Membership would be required for access, conditioned on a willingness by the members to acknowledge that the sanctuary was created for the preservation of wild flora and fauna, not for the caprice of human visitors. Here, within the boundaries of the sanctuary, began a constant process of husbandry, scientific observation, and data collection that, along with similar long-term effort at the Trailside Museum at Bear Mountain, continues to produce a treasure-trove of baseline scientific data, knowledge of the natural world, and an increasingly sensitive understanding of the sweep and majesty of life on fragile planet Earth.

The second-generation commissioner team, led by Perkins Jr., Laurance Rockefeller, Averell Harriman, and Jube, was quick to respond to opportunities to protect the natural environment by expanding the PIPC's holdings. They lacked the fabled ability of Perkins Sr. to contact a few associates and friends and raise millions within a matter of days, but Perkins Jr., Rockefeller, and Harriman were quick to reach for their personal checkbooks to maintain the momentum of building a park system.

One such opportunity came to Rockefeller, who, on learning from A. K. Morgan of the availability of 640 acres on Dunderberg Mountain, just south of Bear Mountain, promptly donated $25,000 to the PIPC to cover the $40-per-acre purchase price.

Unfortunately, some projects were considered just too large or remote to capture the individual or collective attention of the commissioners. Ridsdale Ellis, representing the New York/New Jersey Trail Conference, thought the commissioners might do well to look west past Harriman Park to the 17,000-acre "Sterling Lake tract," which was still owned by the Harriman-controlled Sterling Iron & Railway Company. In a letter to Rockefeller Jr. in October 1951, Ellis and his Trail Conference colleagues Louis A. Sigaud and Paul A. Reynolds suggested that the Sterling Lake property could be purchased for $975,000. In a tactful response, Rockefeller Jr. declined, explaining that his financial participation in the PIPC's activities was already very significant. Ellis's letter was passed on to the PIPC, but it elicited no further action.

Rockefeller Jr. declined the Sterling Forest opportunity, but stepped forward again financially in response to a property matter in New Jersey that threatened to disrupt construction of the Palisades Parkway and wreak havoc with historic Fort Lee. A developer wanted to build five apartment buildings, each fifteen stories high, on a fifteen-acre plot of land located just south of the George Washington Bridge at the site of the old fort. The borough of Fort Lee had gained ownership of the land because of non-payment of property taxes during the Great Depression. The mayor and borough council were eager to unload the land to a developer so that property-tax income again would flow.

The commissioners were alerted to the mood of the borough's elected officials in a letter from Ida W. Certo to Rockefeller Jr.: "Forgive me for intruding on your convalescence. I am the only local citizen who dares to even suggest this land belongs to the people. Letters . . . received by offi-cials . . . are greeted with hoots, catcalls, and wisecracks." The attitude of the Fort Lee mayor was that every possible dollar in property-tax revenue should be squeezed from the Fort Lee site, leaving only an undefined "small parcel for patriotic purposes," probably in the form of a plaque on a rock.

The "second battle" for Fort Lee would extend for more than four years, from 1951 to 1956, demanding the involvement of Governor Driscoll, the commissioners, the *Bergen Evening Record*'s Borg, and an increasing cho-rus of advocates for the preservation of the historic Revolutionary War site. Laurance Rockefeller convinced his father to enter the fray after the PIPC was repeatedly rebuffed in bids for the property that reached $385,000. Rockefeller Jr., through his Sealantic Corporation, proposed to purchase the property for $250,000, donate to the PIPC a crucial 7.7-acre parcel where the old fort stood, and give the remainder to the borough of Fort Lee for development with restricted building-height limitations to ensure that the roofs of any newly constructed buildings would remain below the tree line.

The development strategy favored by the mayor shifted from fifteen-story apartments to ten-story office buildings to be built squarely on the old fortification, fifteen feet from the edge of the Palisades cliffs. A headline in the *Daily News* declared, "Cops to Guard Palisades Zone Hearing," fur-ther reporting that the "bitterest fight in history" was taking place in the borough's council chambers. In the face of growing public outcry to save the fort, the mayor and council refused to budge. A strong majority of out-raged borough residents made their feelings known in the November 1955 election by voting the mayor and his political allies out of office.

Despite the election results, the issue dragged on for another six months. Rockefeller Jr.'s Sealantic Corporation upped its offer to $300,000, with the understanding that a six-acre parcel, the absolute minimum needed to preserve the historic site, would be donated to the PIPC. With height restrictions in place, this deal was approved by borough officials. Subsequently, they sold the borough's seven-acre parcel to the Washington, D.C.–based Marriott Hot Shoppes Corporation for the construction of a hotel and conference center.

The height restrictions then became a point of extended and contentious combat between the PIPC and Marriott. The PIPC prevailed. The proposed hotel/conference center never was built. Historic preservation won the day, guaranteeing that the old fort would not be buried under concrete and parking lots or tucked into the shadow of a multistoried hotel with a revolving restaurant on top.

The effort to preserve the Fort Lee historic site was taking place at almost the same time that another confrontation faced the commissioners in the persons of Dr. Burnett and his wife, Cora. "In barbed-wire isolation on the edge of the Palisades here lies the mystery-cloaked sanctuary of Dr. and Mrs. John Clawson Burnett," reported the *Newark Sunday News.* "Designed by the recluse couple a quarter century ago to shut out the world, the cliff-edge estate, fenced by metal mesh and opaque foliage and guarded around-the-clock by men and dogs, has successfully hidden the Burnetts from the general public since that time."

The problem was that the Burnetts's fifty-four-acre estate stood squarely in the path of the Palisades Parkway. The Burnetts had chosen to be married on the property in 1920 and had developed it into their version of isolated perfection, just across the river from teeming New York City. The property possessed "a variety of fairy-land-like buildings and acres of cliff top woodlands, a pear-shaped swimming pool and a reinforced concrete bomb shelter, carved by hand from the roof of the Palisades, which could accommodate almost all this [Alpine] borough's 644 residents," said William J. Kohm, the reporter fortunate enough to win agreement from the Burnetts to visit the property. Mrs. Burnett, "the former Cora Timken of the multimillionaire Timken roller-bearing family, was in seclusion" when Kohm arrived, but Dr. Burnett took the reporter on a limited tour, once certain rules were understood: "The doctor ruled that no photographs would be taken of him or his wife, and that no personal questions, or questions about the background of either, would be asked."

The house stood near the highest point of the Palisades, ten feet from the cliff's edge. Inside, in addition to a "magnificent Turkish prayer rug" and "a disorderly room crammed with art objects of all shapes and sizes, was a complete hand-carved copy of an Indian temple." The amazed Kohm continued, "A life-sized Egyptian pharaoh stands wide-eyed before a slit-eyed Buddha; two feminine figures from the Middle East relax disdainfully beside the Grecian Sister of Venus de Milo; a Gauguin shouts its colors at the carved silence of an Indian musician. A place of honor is reserved for a plaster-of-paris head of Christ, a copy of an original in the Louvre in Paris."

The Burnett art collection was scattered through nine buildings on the property, but Kohm was not allowed to see "many of the better pieces" in an off-limits house containing guest apartments. In a separate dining building, the reporter saw one room "devoted solely to several floor-to-ceiling food lockers." The dining building included "an organ, one 12-foot statue and several figures of lesser size and a collection of Far Eastern dinner gongs." Nearby was the service building containing enough tools, machines, supplies, and equipment to "launch a small factory," including the estate's independent water and power systems. This building contrasted with a softly elegant Japanese teahouse. It was one of three studio buildings perched right on the edge of the cliff where Mrs. Burnett did her artwork.

The sequestered world of the Burnetts was facing invasion by the Palisades Parkway. Dr. Burnett, who earlier joined Lamont in an attempt to stop the parkway project altogether, was largely on his own in defense of his property. Construction machinery was moving in his direction. He dug in and fought. Burnett hired lawyers, sued, proposed a series of compromises that would force alignment of the parkway farther to the west, away from the major portion of his property, and sought public sympathy for his plight. Against him stood the two states, a project estimated at $46 million in construction costs, the image of a beautiful linear parkway beckoning motorists along a leisurely forty-two-mile drive from the George Washington Bridge to Bear Mountain, the news media, and the PIPC.

Robert Moses, in this instance an ally of the PIPC and a champion of parkways in New York, was leading an effort to acquire necessary land for the Palisades Parkway route along the thirty-mile stretch in New York that would connect to the New Jersey border. Almost every other parcel in New Jersey was already in the hands of the PIPC, with one other exception.

That was Bill Miller's Riviera, an elegant supper/nightclub of yellow stucco and blue trim resembling the stern of a yacht, built on the very edge of the Palisades cliffs. The original Riviera, constructed by Ben Marden, was a booming venue during the prohibition years of the 1930s. According to James Stevenson in a *New York Times* op-ed, "There had been rumors during Prohibition that the Riviera featured a secret chute that went straight down the side of the Palisades. At the first report of an impending raid, bottles could be sent clattering safely into the Hudson." Miller acquired ownership after the war years. From a revolving dance floor, dramatic

Ben Marden's Riviera, Palisades Cliffs (PIPC Archives)

views of a twinkling, nighttime cityscape prompted many of the Riviera's clientele to ardently refer to the place as "the most beautiful nightclub in the world." The Riviera could accommodate 900 customers per night. Sophie Tucker, Joe E. Lewis, Jimmy Durante, Danny Thomas, Milton Berle, Harry Richmond, Martha Raye, Harry Belafonte, Jane Froman, and the Ritz brothers were among the many stars who performed at the Riviera. But, in addition to the Burnett property, the nightclub had to go. It blocked a needed parkway off-ramp that would connect to the George Washington Bridge. (The present-day toll booths for the GW Bridge at the southern end of the Palisades Parkway stand almost exactly on the site of the old Riviera.)

Turning again to Rockefeller Jr., the commissioners won a pledge of $500,000 to pay one-half of the purchase price for the Riviera with the understanding that New Jersey would pay the rest. A side agreement with state officials was that New Jersey would also condemn and purchase the Burnett property for transfer to the PIPC. Newspaper owner Donald Borg, continuing to act as an advocate for the PIPC, negotiated the deal with Miller and Governor Driscoll. On the night that Borg met at the nightclub with Miller to confirm the transaction, he was accompanied by his ten-year-old son, Malcolm, who would one day become a PIPC commissioner. Durante, Belafonte, and Froman were performing at the Riviera that night.

Governor Driscoll's staff was not enthusiastic about the Burnett part of the deal. Each time the PIPC thought it had gained airtight assurance that New Jersey would condemn the entire Burnett property, the commissioner of transportation and his engineers would retreat in the face of furious opposition from the Burnetts. Morgan advised Perkins Jr., "I can see no good reason why we shouldn't ask the Governor to stop this foolishness, and once and for all condemn Burnett and do the job right." Commenting about a meeting with New Jersey highway engineers, Morgan said, "Kilpatrick intimated that the building of the Parkway would be stopped unless the arrangement was made to build . . . around Burnett. He also said that he had a very definite idea that the Highway Commissioner could build the Parkway where he wanted to and spend the money where he wanted to."

Still hesitant to force acquisition of the Burnett property but otherwise sympathetic to PIPCs activities, Governor Driscoll appointed Donald Borg to the PIPC in late 1953. Borg, a Phi Beta Kappa graduate of Yale, whose father bought the *Bergen Evening Record* in 1930, became editor of the newspaper at age twenty-six, setting a standard for independent reporting that, in later years, included sometimes unflattering coverage of himself. Circulation of the newspaper rose from 20,000 to over 150,000 during his tenure, and the *Record* became the most important daily news source in northern New Jersey. In his characteristic style, Borg, replying to an invitation to be considered for the Newspaper Boys Hall of Fame, said, "I regret my disqualification. . . . I have never carried newspapers, but for quite a while, this newspaper has carried me."

His blunt and effective style of communication would serve the commission for twenty-two years. On one occasion, writing to Perkins Jr., who was serving in Paris as the U.S. representative to the North Atlantic Treaty Organization (NATO) while still retaining his office as president of the PIPC, Borg commented on yet another property matter: "The Allison thing is showing signs of life, too. Arthur T. Vanderbilt (a retired Chief Justice for the State of New Jersey) called the trustees into his office last week and told them, in effect, to explain themselves to me. I expect an elaborate rationalization of the previous inaction. As you know, there are two postgraduate larcenists and an honest man, Paul Hudson. I intend to ask them merely what happened to the $700,000 shrinkable between the will probate and their last accounting, also, what they intend to do after they have disposed of all the real estate except Allison Park. As I see it, there are three possibilities; sit on the existing park and exhaust the $2,400,000 corpus in fees and maintenance; buy some more park land, develop it, and go out of business sooner; or get honest and turn the damned thing over to us. I shan't bother you with the blow-by-blow accounts, but I think the long-term prognosis is good, especially with Arthur on our side. If it's all right with you, I'll play it gently for about a month until I get this cast off (Borg had fallen and broken his hip). As things are, I can't fight, and I can't run."

(The reference was to William O. Allison, who, on his death in 1924, left a will with vague language about disposal of the many hundreds of acres of land he had acquired on the Palisades. Probate and lawsuits dragged on for years. In 1967, the PIPC finally succeeded in gaining title to a key Allison property at the highest elevation point on the Palisades, an overlook site that is a popular stop for motorists on the parkway.)

The entrenched Burnetts continued to hold out but admitted that they were thinking of moving to an island they owned in a river in Montana where "we won't have to worry about highway construction." The move never took place. Continuing to resist, Dr. Burnett and his wife faced a new challenge. In November 1954, Robert B. Meyner was elected governor of New Jersey and, soon after, authorized the court-order condemnation of the Burnett property to proceed. Dr. Burnett appealed a condemnation award of $1,245,321 for his home and land. While the appeal was under consideration, Cora Burnett passed away. In May 1956, a court ruling awarded Dr. Burnett $1,585,600, and the embattled recluse vacated his property the following month, almost twelve years after George Perkins Jr. first contacted him seeking a willing-seller, willing-buyer transaction.

Governor Meyner was firm on the Burnett matter, in part because he had been encouraged by New York's newly elected governor, Averell Harriman. Harriman succeeded Thomas Dewey to the governorship in 1954. By then, Harriman had been serving with the PIPC for forty years. Governor Harriman resigned from the PIPC on January 1, 1955, the first day of his term, but wasted no time in appointing his brother, Roland, to the commission.

One immediate benefit was that Robert Moses was attentive to the new governor's wishes. While Dewey still was in office, a frustrated Kenneth Morgan, continually fighting for attention to the parkway project by Moses, wrote to Laurance Rockefeller suggesting that Moses be taken in tow on a visit to Governor Dewey at his Long Island estate. Morgan said, "I favor going to Dewey's residence mainly for the psychological victory it would give Governor Dewey to have Mr. Moses seek out a favor." This tactic failed, but with Harriman in office, the Palisades Parkway project went to the top of State Council of Parks priorities.

Morgan and staff were breathing a sigh of relief about the parkway when two men from Connecticut appeared at Bear Mountain with Geiger counters. The men knew of federal laws, approved by Congress in the 1800s in support of mining activities in the West, that allowed prospectors to stake claims on practically any mineral deposits found on public land. The headline in the *New York Times* declared, "Bear Mountain Uranium Hunters Put Ski Jump Among Filed Claims." Not only was the ski jump claimed, so were 320 acres, including land under the PIPC's administration building and two-thirds of the great lawn between the administration building and the Bear Mountain Inn.

The prospectors were particularly interested in the administration building, where they found the "hottest" reaction on their Geiger counters. Reminding the commissioners that uranium ore was urgently needed for nuclear weapons, not to mention the fact that the men would become wealthy if they could successfully mine uranium at Bear Mountain, the prospectors sought permission to proceed with test drilling. Writing to newly elected Governor Harriman, William Kean reported that "uranium readings at Bear Mountain are higher than anything heard of by the Atomic Energy Commission, even in the west." Attorneys in Albany scrambled to find a legally sound reason that the drilling request could be denied. The prospectors were refused because the PIPC, not the state of New York, owned the land.

The commissioners were presiding over the largest land holdings in the New York/New Jersey metropolitan region, prompting constant inquiries from agencies and individuals for various uses of the land. The borough of Fort Lee in New Jersey wanted Miller's Riviera and eighteen acres for a grade school. The U.S. Army wanted eighty acres for a Nike Missile Base. Civil Defense and the local police wanted radio antennas on the summit of High Tor. Various towns wanted land for fire stations, court buildings, and police barracks. One entrepreneur wanted land for an enclosure where hunters could shoot "wild" African game, for a fee. The uranium prospectors joined a growing list of disappointed people who learned that the PIPC was steadfast in its resolve to protect natural values and history, not to become a vendor for myriad nonpark ventures.

Moses was keen on one project that, in his eyes, was exceedingly worthy. Taking a page from construction activities in his parks on Long Island, Moses began pressing hard in 1957 for construction of "separate but equal" picnic areas, playgrounds, and swimming pools at Hook Mountain. Construction funds for facilities throughout the New York park system were filtered through Moses's State Council of Parks, so even the PIPC could not ignore his command of the purse strings.

Moses wanted separate facilities at Hook Mountain, especially to accommodate visitors from "Harlem and 125th Street," emphasizing that these "groups" should enjoy the same privileges as any other park "group," but in their own designated areas of the park system. As a rationale, Moses argued that access to Hook Mountain by ferry was best for lower-income people, even though the 1955 opening of another giant bridge, this one spanning the Hudson River's Tappan Zee, all but put the ferries out of business.

The ferryboat operators were clinging to a rapidly declining clientele, looking at red ink, and beginning to suggest the need for subsidies from the PIPC just to stay in operation on reduced schedules. Moses seemed unconcerned. He started pushing construction funds into the State Council's budget for the Hook Mountain project. Morgan, writing to Perkins Jr., alluded to other construction priorities of much more interest to the commission and cautioned, "Shouldn't you also see Robert Moses before the Commission meeting? He may blow his top again if the Commission acts contrary to his ideas."

Moses's "ideas" were glaringly out of step with momentous societal changes. In 1954, the Supreme Court, in a unanimous decision, found in *Brown vs. Board of Education* that school segregation is unconstitutional. In 1955, Rosa Parks refused to give up her seat and move to the back of a bus in Montgomery, Alabama. Separate-but-equal at Hook Mountain was not approved by the PIPC. Whether Moses blew his top is unrecorded.

Laurance Rockefeller was interested in the Hook Mountain area for a very different reason. Land at beautiful Rockland Lake, adjacent to the mountain, was for sale. The Knickerbocker Ice Company had gone into business in the 1850s by harvesting ice from the lake, with Mother Nature the production specialist. In winter, large blocks of ice were sawn from the frozen surface of the lake, maneuvered to shore by men wielding long poles, hauled out with giant ice tongs attached to the harnesses of horse teams, dragged uphill to the top of a long wooden chute, and then allowed to slide by the force of gravity to the river's edge. Daring laborers occasionally would leap onto blocks of ice and joyride down the chute. The ice would be transported on the Hudson River by steamboat and barge to supply the cold-storage lockers of hotels, clubs, and fine homes in New York City.

At the height of production, the Knickerbocker Ice Company employed 3,000 workers and could store 50,000 tons of ice for transport to the city. The company thrived for about seventy-five years before refrigeration technology put it out of business. But the crystalline lake remained, and Laurance Rockefeller saw its superb park potential.

He took the lead among his fellow commissioners, seeking to quickly acquire as much land as possible around the lake. Rockefeller's goal was to raise $750,000 in private funds. He turned to his PIPC colleagues Perkins Jr. and Roland Harriman, winning pledges from each for $25,000 to complement his own pledge of $50,000. He then convinced trustees of the Rockefeller Brothers Fund to contribute $400,000 and directed the Jackson Hole Preserve, another Rockefeller charitable trust, to add $250,000.

This was financial incentive enough to convince Governor Averell Harriman and the New York State legislature to match the amount, bringing the total to $1.5 million. Acquisition of land surrounding Rockland Lake was underway. Further purchase opportunities prompted Rockefeller to donate an additional $25,000, Perkins Jr. $33,333.33, and Roland Harriman $66,666.67. The Rockefeller Brothers Fund matched these supplementary donations with $125,000, bringing total private donations to $ 1,000,000.

In 1958, the PIPC succeeded in purchasing the lake and 225 adjoining acres. By 1964, more than 1,000 acres had been acquired, creating an immensely popular park adjacent to 700-acre Hook Mountain. The combined parks are island-like, surrounded by a sea of suburbs. Laurance Rockefeller took the initiative at a moment of opportunity. Otherwise, the land around the lake would have become a developer's dream, and thousands of annual visitors, taking advantage of a variety of park benefits, would be nonexistent. The whole of Rockland Lake and Hook Mountain has proven to be an outdoor recreation triumph.

Rockland Lake State Park (PIPC Archives)

At about the same time, a *Newburgh News* headline announced, "Rockefellers, Harrimans Get Praise at Parkway Dedication." In August 1958, Governor Harriman and a large, enthusiastic crowd celebrated the opening of the last five-mile link of the forty-two-mile Palisades Interstate Parkway. Moses "lauded the generosity of the two families," and Harriman's grandsons, Averell Harriman Fisk and David Harriman Mortimer, cut the ribbon.

Only a traffic circle at one road junction remained to be completed, a task accomplished in 1959. The artful link between city and park, envisioned for decades, was in place. The parkway is a design success, living up to expectations that a major roadway could be constructed in a manner that rested gently on the land, presenting motorists with a green corridor looping along the contours of rolling terrain, passing under stone bridges of elegant style, meandering through a linear park of trees and flowers, and avoiding the typical recipes for road building that too often result in dull troughs.

On the fine August day of the parkway's dedication, Governor Harriman was enjoying the moment. After the ribbon was cut, he hopped into a 1928 Model A Ford, loaded it with his grandsons and other small, lively kids, and proceeded to lead a motor procession toward the Bear Mountain Inn, where he and New Jersey Governor Meyner were scheduled to join other guests at a celebratory luncheon.

An alert police officer, seeing the Model A chugging along in front of the shiny official Cadillac assigned to the governor, assumed that some enterprising spectator had broken into the motorcade. He pulled Harriman over. The problem of identity was quickly resolved. During the luncheon, Harriman spoke to the invited guests, singling out John D. Rockefeller Jr. for special praise in the years-long effort to acquire land needed for the parkway.

In so doing, the PIPC once again was providing common ground for influential people of different political persuasions and ambitions. As he spoke at the luncheon, Democrat Harriman was fighting a vigorous campaign to retain the governorship and keep it from falling into the hands of a formidable Republican opponent, Nelson Rockefeller, son of Rockefeller Jr. and older brother of Laurance. Nelson Rockefeller won in the November election. By March 1959, Averell Harriman had returned to the PIPC, succeeding his brother, Roland, who stepped aside so that Governor Rockefeller could reappoint Harriman to his old commission post.

Harriman rejoined his colleagues shortly before Kenneth Morgan received a communiqué from Lt. Col. J. B. Meanor Jr., Corps of Engineers, confirming a conversation in which Morgan had been alerted to interest by the U.S. Air Force in taking over Storm King Mountain "for a military project of a classified nature." Meanor reminded Morgan that "it is understood that you will advise the Commission of the Government requirement for the land and will avail it of the information furnished you during the discussion. Should the Commission desire additional information, attempts will be made to obtain the necessary security clearance to furnish the data it requires."

Without providing any desired "additional information," Meanor pressed for and gained permission to begin making topographic surveys and subsurface test borings on Storm King Mountain. A representative of the Air Force Air Defense Command, based in Colorado Springs, met with the commissioners in June 1959 to further explain the top-secret project. Even at the meeting, the Air Force representative was vague, but in a follow-up discussion with Perkins Jr., who, through his senior assignments with the State Department and as ambassador to NATO, had all the security clearances imaginable, the Air Force plans unfolded.

The Air Force wanted a fifty-year lease or outright ownership of 600 acres on Storm King Mountain. Two huge tunnels were to be driven deep into the mountain. Under a rocky subterranean roof 500 to 600 hundred feet thick, chambers would be hollowed out to accommodate equipment "vital to the Continental defense" of the United States.

A complement of one thousand military and civilian personnel would be assigned to work around the clock, 365 days a year, within the depths of the mountain. The commissioners were assured that, except for the tunnels, the surface of the mountain would not be disturbed other than for "air vents, exhaust vents, cooling towers, and related structures made to resemble farm buildings."

Prompted by the hard bomb-proof quality of the Storm King Mountain rock and the logistical convenience provided by nearby Stewart Air Force Base at Newburgh, New York, the Air Force wanted to build a Semi-Automatic Ground Environment (SAGE) site inside the mountain. Once activated, the underground air-defense center would watch the skies, connected through a radar network to nine other SAGE installations across the nation, waiting for the first hint of a sneak attack by bombers from the Union of Soviet Socialist Republics. Should an attack be detected, SAGE would alert the Pentagon and White House through the Air Force's North American Air Defense System based in Colorado Springs. Using SAGE radar vectors, Bomarc II rockets would be launched against the incoming enemy aircraft.

The Air Force was anxious to proceed, citing the need for "early completion of this important defense facility." In 1959, the Soviet Union had already placed Sputnik into orbit. Fidel Castro was on the verge of overwhelming the dispirited forces of Cuban military dictator Fulgencio Batista. The temperature of the Cold War was dropping toward frigid.

Nonetheless, the commissioners and nearby residents of the village of Cornwall-on-Hudson were not enthusiastic about this purported national security need. Some of the land wanted by the Air Force included village water-supply reservoirs on the slopes of adjacent White Horse Mountain, just west of Storm King Mountain. PIPC and village representatives asked for proof from the Air Force that other alternatives had been explored, including other mountain locations in the Hudson River Valley.

The fact that local congresswoman Katharine St. George sat as a member of the House Armed Services Committee helped ensure that the Air Force paid attention to provincial concerns. Morgan pointed out that most of the land sought for the SAGE site had been donated to the PIPC by the Stillman family with the condition that, if the land ever were used for "other than park purposes," title would automatically revert to the Stillmans. This might have presented a complication, but it posed no real barrier to the Air Force's plans. If necessary, the land needed, including the PIPC's land on Storm King Mountain, would be condemned and taken by the U.S. Department of Justice.

In fact, the air force was already busy studying other alternatives and fighting its own perpetual battle of the budget. In 1960, Congress cut funding for SAGE, based on the conclusion that Bomarc II missiles would be obsolete even before they were in position to be launched against incoming Soviet bombers. Soviet missiles, not bombers, had become the real threat. The air force's strategy shifted to other projects and locations, and Storm King Mountain seemed secure once again. But an assault on the scenic and natural integrity of the mountain would soon come from an entirely different direction, this time shaking the PIPC to its environmental core.

Perkins Jr. would not live to witness the second Storm King controversy. He had effectively retained the presidency of the commission for fifteen of his thirty-eight-year tenure. When on assignment in Europe, Perkins Jr. corresponded constantly with his commissioner colleagues and Morgan and diligently attended meetings whenever he was in the United States. The combined years of service by the Perkins's, Sr. and Jr., accounted for fifty-eight of the sixty-year existence of the commission. In January 1960, Perkins Jr. passed away at age sixty-five. The eulogy was delivered at the

Perkins Memorial Tower on the summit of Bear Mountain by Commissioner Laurance Rockefeller. After summarizing a public career that spanned responsibilities from army private to NATO ambassador, always with the PIPC as centerpiece, Rockefeller closed by saying, "Those of us who knew George Perkins are also thankful for his simplicity and for his humility of spirit and his sense of service to his fellow man that marked him as a distinguished and outstanding American."

15

▟▟▟

Storm King

ELIZABETH VERMILYE WOULD have been proud. On March 21, 1960, almost sixty years after she was denied official involvement with the PIPC because of her gender, the first woman was appointed to the commission. Linn Perkins, who married George W. Perkins Jr. in 1921, was appointed by Governor Nelson Rockefeller to succeed her husband and, in so doing, to carry on the distinguished tradition of park and historic-site conservation that so typifies the Perkins family.

Linn Perkins was the daughter of George and Friedrike Merck, thereby bringing to her marriage and personal interests the guarantee of access to a portion of the Merck Pharmaceutical fortune. Her fellow commissioners would soon learn that, like her husband and father-in-law, Linn Perkins would willingly open her formidable Perkins-Merck checkbook to provide financial support for the PIPC's objectives once convinced of the merits of specific needs and priorities. She often did so during her eleven-year tenure as a commissioner and for many years thereafter as a commissioner emerita.

Averell Harriman, an advocate for her appointment, stressed in a letter to A. K. Morgan, "We are overdue in having a woman on the Commission. . . . We haven't been keeping up with the times." As so often seemed the case, Robert Moses had a different view. He agreed that a continuing Perkins presence on the PIPC was important, but in a letter to Governor Rockefeller, Moses urged appointment of the son of Perkins Jr.

The commissioners nonetheless recommended Linn Perkins to the governor, as confirmed in a letter from Laurance Rockefeller to his brother highlighting her memberships with the Hudson River Conservation Society, American Forestry Association, National Audubon Society, Save the Redwoods League, and the National Wildlife Federation. The governor did not hesitate; he acted promptly and with favor on the commissioners' recommendation.

Albert R. Jube, who had been appointed to the commission in 1941 by New Jersey governor Charles Edison, was elected to succeed Perkins Jr. as president. In a little-noted move, Laurance Rockefeller, serving as the PIPC's vice president, also agreed to serve as vice chair of Moses's State Council of Parks (SCP). His move to the number-two position with the SCP suggested that cracks were beginning to show in the relationship between Governor Rockefeller and Moses.

Although Moses encouraged Rockefeller to accept the vice chair, he had reacted strongly to a report prepared for Governor Rockefeller by Dr. William J. Ronan, the governor's secretary. The report recommended consolidating and streamlining various departments and agencies within the structure of the state government, including the State Council of Parks. "The Ronan Report is a gratuitous public insult," Moses wrote to Governor Rockefeller. "I have no desire to interfere with your program and certainly do not propose after all these years to argue to remain in office if you have concluded that you can do better under other auspices."

The governor did not accept Moses's veiled offer to resign, but the positioning of his brother as vice chair of the SCP hinted that as far as the governor was concerned, Moses, the all-powerful chair of various quasi-independent state authorities created by the legislature, mostly at his urging and recommendation, and overseer of millions of dollars in contracts for the construction of bridges, roadways, and parks, was not indispensable.

In the realm of influence, Laurance Rockefeller was adding to his national stature as a champion of natural-resources conservation by chairing President Eisenhower's Outdoor Recreation Resources Review Commission (ORRRC), referred to by insiders as "ORK." ORK recommendations led to major action by Congress to establish the Land & Water Conservation Fund (L&WCF) within the Department of the Interior. The L&WCF was structured as a repository for funds flowing into the federal coffers from taxes levied on offshore oil production, fees collected for entry into national parks and wildlife refuges, and proceeds from the sale of surplus federal properties. Estimated inflow into the L&WCF was pegged at about

$900 million per year, all to be reinvested in conservation projects throughout the nation.

Part of the fund would be used to expand federal land holdings, with the rest shared on a matching basis with the states. Even though Congress often fails to appropriate the full amount of money flowing into the L&WCF, hundreds of thousands of acres of park, forest, and wildlife refuge land from Maine to Florida, from Virginia to Hawaii, and everywhere in between, has been protected over the intervening years. Laurance Rockefeller had taken on the family's mantle on the conservation front, beginning with the Palisades and pushing outward to America's most distant horizons.

By 1960, the PIPC had truly come of age. The long-ago battle against the Carpenter brothers had led to ownership of 53,320 acres and control of twenty-four miles of Hudson River shore frontage. The forty-two-mile-long Palisades Parkway was meeting every expectation as a beautiful automobile corridor. In Harriman Park, forty-three group camps, though reduced in number from a prewar high of more than 100, were more refined, manageable, and permanent. Each year, thousands of children were finding a kind of magic and memories in the camps, which still rank the PIPC in a class by itself as a vital link between girls and boys from inner-city neighborhoods and preserved parklands.

Annual visitation to the commission's eleven parks stood at a solid 6.3 million and was steadily increasing year by year. The rustic Bear Mountain Inn was a counterpoint to the elaborate and slick hotels and restaurants more typical of the vast urban market, attracting a steady clientele, including President Dwight David Eisenhower and his classmates from the U.S. Military Academy, Class of 1915, who held their 1960 reunion there.

Publicity about the accomplishments of the PIPC prompted a senior partner at the New York City–based law firm of White & Case to suggest to Commission President Jube that a memorial, perhaps in the form of a plaque, be placed somewhere within the park system to honor the memory of J. DuPratt White. Jube presented the request to his fellow commissioners, commenting that White "was Secretary of the PIPC when it was formed in 1900, and later became its President around 1920. . . . I look upon the suggestion with favor." The suggestion disappeared somewhere in the shuffle of internal dialogue; no action was taken.

Laurance Rockefeller was increasingly interested in Iona Island, nestled in an exquisite portion of the Hudson River narrows just downstream from the Bear Mountain dock. The navy acquired the island in 1900, ironically in the same year that the Palisades Interstate Park was formed. Over the years, 144 buildings had been erected on the island's 118 acres, housing enough ammunition to make any military tactician proud.

By 1958, the navy declared Iona Island excess to its needs, choosing, instead, to disperse its store of ammunition to a variety of more remote locations and hardened bunkers rather than presenting such a single, inviting missile target so close to New York City. When the navy moved out, the federal General Services Administration (GSA), keeper of excess properties, moved in. Despite vigorous interest in the island expressed by the PIPC, the GSA decided to fill up the buildings with "strategic materials," including rubber, cobalt, copper, and aluminum.

Governor Nelson Rockefeller placed the state on record in support of the PIPC's initiative to acquire Iona Island. On a trip to Washington, D.C., Governor Rockefeller gained endorsement from the New York congressional delegation for this purpose, but the GSA would not budge, claiming that the tons of natural rubber stored on the island were essential to national defense, even though synthetic rubber had clearly made the natural product obsolete. For the moment, the Rockefellers were stalemated by entrenched GSA officials who were supposed to dispose of excess federal property, not hold it for warehousing purposes. Even the collective clout of the Rockefeller brothers could not budge the GSA, but while the question of stewardship for Iona Island remained unresolved, the PIPC found itself being pressured to assume ownership of another island, this one a symbol of America's invitation to the world to "give me your tired, your poor, your huddled masses yearning to breathe free."

The same Dr. Ronan who had so irritated Moses by suggesting realignments and adjustments within the structure of the state government proposed that the PIPC "take over" Ellis Island in New York Harbor. Ronan accurately contended that ownership of Ellis Island was in dispute between New York and New Jersey. He argued the logic of vesting ownership of the island in the Palisades Interstate Park Commission to settle the dispute and ensure that both states would be represented in its future care.

At a meeting in February 1961, the commissioners considered Ronan's surprising notion. After due deliberation, they declined to seek jurisdiction over Ellis Island. Four years later, by Act of Congress and with the signature of President Lyndon B. Johnson, the island became part of the Statue of Liberty National Monument administered by the National Park Service.

Commissioner Donald Borg probably had forgotten about the PIPC's brief flirtation with Ellis Island when a letter arrived on his desk from J. Willard Marriott Jr. "Please accept this quarter, Mr. Borg, as a 'token payment' for taking five minutes of valuable time from your busy schedule to do me a favor." Marriott Hot Shoppes Inc. was making a market survey for the hotel it proposed to build next to the commission's Fort Lee Historic Site.

Borg was just one of many businesspeople in the Fort Lee area who received the Marriott inquiry and a shiny quarter, but the details of the marketing questionnaire caught his eye, especially since Borg had met personally with Marriott to discuss the building-height restrictions that the PIPC was determined to enforce at the hotel site. He was under the impression that Marriott was willing to abide by the height restrictions that would allow for a four-story structure, but the questionnaire described a "10-story building offering panoramic views of the Hudson River, two restaurants, one located on the roof of the hotel, and 400 attractive guest rooms available at $10 for a single and $15 for a double."

The estimated construction cost was $6 million. Borg answered the two questions asked of him: Would you use the facility? "No." Would you use the facility for conventions or meetings? "No." Encouraged to add a comment in a space provided at the bottom of the questionnaire, Borg wrote, "My personal interest is chiefly the height of the proposed structure and its gratuitous defacement of the Palisades." Alerted by the market survey, Borg and the commissioners began to prepare for a lengthy court fight, convinced that their effort to reach a compromise with Marriott had failed.

Farther north within the PIPC's park system, there was better news. Nine hundred people gathered at Lake Welch in May 1962 to participate in the official opening of a gigantic facility designed and constructed to provide for 30,000 visitors per day, most of them attracted by a half-mile beach of Long Island sand that had been barged up the Hudson River and then trucked to the site. Parking was available for 3,200 cars and scores of buses. This was a scale of development that fit nicely into Robert Moses's concept of park development, and he was there along with Governor Rockefeller and several of the PIPC commissioners to speak at the celebration. The weather was not very obliging, bringing blustery winds and a steady rain. As chronicled by the *New York Times*, "A ten-foot flagstaff was toppled by the wind and the sharp beak of a brass eagle atop it nearly struck Mr. Moses on the head . . . just after he had said some caustic things about people 'who sit in ivory towers, stuffy clubs, and cellar bistros' to judge park problems 'rather than explore the wilderness for a first-hand view.'"

Engineers for the Consolidated Edison Company of New York, Inc. (Con Ed) were, in fact, out exploring the PIPC's wilderness. Con Ed was the giant vendor of electric power serving 8.5 million customers in New York City and the suburbs of Westchester County. In September 1962, Kenneth Morgan made a note to the files recording a visit to his office by a Con Ed representative who told him that the company wanted to build a "pumped storage power generating plant" on commission property. Morgan responded that the commissioners would take an exceedingly dim view of a power plant on parkland but probably could accommodate the utility easements needed as part of the project's infrastructure. The PIPC from time to time issued permits to accommodate utility services for the surrounding communities. Morgan saw the Con Ed proposal as just the latest such request.

Three days later, a news headline heralded, "Huge Power Plant Planned on Hudson." Con Ed was proposing to build the largest pumped-storage hydroelectric plant in the nation on Storm King Mountain. Morgan, under the assumption that the plant would be placed north of PIPC property, followed up with a memo to L. L. Huttleston, the New York State parks director in Albany, stating, "The only thing they need from us is a permit to build a tunnel under our land"—implying that this was yet another routine utility-easement matter.

Immediately north of Storm King Mountain, the little village of Cornwall-on-Hudson found itself being courted by Con Ed. The power company knew that it must persuade the local voters to give up village-owned land needed for the massive project. Not yet knowing many details, a first reaction by village residents was to wonder whether the hydroelectric plant might disrupt television reception. Company representatives described the project as like a "large storage battery" and said the village would be caused "as little inconvenience as possible." Very convenient might be huge economic benefits.

Even so, one of the first cautious expressions of opposition to the venture came from members of the Cornwall-on-Hudson Garden Club. Meeting at the home of Mrs. Dale Bouton, the women voted unanimously to oppose "the transmission of power by an aerial cable across the Hudson River from Storm King," then settled in for a discussion led by Mrs. Marie Hand on the sundry uses of herbs for Christmas decorations. Once again, in the very early stage of what was to become a stupendous environmental battle, garden-club women were among the first to fire a shot across the bow of industrialists bent on exploitation of the Hudson Highlands.

Professor Calvin W. Stillman was also watchful. Those many years ago his father, banker James Stillman, had introduced Perkins Sr. to Wall Street, thereby opening the door for Perkins to join J. P. Morgan & Company. Much of the land owned by the PIPC on Storm King Mountain had been donated by James Stillman or by members of his family in his memory.

When Calvin Stillman first heard of the Con Ed initiative, he promptly stopped by to visit company officials in New York City, advising them that the PIPC would never allow a pumped-storage power plant to be constructed on its land, a message already delivered by Morgan. Stillman echoed, too, the concern of the garden club's members; stringing power lines across the "Wind Gate" of the scenic Hudson River narrows would not do. If Con Ed wished to succeed, Stillman counseled, the company must find an alternate site to Storm King Mountain.

Company officials listened somewhat to Stillman's advice, but the rocky structure of Storm King beckoned. As reported years later by Ron Britzke in the *Cornwall Local*, "Con Ed had a problem in the early 1960s." The problem was demand for peak power in New York City. The coal-fired and nuclear-power generating plants operated by Con Ed were barely enough to meet the city's normal demands for power. These demands peaked in the mornings and evenings when millions of people were home turning on lights, using appliances, setting thermostats to kick on furnaces or air conditioners, and otherwise expecting "unlimited power at the flick of a switch," as Britzke put it.

When too many switches were flicked too quickly, New York City experienced brownouts, menacing dips in the flow of electric power that dimmed lights, reduced cooling and heating capabilities, impacted emergency medical, law enforcement, and fire services, slowed pumps that delivered drinking water, and other events that threatened to push the electric grid into blackout, the final stage of collapse that would force the power-dependent city to its knees. Brownouts were becoming all too common. A major brownout in 1961 captured world headlines by shutting down five square miles of Manhattan.

The "large storage battery" Con Ed's executives and engineers had in mind for Storm King Mountain consisted of eight enormous "reversible" electric-power-generating turbines, each capable of generating 250,000 kilowatts of electric power. They were to be located at the base of the mountain on the shoreline of the Hudson River, just across from the location where Henry Hudson gave up his quest in 1609 for discovery of the fabled passage to Asia and turned back toward Europe.

When reversed to gulp water (and fish) from the river, the turbines would pump 12 billion gallons of water upward through a tunnel forty feet in diameter and two miles long, to be "stored" in a 260-acre reservoir perched more than one thousand feet above the river on land to be acquired from the village of Cornwall-on-Hudson. Pumping to fill the reservoir would occur late at night and in the predawn hours when the city's power demands were low. When demand moved toward peak, billions of gallons would be released from the reservoir to be sent gushing back down through the tunnel, spinning the turbines with an immense hydro impact.

Ironically, more electric power would be required to pump the water up to the reservoir in the first place than would be produced when the water came coursing back down through the turbines. But that was not the point. Con Ed could leisurely use excess nighttime power to pump the water uphill. The whole purpose was that anywhere from 900,000 to 2 million kilowatts of extra electrical energy could be injected into the Con Ed power grid in a matter of minutes by the down-rushing water, depending on how many of the eight turbines were used at any given moment.

Customers in the city could flick switches with impunity during the approximately 4 percent of each twenty-four-hour period when peak demand otherwise threatened to override Con Ed's regular power service. A seemingly inconsequential factor was that fish, first sucked up into the reservoir through the reversible turbines, then blasted back down through the tunnel hours later, would swim no more.

Con Ed emphasized the benefits of the project to Cornwall-on-Hudson residents. In a four-color brochure that contained no technical or design information about the pumped-storage plant but featured several photos of derelict buildings near the river's edge, Con Ed promised to "transform nearly a mile of dilapidated waterfront property north of the plant into a park and turn it over to the Village of Cornwall. This will serve several important purposes. It will beautify the waterfront; it will provide a major recreation area on the river where none now exists; and, perhaps most important, it will preserve for all time an extensive segment of waterfront that otherwise might be redeveloped for industry or other purposes."

Con Ed's tag line proclaimed, "From electric power to serve southeastern New York State will come the means not only to conserve but to add to the beauty of the Hudson River."

With the PIPC standing on the sidelines on the assumption that its primary concern was the matter of issuing a permit to allow for the tunnel to be carved deep under Storm King Mountain, the power company offered a tax treasure chest to residents of the village of Cornwall. If the giant power company became a landowner, power-plant developer, and taxpayer in the village, the property-tax burden on village residents might plummet by two-thirds or more, perhaps even to zero. The heady thought of letting Con Ed take over most of the property-tax burden for the village was, according to reporter Ron Britzke of the *Cornwall Local*, like "a rainbow with the mythical pot of gold."

Jobs, too, were dangled as an incentive, perhaps as many as 400 during the construction phase. The fact that almost no one in Cornwall had any experience in hard-rock tunneling or power-plant construction was only a minor concern. Most of the village's residents began thinking that these incentives were just fine and started rolling out the welcome carpet.

But a few villagers were increasingly skeptical. Britzke wrote, "At the heart of it all was the mountain. Visible from virtually everywhere in the vicinity, it brooded over the river. For the first time, the mostly unspoiled Highlands would be invaded by man and his commerce. Storm King would become a symbol of ecology vs. industry." Among friends and neighbors, elected officials and family members, lines were beginning to be drawn. No one understood the full scale of controversy that would develop. No one assumed that a company with the influence and brawn of Con Ed could be stopped. About the best the skeptics could hope for was that the environmental impact of the power plant might be minimized.

At Bear Mountain, a member of the PIPC's staff alerted Kenneth Morgan that Mr. and Mrs. George Brooks of Cornwall were willing to give their twelve-acre property to the PIPC if the proffered donation would somehow help stop Con Ed. Morgan responded, "The Commission will not get involved . . . if this entails opposition to Con Ed. . . . The Commission usually works with utilities. . . . [We] suggest that you make the best possible deal and sell to Con Ed." Only a short time later, Morgan learned that Con Ed might want to dump rock spoil from the tunneling operation at the PIPC's old CCC campsite on Storm King Mountain. He responded by speculating that this "might be a good way to fill up and make usable this land."

A seemingly unrelated transition occurred in Albany in January 1963. With the informal encouragement and consent of the governor, Laurance Rockefeller replaced Moses as chairman of the State Council of Parks. Strain between the Rockefellers and Moses had grown. Moses's departure from the SCP was an indication that his iron-fisted control of so many New York State construction activities might not survive the Rockefeller administration. He turned over the reins of the SCP to his successor, leaving behind the "advisory" organization Moses founded in 1924 to grab the budget purse strings from those he characterized at the time as "the old park men." Moses's departure from the SCP did not hinder him from commenting on the Storm King Mountain issue. In response to a request to help stop Con Ed, Moses, wearing his hat as chair of the Long Island Park Commission, said, "I have been aware of this proposed development from the beginning, and I feel that the Consolidated Edison Company has used every reasonable effort to preserve the natural scenery."

Con Ed's 1962 stockholders' report brought graphic definition to the company's interpretation of what was meant by the phrase "preserve the natural scenery." Shown on the cover of the report was an engineer's rendering of exactly what Con Ed had in mind for the base of Storm King Mountain. As Britzke described it, "The rendering (by an unknown artist who should be enshrined in the ecological hall of fame) showed a rectangular chunk the size of several football fields blasted out of the foot of the mountain." The enormous notch would house the transformer and switchyard for the eight underground turbines.

People who had been generally aware and somewhat concerned about the Con Ed proposal were suddenly galvanized into action after taking one glance at the cover of the annual report. To the still small cadre of opponents, the drawing was a bombshell. Immediately realizing the consequences of its public-relations gaff, Con Ed announced that the riverfront portion of the plant would be nestled underground. Cost for the project, originally estimated at about $100 million, jumped to $150 million.

The New York/New Jersey Trail Conference, who counted Maj. William Welch among its founders and was long allied with the PIPC, stepped forward to voice opposition to the Con Ed project. In a May 1963 letter to Dr. William J. Ronan, an aide to Governor Rockefeller, Trail Conference

CON EDISON'S PROPOSED HYDROELECTRIC PROJECT, CORNWALL, NEW YORK

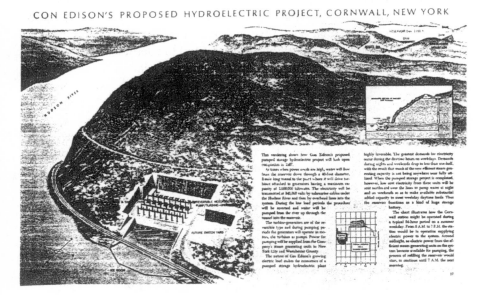

Proposed Consolidated Edison Power Plant, Storm King Mountain (PIPC Archives)

Conservation chairperson Leo O. Rothschild cautioned, "The threat to the northern gate of the Hudson Highlands has become far more ominous in the last few weeks."

Rothschild was reacting, in part, to a declaration by another power company, the Central Hudson Gas & Electric Corporation, that it intended to build its own new power plant on Breakneck Ridge, on the east side of the Hudson facing the Con Ed pumped-storage plant. Central Hudson served the Hudson River Valley north of Con Ed territory, and it saw an opportunity to follow along with the larger Con Ed leading the charge. Ronan sent Rothschild's letter to Kenneth Morgan for comment. A fault line between the PIPC and its traditional environmental constituents was becoming increasingly visible, as confirmed by Morgan's response to Ronan that the PIPC must "determine whether their [Central Hudson's] installation will be as satisfactory as that of Con Ed is expected to be."

As in the days when the commissioners were trying to sell land on the summit of the Palisades to meet budget needs, the PIPC seemed to be losing contact with its purposes and roots. Morgan was seemingly content to view the pumped-storage power plant only from the perspective of a tunnel passing under Storm King Mountain, not in the larger context of protecting the scenery of the Hudson River narrows.

The fact that the PIPC did not own the property on which Con Ed planned to build the turbines and reservoir became a factor in the PIPC's hesitancy to challenge the project, even though the commission never hesitated to challenge quarry companies up and down the river or to use aggressive acquisition tactics if necessary to stop the desecration of natural beauty. Somehow, the PIPC viewed the pumped-storage power plant, potentially the granddaddy quarry of them all, more kindly.

Alerted by Trail Conference members, the *New York Times* broadcast the first of many succeeding articles: "Power Plan Stirs Battle on Hudson." In the article, a direct remark by Rothschild to Governor Rockefeller was quoted: "Knowing your interest in the scenic and historic landmarks of our State, we hope you will do everything in your power to preserve the Highlands. We would appreciate hearing from you."

Reacting to the *Times*'s article, Robert A. Roe, the New Jersey conservation commissioner, asked of Morgan, "What does the PIPC recommend?" Morgan responded, "We are fully aware of these two developments and have made an exhaustive examination of the probable damage to scenery. . . . The projects will not severely damage the scenery. . . . It is not the Forest Primeval. . . . The area is occupied by many old buildings. . . . The forestalling of all enterprise in this area is neither practical nor desirable. . . . The Commission will not object to either one."

Roe must have been pondering Morgan's advice when an editorial, "Defacing the Hudson," appeared in the May 29, 1963, edition of the *New York Times*: "If any utility proposed to construct a plant in the middle of Central Park, the absurdity of such a defacement of precious natural (or nearly natural) surroundings would be immediately apparent. It is almost as bad to plunk down a couple of power installations right in the heart of one of the most stunning natural regions in the Eastern United States: Storm King Mountain and Breakneck Ridge. . . . All of us who have hiked and played in the Palisades Interstate Park know what a beautiful backyard exists 50 miles north of New York City. Is it too close to home to appreciate? 'This is very good land to fall with and pleasant land to see,' said one of Henry Hudson's officers, going up the river under these high blue hills. Baedecker (publisher of a popular travel guide) found the Hudson's scenery 'grander and more inspiring' than the Rhine. The proposed power plants . . . would desecrate great areas of the natural and historic heritage of our country that are still largely unspoiled and should remain that way."

Answering for the PIPC, Laurance Rockefeller responded in a letter to the editor that had been drafted by Morgan and approved by the other commissioners. Expressing regret that Storm King and Breakneck Ridge had been chosen for the siting of new power plants, Rockefeller said, "We have recognized that additional facility developments of this type must be made available to meet the requirements of a growing population." Then, reciting the history of the PIPC as "a strong protector of the natural beauty of the Hudson," he continued, "For some time now the Commission . . . has been working with Consolidated Edison to make sure that the greatest consideration possible is given to scenic and other values in the construction of this power plant. . . . The plant will be at river level, facing north. Most of the plant will be below ground level and it has been designed to blend into the hillside and will be landscaped by an outstanding landscape architect."

Repeating Morgan's contention that the proposed site "is not forest primeval," Rockefeller stated that "on balance the Commission prefers to take the positive approach to ensure that the esthetic, historical, and recreational values of the area are protected as much as possible."

The PIPC's staff was alerted that Con Ed might dump 720,000 cubic yards of rock spoil from the tunnel onto the old CCC site at Storm King, creating "about 28 acres of level land," and pay the PIPC twenty-five to fifty cents a yard for the privilege, producing income for the PIPC of from $430,000 to $860,000. Con Ed sweetened the offer even more by suggesting that it would donate about 160 acres on Storm King Mountain to the PIPC that would be excess to the company's needs once construction was completed.

The PIPC, champion of scenic preservation, but now advocating compromise with the power companies, was feeling growing pressure to defend its position. Laurance Rockefeller, writing in July 1963 to Con Ed's chairman, Harland D. Forbes, and to Lelan F. Sillin, Forbes's counterpart at Central Hudson, inquired, "One conservation organization, The Nature Conservancy, has assigned a representative who is on a fact-finding mission to find among other things the necessity of additional facilities in this area. . . . Would it be possible to obtain some relative facts and figures that would illustrate your reasons for selecting this site as well as this particular method of generating power against other methods?"

Forbes was quick to reply, and he referred to the "world's largest pumped-storage power plant" at Vianden, Luxembourg, as proof of proven technology, citing the "highly advantageous" economic benefits of the proposed Storm King plant, while explaining that "considerable new power capacity . . . would be ready at a moment's notice." Forbes listed various alternatives, all sites along the Hudson River of lesser elevation than Storm King and the Cornwall reservoir, and discarded the choice of depending on old coal-fired steam plants, which would cost about double the hydro option. Forbes did not mention other options for producing peaking power, such as using natural gas power plants, energy conservation, or the import of electric power from other vendors.

Carl O. Gustafson, executive assistant to Laurance Rockefeller, was appointed to the PIPC in 1961. A strong conservationist in his own right, Gustafson nonetheless was comfortable with Forbes's response. "The next time the Nature Conservancy boys draw a bead on me, I'll be able to fire back," he wrote to Kenneth Morgan.

In the meantime, Rothschild communicated again with Governor Rockefeller's aide, Dr. Ronan, challenging the content of the letter to the *Times* that Rockefeller signed. "That the Consolidated Edison plant would supply electricity in an efficient way, I assume is beyond dispute," he said. "What I do dispute is the damage to be done to the river, and my position is that the defacement of one of the finest river stretches in the world cannot be excused because an economical source of power would be developed. The proposal is on a par with a plan to dam the Grand Canyon."

In further contact with Kenneth Morgan, Gustafson said of the Rothschild letter, "It is my opinion that there is enough nonsense in it so that we can at least prepare a response to Ronan, pointing out where Rothschild is in error." This was not to be. Ten days after communicating with Morgan, the forty-two-year-old Gustafson, his wife, and daughter tragically died when Gustafson, a skilled World War II and Pan American World Airline pilot, failed in an attempted emergency landing of his Cessna 190 near Teterboro Airport, New Jersey.

In the village of Cornwall, a large majority of residents were persuaded that the Con Ed project would, in addition to being a boon for the village tax base, bring improvement to the neglected waterfront through creation of the proposed fifty-seven-acre shoreline park built out of spoil material from the tunnel and reservoir excavations. Con Ed also was pledged to purchase and expand the village's reservoir, allowing Cornwall to reap the benefits of upgraded utility systems at company expense.

According to the *Cornwall Local*, about 1,700 residents signed a petition favoring the development. Only 100 were opposed. Two of the Cornwall residents who stood in opposition were Stephen and Beatrice Duggan. They understood that any hope for success against Con Ed must be found outside the village limits. In November 1963, at the home of Carl Carmer, Mrs. Duggan joined a dozen other people to form the Scenic Hudson Preservation Conference. Stephen Duggan, an attorney, also became a founder of the Natural Resources Defense Council (NRDC), another organization sparked to life by Con Ed. The NRDC eventually would gain impressive national credentials as a highly respected champion of sound environmental law.

What the dozen founders of Scenic Hudson did not anticipate at the time was that the Storm King Mountain controversy would span seventeen years. The conflict and results would profoundly change the way the United States conducts its environmental business.

As the volume of the Con Ed debate increased, Kenneth Morgan alerted the commissioners to a startling inquiry from a deputy director of the Atomic Energy Commission (AEC) that had nothing to do with pumped-storage power. "The AEC is considering the establishment of a tunnel from the river into the rock north of West Point for the purpose of servicing atomic submarines," implying that the AEC was unaware of the Con Ed proposal. The vision of submarines floating on a millpond of black water beneath a stadium-sized domed-rock ceiling, armed guards on alert, and technicians working by fluorescent light surpassed in creativity the 1920s proposal to bore a tunnel through Hook Mountain to form an artificial waterfall plunging into the Hudson River.

"I told him I thought that the conservationists had taken about all the defacement of the mountains on either side of the river in this vicinity that they would take, and strongly urged him to leave our property alone and go somewhere else." The AEC listened. The atomic subs went to Groton, Connecticut.

Many other issues remained actively on the PIPC's agenda. In Orange County, the PIPC acquired 1,600 acres to create Highland Lakes State Park. With access to New York State bond funding, the PIPC added almost 7,500 acres to its holdings in 1963 and continued a steady pace of acquisition through the 1960s. Among the acquisitions were 34.5 acres donated to the PIPC by T. Dwight Partridge, son of Dr. Edward L. Partridge.

At High Tor Park, the PIPC concluded that the seventy-four-room mansion, donated years earlier to the PIPC by Archer M. Huntington, was "rambling, disjointed, unorganized, could not be used as a hotel, and had no historical value or charm." After attempting on several occasions to find a suitable use for the structure, the commissioners voted unanimously to demolish the mansion to make way for a scenic view out over the Hudson River toward the city. Demolishing buildings was not unusual for the PIPC. Even the PIPC's own New Jersey office building, which was constructed by CCC workers during the Depression, fell to wrecking crews after World War II to make way for the Palisades Parkway.

An occasional outcry could be heard as some buildings disappeared, but the commissioners were not prepared for the group of women who threatened in midwinter 1966 to "lie down in the road" to save the Huntington mansion. Mrs. John R. Sarno Jr. and thirty like-minded women thought that the removal of the English Tudor-style mansion was a very bad idea. When contractors arrived to begin demolition, they were met by a group of determined women who literally placed themselves in the path of the wrecking ball. To outflank the women, the workers climbed up on the mansion to begin ripping off roof tiles.

The workers' maneuver was short-lived. Reacting to adverse publicity, Governor Rockefeller directed that the PIPC commissioners halt the demolition and make one more good-faith attempt to find a worthy use for the building. The PIPC held a hearing at which the protesters and local political officials were invited to produce a sufficiently funded organization capable of guaranteeing maintenance and public use of the Huntington house. No satisfactory organization was identified. Despite pleas for more time, the commissioners ruled, "The house is coming down." Two months after the confrontation at High Tor, all that remained of the mansion was a flat, open site outlined by low walls, a lawn, a disconnected fountain, and a sunken terrace. Mother Nature was already in the process of reclaiming the site.

Building demolition of a much larger scale became possible when the General Services Administration advised Kenneth Morgan by letter in August 1965, "Your application to acquire Iona Island is acceptable to this

Agency." The PIPC was invited to close the transaction for $290,000, one-half of the appraised value for the 118-acre island. The commissioners wasted no time. Within a matter of weeks, the island and its 144 buildings, totaling 469,000 square feet, were in the PIPC's ownership. The island had been variously considered by the GSA for industrial development, a university campus, a prison, a mental-health institution, and a resort. Most of the commissioners and Morgan considered the island suitable as a site to handle "overflow" crowds from Bear Mountain. A 1,500-car parking lot was constructed on the north side of the island. An Olympic-sized swimming pool, picnic areas, tennis courts, and baseball fields were planned for the rest of the island at an estimated cost of $20 million.

Despite the existence of the new parking lot, commissioners Frederick Osborn and Conrad Wirth (former director of the National Park Service) were not convinced of the wisdom of replacing the navy infrastructure with more development. The National Audubon Society also took a dim view of the PIPC's plans, cautioning that restoration of bird habitat should take precedence. The dissenting commissioners and Audubon members were joined by an unlikely ally, the famous Iona Island "sinking road" that the navy had built straight across a portion of the Iona Island marsh to provide access. Navy truck traffic kept pushing the road base further down into the muddy bottom of the marsh. The navy would have to keep adding more layers of gravel and blacktop to keep the road surface above water.

The PIPC would be faced with the same problem. Thousands of visitor cars traveling to and from the parking lot would just keep sinking the road. Added to this costly maintenance problem was that an at-grade crossing of Penn-Central railroad tracks became a safety concern. The new parking lot, rushed too quickly to completion, was just as promptly abandoned. PIPC contracts were approved to demolish the buildings, tunnels, web of roads, and ammunition bunkers left behind by the U.S. Navy. Only six of the old brick-and-concrete buildings remain on the island. The largest of these, the marine corps barracks, is a drafty, empty relic. The PIPC uses the others for warehousing and storage purposes. Abundant wildlife and plant species, including bald eagles, have returned to the recovering island. Botanists have found rare plant species growing in the unused parking lot.

In New Jersey, Commissioner Donald Borg was keeping track of the legal maneuvers required to convince the Marriott Company that it could not ignore building-height limitations on the Palisades. Borg, Laurance Rockefeller, Kenneth Morgan, and attorney Henry Diamond, a member of Rockefeller's staff, met with New Jersey governor Richard J. Hughes to successfully enlist his support for the development of the Fort Lee Historic

Site and, coincidentally, to join in the effort to persuade Marriott to back off. The commissioners promised to provide $250,000 of the estimated $500,000 cost of developing the historic site.

Con Ed, though, remained at the center of the commissioners' attention. The company dropped off a model of the proposed project at Morgan's Bear Mountain office. The model was 4 x 10 feet, came with its own aluminum platform, and was delivered by truck. Morgan had to recruit members of the PIPC's maintenance staff to help off-load the model.

Opponents to the pumped-storage project continued to align with the Scenic Hudson Preservation Conference, including the city-based Regional Plan Association, the Nature Conservancy, and the New York/New Jersey Trail Conference, all traditional PIPC allies. The twelve people who formed Scenic Hudson found themselves joined by thirty-three affiliate organizations. Before the controversy ended, Scenic Hudson would receive small and large donations from more than 20,000 people, including many Con Ed stockholders.

In written statements to the Federal Power Commission (FPC), Con Ed's attorneys began referring to the Scenic Hudson advocates as "extreme conservationists." Dale E. Doty, a Washington-based attorney retained to represent Scenic Hudson in the FPC proceedings, admitted in response that his clients were "conservationists," but added that whether they were "extreme" would be left to the judgment of history.

With the consent of the commissioners, Kenneth Morgan journeyed to Washington to appear before the Federal Power Commission. This federal agency was in the eye of the Storm King controversy because it had legal authority to accept or deny Con Ed's application to build the power plant. In his testimony, Morgan contended that "the project, when completed, would not damage the scenery," that "dual use would occur, [with] the tunnel below [and] recreation above," and that "it would be unreasonable to oppose the project provided that the company does everything within reason to minimize impact on the scenery." Morgan's comments were confined to possible impact on PIPC land. Even though the Con Ed turbines and reservoir dams would be only a stone's throw to the north of the commission's property lines, the PIPC continued to break with its own decades-old tradition by hunkering down behind its own boundaries.

The FPC hearing concluded in March 1964 with a finding that the Storm King plant "would have little adverse effect on the scenic beauty of the Hudson River Valley." A license was granted by the FPC to Con Ed to allow the project to proceed. But opposition was becoming more vocal by the day. The most constant PIPC allies, the garden clubs, now represented by the Garden Club of America, found that a power plant with 700 feet of frontage on the river, quarried 900 feet into Storm King, standing at a height of 122 feet, and backed by five earth and rock dams, each 250 feet tall, the largest of which would extend laterally for 2,000 feet, would indeed have horrendous scenic impact on the Hudson River Highlands.

Writing to Robert Moses, who was listed on the Garden Club of America letterhead as an adviser to the conservation committee, Mrs. Alexander Saunders asked for his advice on how to stop Con Ed. Moses was consistent in his reply: "I think you and your Conservation Committee should reexamine your decisions." Then, quoting Morgan's testimony before the FPC, he said, "No eyesore or desecration will result," adding that there is an "urgent need for this additional power at a cheap price."

Randall J. LeBoeuf Jr., of the law firm LeBoeuf, Lamb & Leiby, was representing Con Ed in hearings before the FPC. His firm's letterhead listed eighteen lawyers. Doty, representing Scenic Hudson, had a letterhead with a single name, his own. LeBoeuf alerted Kenneth Morgan that the FPC, feeling pressure, was planning to hold further hearings, and that Doty specifically would request Morgan's appearance. A letter from Doty to Morgan confirmed the warning: "I feel that the present position of your Commission is not clear and that your testimony would be of benefit not only to the FPC but to better public understanding of the PIPC's position in this matter." Morgan drafted a written statement and sent an advance copy to LeBoeuf, asking for his comment and guidance on the text of the statement, but did not extend the same courtesy to Doty.

In his further statements and correspondence, Morgan took the position that the PIPC had won major concessions from Con Ed. Construction scars and the power plant's infrastructure would be professionally landscaped, power cables would pass under the river, and 1.5 miles (of a projected 23.5 miles) of power cable on the east side of the Hudson River would be buried underground. "I feel that this type of cooperation, rather than opposition, has been worthwhile," Morgan claimed. As an increasing number of letters reached his desk, asking that the PIPC intercede more vigorously against Con Ed, Morgan gave a standard response: "Contact the FPC."

But the political ground was shifting under the feet of those who had assumed that the best hope for protecting scenery in the narrows of the Hudson River was to win as many design and operating compromises as possible from Con Ed. In a legal brief submitted for the record to the FPC, Doty described the imposing dimensions of the power plant and contended, "To argue that such a plant will not mar Storm King and the scenery of the area is nonsense. . . . To speak of landscaping such a monstrosity is sophistry."

The Hudson River Conservation Society, long existing as a stately forum for dialogue among some of the most influential families in the lower Hudson Valley, found itself struggling with the controversy. The society's founder was William Church Osborn, father of PIPC commissioner Frederick Osborn. Frederick's brother, William H., served as president of the Society. Mrs. Frederick Osborn served on the society's board, as did Mrs. Leroy Clark, president of the Palisades Nature Association (Greenbrook Sanctuary), Laurance Rockefeller, Calvin Stillman, Mrs. Lila Acheson Wallace (a founder of *Reader's Digest*), and Carl Carmer, one of the founders of Scenic Hudson.

Hearing rumors that the Con Ed plant, once constructed, might be expanded, the Society Board sent a communiqué to its members, confirming its original position that "the best way to protect the river was not to oppose the entire project in a battle we were bound to lose . . . [but] if the protests piling in on the FPC succeed in completely stopping this project, our approach has of course been proved wrong." The board then passed a resolution that, in part, stated, "The plans of Consolidated Edison . . . will cause objectionable damage in our judgment to the natural scenic beauty."

Chairing the State Council of Parks, Laurance Rockefeller found another reason for concern. High-voltage power lines from the project were shown to pass over the Clarence Fahnstock State Park on the east side of the Hudson River. By unanimous vote, the State Council voted to oppose the alignment of the high-voltage lines. In a statement to reporters, Rockefeller announced opposition, and added, "New York is leading the nation in park development," citing $100 million made available for land conservation during his brother's tenure as governor. Fahnstock was the immediate concern, but perhaps the real message was that Rockefeller was shifting toward the Con Ed opponents.

The *New York Times* captured the contentious issue in an editorial headlined, "Preserving the Hudson Highlands": "If the area is to be fully protected for future generations, it cannot be most efficiently used now for a power plant or a dam, for mineral exploitation or for grazing: If it is to be used for such purposes, its particular esthetic or scenic qualities, its beauty and its silences, will be lost forever." Though the *Times* editorial spoke of Storm King, in a much larger sense, it captured the essence of park battles yet to be fought throughout the nation.

Nationally prominent *Life* magazine joined the *Times* in editorial opposition to the pumped-storage power plant, posing the question in its headline "Must God's Junkyard Grow?" Dr. Nathan M. Pusey, president of Harvard University, did not think so. Harvard had a stake in the natural integrity of 1000-acre Black Rock Forest, just west of Storm King Mountain. The forest had been donated to the university by the Stillman family. Pusey expressed personal opposition to the pumped-storage power-generating project.

Kenneth Morgan, in response to an inquiry from Laurance Rockefeller, confirmed in June 1964 that Con Ed's plant likely would be expanded once in place. He said that, based on recommendations from the staff of the FPC, Con Ed would consider raising the dams by fifty feet and adding more turbine generators at the river's edge. Rumors of possible expansion spurred Con Ed's opponents to search even more vigorously for any means to block the project. Morgan shared engineer-to-engineer camaraderie with Earl Griffith, his principal contact at Con Ed. When Griffith alerted Morgan that opponents might attempt a "taxpayer's law suit" against the PIPC to enjoin it from issuing a permit for construction of the tunnel under "Stillman gift lands," Morgan wrote a note to the file: "I told Earl that the Stillman gift land was not above the area expected to be used for the tunnel. We both got a big laugh out of this."

A flotilla of fifty boats moved up the Hudson River in September 1964, led by the *Westerly*, the seventy-nine-foot flagship of the New York Yacht Club. Behind the *Westerly* came other motor-powered yachts, sailboats, outboard motorboats, two houseboats, and four kayaks. Under the shadow of Storm King, three teenage boys dressed in Revolutionary War uniforms rowed ashore from the flotilla, invaded the Con Ed site, and, playing word games with Con Ed's motto, "Dig We Must," planted a sign, "DIG YOU MUST NOT!" Cheers and boat horns echoed in the narrows.

Dig Con Ed would not. This was the reality facing the giant power company, but no one knew it in November 1964 when 107 people appeared before a New York Joint Legislative Committee on Natural Resources that held a two-day hearing at the Bear Mountain Inn. The purpose of the hearing was to examine the prospect for strengthening state control of uses along the Hudson by establishing a "Hudson River Valley Authority," but the spotlight shone unblinkingly on the pumped-storage power-plant project. Predictably, speakers delivered mixed messages. Dr. Michael J. Donahue, mayor of the village of Cornwall-on-Hudson, expressed his continuing strong support for the project, as did the mayor of Newburgh, New York, several labor-union representatives, a handful of citizens who introduced themselves as "taxpayers," and a spokesperson for the power company.

Standing against the project were Scenic Hudson, the Sierra Club, and residents from throughout the lower Hudson River Valley who urged protection of the scenic beauty of the river. John J. Tamsen, retired superintendent of Bear Mountain State Park, and once a contender for the position occupied by Kenneth Morgan, stepped forward to claim a "change in philosophy of the commission that creates a serious void in the forces which were formerly dedicated to the preservation of the Hudson." Citing a Greek adage, Tamsen reminded those in attendance at the hearing that "life is the gift of nature; beautiful living is the gift of wisdom."

Poetry and philosophy, however elegantly expressed, were not going to win the day against the strong advantage of the FPC license held by Con Ed. The hearings at the Bear Mountain Inn were more allegorical than substantive. Still, proponents and opponents alike should have been paying close attention when Robert H. Boyle, a writer for *Sports Illustrated*, strode to the microphone to present his allotted five minutes of testimony. Boyle represented a little-known local organization identified for the record as the Cortland Conservation Society. In his brief comments, Boyle said, "The very life of the Hudson, itself, is at stake." Calling the lower Hudson River "a remarkable marine nursery and spawning ground," Boyle implored that "such vital resources should not be hastily overridden by the ichthyological illiterates of the power commission."

Most of those opposed to the power plant reacted to the potential negative impact on natural scenery that would occur along the shoreline of the Hudson River and the power line that would cut a swath through the forest for miles from the river to a Con Ed substation in Westchester County. Boyle was raising a separate issue: the impact of the power plant on fishery resources in the river.

Rising in the Adirondacks, the Hudson is 315 miles in length, but, incredibly, the drop in the river's surface elevation for the last 150 miles, from Albany to the tip of Manhattan, is less than five feet. Starting as a tumbling mountain brook, the great river slows at Albany, where it is met by the tidal flow from the distant Atlantic Ocean. From this point downriver, the Hudson essentially is a fjord, a glaciated arm of the sea, that allows for the wash and mix of salt and fresh waters.

This intermixing of nutrients and salinity creates the vibrant "nursery" that Boyle referred to on that November day at the Bear Mountain Inn. Shad, striped bass, and many other fish species would swim from spawning grounds in the Hudson River to the ocean. Boyle took a dim view of the power plant colossus that promised to chop the fisheries' resource unmercifully. The intake to the pumped storage plant was to be located just where the broad upper river was constricted by the Highlands. The natural flow would push migrating fish right into the mouths of the power plant's turbines.

Boyle's five minutes at the microphone likely became just part of the numbing verbal tabloid woven by more than one hundred speakers over two days of public hearings, but his fishermen's association would not be buried quietly in the pages of a corpulent hearing record. Scenic Hudson and the Hudson River Fishermen's Association had legally intervened in the FPC licensing process, contending that negative impacts on scenic beauty and natural resources should be considered and that the voices of concerned citizens who might be adversely affected should be heard. The FPC had rejected these pleas, proceeding, instead, to issue the license to Con Ed. Scenic Hudson and the Fishermen's Association appealed this decision, taking their legal arguments to the U.S. Court of Appeals for the Second Circuit.

And then came the legal bombshell.

In December 1965 the Court of Appeals announced its decision. "The Commission's [FPC] renewed proceedings must include as a basic concern the preservation of natural beauty and national historic sites, keeping in mind that in our affluent society, the cost of a project is only one of several factors to be considered." Then, going further, the court stated, "On remand, the Commission [FPC] should take the whole fisheries question into consideration before deciding whether the Storm King Project is licensed." In this ruling, the court set aside Con Ed's license.

The Circuit Court ruling on the Storm King license was momentous. In dry legal wording, it established legal precedent that would sweep through government and across the nation, leading to approval by the U.S. Congress of the National Environmental Protection Act (NEPA) and establishing citizen rights of legal standing in federal, state, and local government proceedings that threaten environmental damage. As stated in a Marist College review, the Appeals Court decision on Storm King "is credited with launching the modern environmental movement."

For Con Ed, the licensing process would begin anew, this time by formally allowing into the debate the organizations and citizens who saw in the pumped-storage plan for Storm King a potential act of severe environmental abuse. A telling signal that political opposition to Con Ed was strengthening came a month after the court ruling when Governor Rockefeller issued an executive order creating the Hudson River Valley Commission (HRVC) and appointed Laurance Rockefeller as its chair. The governor was responding to a bipartisan initiative by Democratic congressman

Richard Ottinger, Republican senator Jacob K. Javits, and Democratic senator Robert F. Kennedy Jr. to designate the lower Hudson River Valley a federally protected conservation zone.

For obvious reasons, the PIPC endorsed the newly formed HRVC, and though still publicly maintaining the position that "dual needs" could be served at Storm King, the commissioners were beginning to move cautiously toward opposition.

About a year after the Bear Mountain hearings, the newly formed Hudson River Valley Commission came forward with its recommendation about the Storm King Mountain pumped-storage electric project. The signal to Con Ed was unmistakable. In addition to Laurance Rockefeller, the newly created Hudson River Valley Commission included Averell Harriman, Ford Foundation president Henry Heald, famed journalist Lowell Thomas, IBM's Tom Watson, and Marion Heiskell of the *New York Times* among its roster of prominent corporate, community, and educational leaders.

"The HRVC strongly believes scenic and conservation values must be given as much weight as the more measurable economic values and that we should not necessarily destroy one to create another," said the HRVC. Then, narrowing its comment, the HRVC added, "The immediate case in point is the plan of Con Edison to build a pumped storage plant at Storm King Mountain. The HRVC believes that scenic values are paramount here and that the plant should not be built if a feasible alternative can be found."

In response, the alternative proposed by Con Ed was to pursue the Storm King site, but somehow place the huge pump turbines underground. This prompted the Hudson River Fishermen's Association to intervene again, motivated by the obvious conclusion that immense volumes of water sucked into the power system intakes from the river would include fish, regardless of whether the turbines were above or below ground. In October 1966, the fishermen and all the other opponents to the Con Ed plan gained a strong ally when Secretary of the Interior Stewart Udall recommended that the plant "not be built."

More hearings were held, and the case dragged on until August 1968 when, to the amazement and regret of the Storm King opponents, an apparently stone-deaf Federal Power Commission hearing officer again recommended that a license-to-build be issued to Con Ed. Then, surprisingly, New York City, the potential recipient of "desperately needed" peaking power according to Con Ed, intervened in the proceedings when city officials realized that placing the reversible turbines underground would require the removal of 580,000 tons of rock to create three underground chambers, the largest of which could accommodate a fifteen-story building.

Blasting for the underground chambers would threaten the city's Moodna Tunnel, hidden in the adjacent rock formation only 140 feet from where thousands of dynamite blasts would be triggered. This was not just any tunnel. Constructed through Storm King Mountain before World War I, the Moodna Tunnel delivered more than 40 percent of New York City's drinking water from the distant Catskill Mountains into the city's vast reservoir and plumbing system. When the tunnel was constructed more than fifty years before Con Ed arrived on the scene, unstable rock formations presented an engineering challenge.

Peaking power or no peaking power, the city did not want Con Ed's explosions to jeopardize its crucial water supply tunnel, claiming possible "catastrophic consequences" for 8 million people if blasting for the turbine chambers either cracked the Moodna Tunnel or caused sections to collapse. The FPC, by now notorious for its disdain of environmental arguments advanced by Scenic Hudson and its allies, could not ignore this new heavyweight contender for a voice in the proceedings, nor could the power company.

In response to concerns raised by city officials, Con Ed attempted to reconfigure the pumped-storage power plant by reverting to an earlier concept that would move the plant away from the city's Moodna Tunnel and onto PIPC land, still with the idea of digging the enormous underground chambers. It justified this stratagem by saying that it would build the power plant "underneath" the park, as if building economically feasible industrial sites "underneath" Yellowstone, Mount Rushmore, or any other park were logical beyond dispute.

It was as if parks were icing on a geologic cake, good for decoration, but of little consequence in the exploitation of any profitable underground resources. In an "action letter" signed jointly by Scenic Hudson, the New York/New Jersey Trail Conference, the Sierra Club, and the New Jersey Conservation Foundation, Con Ed's reasoning was robustly attacked. "To concede the rights to build a great power plant under any part of the park . . . would not only set a grave precedent for the Palisades Park but would in effect endanger all parks. . . . Such a precedent could range far and wide across parks of many states. Which then would be immune to industrial invasion if the land suited some private purpose?"

As a result of Con Ed's gambit and the action letter, the PIPC and its commissioners were flooded with demands to unequivocally join in the fight to stop the power company. By this time, the commissioners needed little convincing. In a general letter of response, PIPC President Jube said, "The Commission is taking the appropriate steps to make its opposition to the location of the proposed power plant on park land known to the Federal Power Commission formally. If the FPC opens hearings on this subject, I can assure you that the PIPC will present its opposition vigorously." The PIPC was ambivalent no more. It legally intervened in the Storm King proceedings in December 1968 to ensure that its "vigorous opposition" would go on record.

In March 1969, Commissioner Conrad Wirth was seated before the FPC's hearing officers in Washington, D.C. He testified that the PIPC was unalterably opposed to any uses of preserved parklands for industrial purposes except for essential utility easements. Wirth was distinctly qualified to deliver this message on behalf of his PIPC colleagues. He brought to the hearing a distinguished professional record in landscape architecture and resource management, marked most prominently by his service as director of the National Park Service from 1951 to 1964. When questioned at the FPC's hearing, Wirth spoke for everyone who cherishes precious natural places.

Question: "What are the general criteria used in what you call 'environmental planning?'"

Answer: "To be a good environmental planner, one must recognize the needs of all the social, cultural, and economic values and be able to place them in balance. To do this is one of the most difficult of all tasks. One is confronted with things that are described in dollars and cents but that must be compared to human values, values that affect and govern our social and cultural growth, and, in fact, our very civilization. These are not new values. They are basic values that are fundamental in the establishment of this nation. They are references to the freedoms and the right of people not only to work for a living but to enjoy life itself. They control those things that contribute to the enjoyment of that life as described in our Bill of Rights, Declaration of Independence, and the Constitution."

Question: "How would you apply these criteria or values to the subject of parks?"

Answer: "Well, with respect to parks, this balance has already been struck. Parks are set aside for the enjoyment of people and to preserve some of our outstanding natural scenery as a heritage to be passed to future generations. Things of this kind are found only where the good Lord placed them, and man cannot create them. He can only destroy them in order to gain monetary value, usually for only a relatively few years. Therefore, we in this country have decided that our outstanding natural heritages must be set aside for their true value to all the people."

After more months of deliberation, in an amazingly fallacious decision, the FPC determined that the original site for the power plant near Cornwall-on-Hudson was fine. After eight years of legal contest, Con Ed was relicensed to proceed. After eight years of citizen struggle, and probably a large dose of exhaustion, Scenic Hudson, the fishermen, the City of New York, and now, the PIPC appealed. In the following years, the grass-roots effort to save Storm King Mountain seemed to verge on hopelessness as courts and regulatory agencies seemed to tilt in favor of Con Ed, but there were always the fish.

In March 1974, Con Ed began construction, but in May, the U.S. Court of Appeals ordered further examination of the neglected fisheries issue. Hard-rock miners, mostly from the West, were moving into Cornwall-on-Hudson neighborhoods. When the ruling came down, construction was halted. One miner from California, who had moved into his newly purchase Cornwall house one day, put a FOR SALE sign on his lawn the next.

The Environmental Protection Agency (EPA), established in 1970 by President Richard Nixon while the Storm King controversy was boiling, became a crucial regulatory player. With Scenic Hudson and the Fishermen's Association actively in the mix always, the back-and-forth of legal proceedings continued, with the Federal Power Commission taking a backseat to the EPA on the matter of fish. In 1977, the EPA dropped a broadly applied bombshell. It concluded that "closed-cycle" cooling towers should be built at existing or proposed Con Ed, Central Hudson Gas & Electric, and other Hudson River power plants to protect fishery resources. In other words, countless fish no longer should be sucked into power plant plumbing systems as victims of cost-saving engineering design. Instead, the power vendors would have to build massive water recirculating structures to protect fisheries' resources.

Con Ed had started with the idea for a new pumped-storage power generating plant at Storm King, only to find after years of contentious debate and 12,542 pages of testimony on the record that it faced an unanticipated and very costly challenge to retrofit its downriver Indian Point Nuclear Power Plant with closed-cycle cooling towers. Company officials decided to settle with their opponents and walk away from Storm King. Legal combat was replaced by negotiation.

In 1980 Mrs. Willis (Franny) Reese, chairwoman of Scenic Hudson, was among the fatigued but euphoric and proud signatories to a settlement agreement with Con Ed. Albert Butzel, by then Scenic Hudson's battle-tested counsel, looked on. Con Ed surrendered its FPC license for the

Storm King project, vowing not to attempt any similar project for at least twenty-five years, and agreed to pay $12,000,000 to fund independent fisheries research in the Hudson River. It paid another $500,000 to offset the legal expenses accumulated over many years by Scenic Hudson, the NRDC, the Hudson River Fishermen's Association, and other litigants. Land on the river at Storm King, acquired for the main power-plant infrastructure, was deeded to the PIPC. Land acquired at the Cornwall-on-Hudson reservoir site went to the village.

Central Hudson Gas & Electric abandoned its attempt to build a cross-river power plant at Breakneck Ridge, and it agreed to sell the site, more than 600 acres, to the state of New York. In return, the EPA agreed that Con Ed, Central Hudson Gas & Electric, and other companies would not be forced to build closed-cycle cooling towers at their existing Hudson River power plants.

Con Ed had invested more than $35,000,000 in the Storm King venture and came away empty-handed.

The settlement greatly strengthened conservation organizations throughout the United States. The environmental impacts of new development now must include identification of at-risk flora, fauna, and natural resources, including data collection and research, independent analysis, identification of alternatives, and means of possible mitigation. If the impacts are too great and mitigation is not feasible, federal and state laws require that projects must be modified, or even dropped, to protect environmental values.

For the PIPC, Storm King Mountain was a hard lesson. The PIPC's reputation was bruised by its initial dual-use approach to Storm King before finally joining vigorously in opposition. But the pace and good purposes of the PIPC's other activities continued. The day after Robert Boyle spoke about fish at the Bear Mountain Inn, Laurance Rockefeller, Borg, Jube, and Kenneth Morgan were in New Jersey with Governor Hughes to witness the firing of a vintage cannon from historic Fort Lee for the first time in 188 years. They were determined to build a wonderful museum on the site where history was made in 1776.

Another hint of the path ahead for the PIPC came in early 1969 in the midst of the Storm King controversy when Commissioner Frederick Osborn mentioned at a commission meeting that "a 7,000-acre tract believed to be owned by real estate people" might be available for purchase on the Shawangunk Ridge, just west of New Paltz, New York. The tract was near the famous Cliff House, a Victorian-style hotel, vintage 1879, perched on the edge of a cliff at Lake Minnewaska, a so-called "sky lake." Osborn said the property was "scenic and might be of interest to the PIPC." With the consent of his fellow commissioners, Osborn asked the PIPC's staff "to try to find out what is proposed to be done with this tract."

Kenneth Morgan was not at the meeting when Osborn recommended that the Shawangunk Ridge be investigated. In response to inquiries by the commissioners and staff, J. O. I. Williams, the PIPC's assistant general manager, advised, "Ken is feeling better and thinks he might be back in circulation soon," but Morgan's deteriorating health prevented him from pursuing the Minnewaska opportunity as recommended by Osborn. After twenty-nine years as the PIPC's chief engineer and general manager, health problems forced Morgan to retire on September 1, 1969.

16

***

Minnewaska

ANTICIPATING CHANGE AT the top of the PIPC's staff organization in 1969, Laurance Rockefeller was thinking about exactly the right person to succeed A. K. Morgan. On his many trips to Washington, D.C., to chair the Outdoor Recreation Resources Review Commission and, later, to chair the White House Conference on Natural Beauty, Rockefeller met Nash Castro. Castro was serving as regional director for the National Park Service (NPS) in the nation's capital. NPS holdings in Washington, D.C., include a treasure-trove of cherished historic sites, among them the Lincoln and Jefferson memorials, the Washington Monument, the Mall and Vietnam Memorial, and the President's Park (the White House).

Unlike more traditional NPS operations that are focused on natural areas, historic and archaeological sites, battlegrounds, and outdoor recreation areas usually far removed from city centers, Castro was responsible for a full-blown urban-park operation. At age forty-nine, he was near the pinnacle of his profession, standing among a select few who exercised senior administrative and management influence over NPS policies, priorities, and style.

A naval aviator during World War II, Castro had a wide range of responsibilities, including service as executive vice president of the White House Historical Association, an organization he helped found during the Kennedy administration. Castro's continuing work with the historical society led him to assist Mrs. Lyndon Baines "Lady Bird" Johnson, who was advancing beautification projects throughout the nation. In this initiative,

Johnson had enlisted the celebrity and kindred interest of Laurance Rocke-feller and the administrative skills of Castro.

While traveling with Lady Bird Johnson to visit beautification projects, Rockefeller and Castro "came to know each other quite well," as Castro later recalled. Rockefeller obviously liked what he saw in the person and competency of Castro, confirmed in 1966 when Rockefeller, wearing his hat as chair of the New York State Council of Parks, attempted to lure Castro to New York to become director of the New York State Park System.

Castro declined, citing his work with Mrs. Johnson as well as other professional commitments to the NPS. Nonetheless, Rockefeller remained determined. Soon after the inauguration of President Richard Nixon in January 1969, Castro received a message that Rockefeller would be in Washington, D.C., and wanted to meet with him. "I speculated that he might wish to underwrite one or more unfunded beautification projects," Castro said. "Instead, he surprised me by inviting me to succeed Ken Morgan at the PIPC. L.S.R. said to me, 'This time don't say no. Come to the Palisades and take a look.' Laurance Rockefeller was aware that Kenneth Morgan planned to retire in a few months and wanted a successor who could step quickly into the position."

Accepting the Rockefeller invitation "more as a courtesy than anything," Castro toured the PIPC's parks and historic sites. "I did a lot of soul-searching subsequently, remembering that I filled a very satisfying niche in the National Park Service. . . . To offset this, I dwelled considerably on L.S.R.'s most commendable environmental work of 30 years in which he more than distinguished himself."

Still, Castro hesitated. Even after the Johnson administration was succeeded by the Nixon administration in January 1969, Mrs. Johnson continued to make frequent return trips to Washington to continue her advocacy for beautification. Along with Laurance and Mary Rockefeller, Castro and his wife, Bette, were included in various social events when the former First Lady was in town. Rockefeller would take advantage of these gatherings to again discuss the PIPC opportunity with the Castros. Bette Castro fondly remembered Rockefeller as being "very persuasive."

In June 1969, Rockefeller dispatched his Citation jet to Washington to transport the Castros to the Hudson River Valley. Aboard Rockefeller's boat, the *Dauntless*, a converted Canadian minesweeper, the Castros were treated to a waterborne view of the Hudson River from Tarrytown to Bear Mountain. Rockefeller escorted the Castros to lunch at the Bear Mountain Inn, where they were joined by commissioners Linn Perkins, Averell Harriman, Frederick Osborn, Conrad Wirth, and Donald Borg.

Not leaving anything to chance, Rockefeller also included Commissioner-emeritus Horace Albright. Both Albright and Wirth were former directors of the National Park Service. Two weeks after the visit, Castro agreed to bring his NPS career to conclusion and to join the PIPC as its general manager. "Bette and I considered all the pros and cons about making such a big change in our lives," Castro remembered. "Every discussion recalled the prospect of working with a real star of the conservation fraternity, Laurance S. Rockefeller." In September 1969, in a smooth management transition, Castro began a second career, one that would extend over two decades.

Castro had been on the job for only a few days when Commissioner Osborn telephoned him at the Tarrytown Hilton, where he and his family were temporarily lodged. Osborn's call was prompted by an article in the *New York Times* describing the Shawangunk Ridge and its classic Victorian-style hotels at lakes Minnewaska and Mohonk. In the 1870s, twin brothers Albert and Alfred Smiley purchased a tavern on the shore of Lake Mohonk, a "sky lake" on a knife-edged promontory near the northern end of the thirty-mile-long ridge. Following their Quaker beliefs, the Smileys forbade the drinking of alcoholic beverages, card playing, or dancing in their newly acquired tavern, named the Mohonk Mountain House.

Despite these restrictions, the brothers began to transform the mountain house into a successful resort for guests, primarily from New York City. The guests would journey by train to a railroad stop near New Paltz, New York, then would be transported by horse-drawn carriages to Lake Mohonk. Inspired by a carriage trip to nearby Coxing Pond, the other "sky lake" on the ridge, Alfred Smiley determined to develop a second resort hotel at the site. He renamed the pond Lake Minnewaska, borrowing a word associated with the Native American language of the region. Alfred's Minnewaska Mountain House (later renamed the Cliff House) was constructed just beyond the edge of the cliff that surrounded the lake and accommodated 225 guests. Within a decade, the Wildmere, a second hotel that accommodated 350 guests, shared the locality with the Cliff House.

The *Times*'s article confirmed that the geologically elaborate Shawangunk Ridge was a phenomenal nursery for thousands of plant and wildlife species. Osborn said to Castro, "Let's drive up there and take a look." Enlisting Commissioner Wirth to the reconnaissance team, the three men confirmed that the Empire National Bank was in fact threatening to foreclose on thousands of acres on the ridge owned by Kenneth E. Phillips Sr. In 1955, Phillips, an ardent spokesperson for the sensational beauty of the Shawangunk Ridge, had acquired the Cliff House and Wildmere Hotels and 10,000 acres of land surrounding Lake Minnewaska for a reported $75,000 from the heirs of Alfred Smiley.

The hotels retained their outward charm when the Osborn team visited in 1969, but weather and wear and tear had taken their toll on the old wooden structures. The cost of maintenance was spiraling upward. Travel patterns were changing. City dwellers who once provided a dependable leisure clientele for the Shawangunks, Catskills, and Adirondacks were routinely traveling to more distant places. Ownership of the old hotels, combined with a property-tax burden on so much land, was proving to be beyond the limits of Phillips's pocketbook in a faltering market.

Maneuvering to financially reinforce the Lake Minnewaska operation and avoid foreclosure, Phillips signaled a willingness to sell 7,000 acres of land while retaining 3,000 acres nearest to the lake. Resting along the ridgeline south of the lake, the land offered for sale was a stunningly beautiful domain of cliffs, forests, and narrow valleys. Osborn, Wirth, and Castro needed no convincing that such a place should be preserved. They quickly learned, too, that the PIPC had a readily approachable ally in the Nature Conservancy. The conservancy was known for its businesslike approach to the acquisition of natural areas noted for biological diversity and vitality. The fact that Pat Noonan, the Nature Conservancy's executive director, was interested in the Shawangunk Ridge ecosystem was a huge plus for the PIPC.

In November 1969 the PIPC went on record in support of the initiative to acquire 7,000 acres from Phillips at a cost of $1.5 million. Castro affirmed that the Nature Conservancy was positioned to handle the transaction and was willing to transfer title to the PIPC on the handshake expectation that the commissioners would succeed in arranging for enough public funding to reimburse the conservancy. If successful, the acquisition would stand among the most meaningful on the PIPC's ledger. Phillips's acreage was of sufficient scale to create a stand-alone park. Miles of carriage roads and hiking trails were already in place. A forest of rare dwarf pitch pine trees, abundant wildlife, sweeping views, and two wilderness lakes awaited future park visitors.

Just south of the land being sought by the PIPC, deep cave-like fissures in the striated cliffs sheltered ice formations that grew from trickling water during the winter months and were permanently sheltered from the summer heat in naturally refrigerated alcoves. Only a handful of ice caves exist in the United States, and very few worldwide. Just north of Phillips's acreage, rock climbers had discovered the "gunks" (shorthand for the Shawangunk cliffs) in the mid-1950s and added the hard rock cliffs to their informal network of prized climbing destinations. Climbers often train on the gunks in anticipation of later adventures on the big walls of Yosemite.

Working out the details to purchase a portion of the Phillips property put the PIPC and the conservancy in a race against the Empire National Bank. The first challenge was to determine exactly how much land Phillips was selling. His estimate was 6,725 acres. A survey and title search boosted the figure to 7,100 acres. Phillips did not quibble; his price was firm regardless of the acreage discrepancy. With support for the transaction expressed by Governor Nelson Rockefeller, the conservancy closed the deal in June 1970, only days before the bank was prepared to foreclose. With signature affixed to the legal documents, the PIPC assumed management of a major new park of rare merit. Reimbursement to the conservancy took months of concerted effort by the commissioners and Castro. Half of the $1,500,000 purchase price was slated to come from the federal Land & Water Conservation Fund on condition that New York State would match the amount.

Two years after Phillips sold to the conservancy, the commissioners commended Castro for the "infinitely painful and highly effective effort in the matter of the Minnewaska acquisition." The pain came from the after-the-fact need to convince federal and state officials that the PIPC's commitment to the Nature Conservancy was valid, justifiable, and undeniably beneficial to the public. Even with the weight of Governor Rockefeller's support, complemented by the national standing of Laurance Rockefeller as a conservation champion, the many keepers of the public purse still had to be persuaded. New to his PIPC job, Castro used diplomacy and his personal contacts in Washington, D.C., to lock in the $750,000 federal investment in the Shawangunk Ridge. The New York matching amount followed.

Commissioner Osborn was in a lame-duck position with the PIPC when the transaction for Phillips's land was completed. His five-year term had expired in 1969. Commissioners with expired terms could continue to serve until they or their successors were appointed to a new term. Osborn learned that Governor Nelson Rockefeller did not intend to reappoint him. The governor's reasoning was that Osborn, at age eighty-two, was too old to be reappointed. On learning the news, Osborn commented that "this association has been one of the happiest of my life." Out of courtesy, Governor Rockefeller hesitated to appoint a successor, and Osborn went right on with his volunteer work for the PIPC, including pursuit of Phillips's property. Poor health finally caused him to submit his resignation in 1972. Osborn had served with the PIPC for forty-four years. Speaking on behalf of the commissioners, Castro credited Osborn with a "triumph of the first order" on the Shawangunk Ridge.

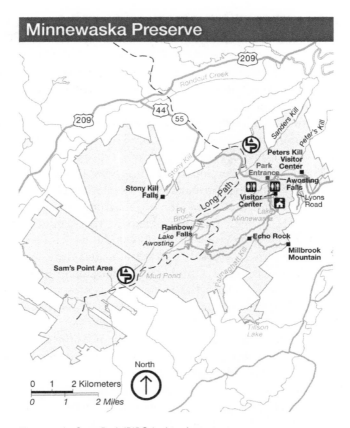

Minnewaska State Park (PIPC Archives)

Castro moved quickly to begin developing a management plan for the duly named Minnewaska State Park. One of the immediate challenges was that a rustic camp was located at remote Lake Awosting consisting of several "rather marginal buildings" and "more especially, a problem from a sanitation point of view," as Castro delicately advised the commissioners. Practitioners of yoga used the camp for prayer and meditation. The campers promptly launched "a highly effective letter-writing campaign" to retain use favors in the center of the new park.

Castro knew well this type of demand for privileged use of parklands. Groups and individuals not necessarily involved in the hard-fought political efforts to create new parks frequently were first in line to insist on entitled access. Usually claiming that "our tax dollars created the park," groups and individuals motivated by singular agendas often seek to stake out territory by means of whatever political clout may be available to them. The yoga campers at Lake Awosting were no exception. They wanted not to budge or to be disturbed. Close after them, hunters, trappers, and snowmobilers made known their expectations that the new state park would be immediately open for their purposes.

To find answers for the future stewardship of Minnewaska, Castro turned to a planning technique routinely used by the NPS. He imported a team of NPS experts and opened the planning process to public debate. A strong consensus quickly developed for preserving the wild character of Minnewaska. The minority, particularly representatives of nearby communities who wished maximum economic benefit from tourism, expressed a contrary view. They urged rapid development that would include bus access, use of the carriage roads by automobiles and motorcycles, motorboats on Lake Awosting, off-road vehicle trails, and the construction of tennis and basketball courts, swimming pools, and food stands. Snowmobilers aligned themselves with the community representatives, seeking strength in numbers as a springboard to influence the PIPC and elected officials in Albany.

Claiming that the park should not be "locked up for the hardy few" who enjoyed the solitude of wilderness trails, the proponents for development of Minnewaska discovered that their voices were overmatched by the "hardy few" whose numbers far surpassed their own. Spokesperson after spokesperson appeared before the planning team to urge strict limits on development and motor-vehicle access in favor of park uses including hiking, nature education, cross-country skiing, sledding, ice skating, horseback riding, bicycling, bird-watching, photography, fishing, swimming in the lakes and streams, canoeing, rock climbing, and the simple pleasure of experiencing the sights and sounds of a natural place unencumbered by blacktop, buildings, and combustion engines.

The planning team found in favor of the "hardy few" by recommending to the PIPC that snowmobiles (later allowed), motorcycles, motorboats, and automobiles not be permitted access to the carriage roads and lakes in the new park. In varying degrees, the team reasoned that these activities might severely limit other uses, could be dangerous, and would be destructive to the natural integrity of the park. The team concluded that these "artificial" forms of recreation were not appropriate in such an exquisite and complex natural ecosystem.

Hunting was another matter. The NPS planning team recommended against hunting at Minnewaska State Park, taking a page from federal rules

Minnewaska State Park, Lake Awosting (Photo by Angela Ciao)

that deny hunting in national parks. But the tradition of hunting on most public lands in New York runs deep. Despite the pleas from many who participated in the planning dialogue that the sound of gunfire and bullets whizzing through the trees would be intrusive and dangerous, Castro judged that a compromise on hunting was necessary. He recommended that the commissioners approve hunting limited to the use of shotguns only in a 1,600-acre section of the new park.

At about the time that the Minnewaska plan was being settled, the PIPC found itself affected by a nationwide economic recession. In late 1970, Castro alerted the commissioners that Governor Rockefeller was imposing an austerity program on all state agencies, including the PIPC. A hiring freeze was imposed, and all agencies were directed to reduce operations by 40 percent. This economic squeeze caught up with the last of the side-wheel steamers still operating on the Hudson River. The *Alexander Hamilton* made its final voyage from New York City to Bear Mountain on September 6, 1971. When the steamer turned downriver from the Bear Mountain dock, the grand era of steamboat service on the river came to an end.

While grappling with the challenges of the economic recession, Castro was also looking at the details of the operation. At his direction, the staff began changing the language on the signs that greeted the host of visitors to the parks and historic sites. Message changes included:

Old Sign: "Persons Discarding Rubbish Will be Arrested; No Skating; Stay on Walks; No Picnicking."

New Sign: "Please Keep Your Park Clean; Sorry, Ice Unsafe for Skating Today; Please Use Walks; Please Picnic in Designated Areas Only."

User-friendly initiatives at the grassroots level, combined with Castro's bid to ensure that the public was heard on the questions of use at Minnewaska, brought praise to the PIPC. Commissioner Conrad Wirth echoed this sentiment when he decided to conclude his volunteer service with the PIPC in 1972. Writing to Castro, the former NPS director said, "You are demonstrating, as I knew you would, what a fine professional park man can do. One of the most important aspects of administration (it may be the most important) is understanding people." By attempting to improve contact between the PIPC and its patrons, Castro was inviting the public to join in the care of their parklands.

This invitation, though, had its limits. Park visitor Christopher J. Schubert, a geologist on the staff of the Museum of Natural History, confirmed that crystals found near Bear Mountain "are Jeffersonite, a manganese-zinc bearing pyroxene. These have respectable market value and word of their occurrence should be kept to a minimum." But, he added, "it would be most appropriate to try to incorporate these pockets into a natural trail exhibit. So rarely does a student have the opportunity to see museum quality crystals in their natural setting." Castro only heeded Schubert's first advice. The location of the crystals remained undisclosed. Even in a user-friendly park, to do otherwise almost surely would guarantee that the crystals would disappear, either as curiosities and mementos or at the hands of commercial raiders.

While Castro was encouraging visitor cooperation in the care of the parks, he found himself on a collision course with the PIPC's own police force in New Jersey. The police officers were unhappy with their level of pay. They had been lumped into a general state of New Jersey pay category with game wardens and prison guards even though their duties closely paralleled those of the more highly paid State Highway Patrol. The complaint by the Palisades Parkway policemen was directed at New Jersey's union pay structure, not at the PIPC, but their leverage was the parkway. Urged on by the Patrolmen's Benevolent Association, the Parkway police began stopping rush-hour motorists to check driver's licenses and automobile registrations. The officers were careful and took their time, so much so that the result was a ten-mile-long traffic jam on the Palisades Parkway involving an estimated 9,200 vehicles occupied by infuriated commuters.

The "job action" produced exactly the opposite result of that sought by the police officers. They generated so much bad publicity for themselves that state officials stiffened their resistance to any change in the union status for the Palisades Parkway police. The commissioners and Castro, acting to stem an instant erosion of public support, directed that key leaders of the "job action" be reprimanded and reassigned to foot-patrol duties at the Englewood and Alpine boat basins.

Nor had the user-friendly approach prevailed in the lengthy dialogue between Commissioner Donald Borg and representatives of the Marriott Corporation. After almost fifteen years of effort, during which the PIPC had sought to reach a compromise with Marriott to preserve the character of the Fort Lee Historic Site, the corporation was finally forced from its 6.43-acre development site through condemnation proceedings by the state of New Jersey. Marriott received $1,012,500 as payment for the acreage. The PIPC's share of the payout to Marriott was 25 percent, about $253,000. Marriott hotels subsequently were built nearby on the New Jersey side of the Hudson River, but not in a manner to dominate the PIPC's thirty-three-acre Fort Lee Historical Park, including reconstructed fortifications and a visitor center.

Borg and his fellow commissioners might have been forgiven a sigh of relief at the departure of such a stubborn foe. They would find, though, that their contest with Marriott at Fort Lee would pale by comparison to a conflict yet to come. Marriott was about to arrive on the scene at Minnewaska.

"Dear Laurance: My resignation as a member of the Palisades Interstate Park Commission has been accepted and I am writing to tell you and the other members how very much it has meant to me to serve with you gentlemen." With this note, dated September 18, 1973, Linn M. Perkins, the first woman appointed to the PIPC, signaled the beginning of a transition among the commissioners equal to major changes that occurred in the late 1930s. Less than three months after Mrs. Perkins resigned, Donald Borg stepped down because of health reasons. Then, in March 1974, former governor and ambassador Averell Harriman followed suit after more than a half-century of involvement with the PIPC.

These resignations, coming so closely in line, might have opened the PIPC to political pressures of unaccustomed scale. Instead, the resignations served to reinforce family commitments to the PIPC. Governor Rockefeller appointed Mrs. Perkins's son, George Perkins Jr., to succeed her. In late spring 1974, Malcolm Wilson, who became governor of New York when Nelson Rockefeller was affirmed as vice president of the United States in the Ford administration, appointed Mary Fisk, daughter of Averell Harriman, to take her father's place. A few weeks later, New Jersey governor Brendan Byrne appointed Malcolm A. Borg to succeed his father.

These family commitments demonstrate remarkable dedication to the purposes and activities of an organization that otherwise is only vaguely recognized in the New York/New Jersey metropolitan region and among established national conservation organizations, if recognized at all. The incentive for such constancy lies in the product of the commissioners' work. They can see, touch, measure, and cherish the results. Successful conservation of land and historic treasures is tangible. This seems reward enough for the PIPC baton to be passed from one family hand to the next.

During the transition among the commissioners, the 3,000-acre portion of property at Minnewaska still owned by Kenneth L. Phillips Sr. was at risk. At a PIPC meeting in April 1974, Castro reported that the *Daily Freeman*, a Kingston, New York, newspaper, carried an article confirming that Phillips "has not arranged financing" and that he was planning to demolish the old Cliff House and Wildmere Hotels in favor of subdividing the land to accommodate motels, condominiums, and private homes.

George R. Cooley, a local Hudson River Valley member of the Nature Conservancy, wrote to Castro later in the year: "Word comes to me that Phillips is getting slow in making his monthly payments to the First National Bank of Highland on the $750,000 first mortgage which the bank owns. Is the ultimate acquisition of the hotel property and its 2,895 acres something which interests you and should the Conservancy step in at the right time?"

A headline in the *Sunday Times Herald Record*, Middletown, New York, announced, "$18 Million Project Proposed for Minnewaska." According to the article, a "1,136-unit condominium project that would transform a large tract of wilderness land adjoining the Minnewaska State Park into a four-season resort has been proposed to Town of Rochester and Gardiner officials." Included in the development plan, said reporter Peter Kutschera, were "commercial tracts that would include a theater, motel, restaurants, service stations, retail shops and professional offices." He added that the plan also included "an equestrian ranch, tennis courts, a children's recreation area, hiking trails and an expanded golf course."

Subdivided lots for 110 private homes were part of the development concept. Kutschera speculated that "as word of the proposal spreads through the region, it appears that a battle may be in store." Quoted in the same article, Frederick Faerber, president of the Ulster Sportsmen's Federation, confirmed that his organization would oppose the plan, contending that the proposed development "would further threaten a continually diminishing supply of available game land. A high-density development next to a forever wild park makes absolutely no sense to us." The Rondout Valley Sportsmen Club echoed Faerber's view. Its representative sent a letter to Ogden Reid, New York's newly appointed commissioner of Environmental Conservation, recommending that the Phillips property be annexed to the park. "That's the only way we can see something like this being stopped cold," contended club member Steve Schwartz.

Within days of the *Times Herald* article, Phillips and Castro were meeting at Bear Mountain to discuss a possible willing-seller, willing-buyer transaction. As he had done once before, Phillips suggested sale of one-half of his remaining land holdings. "This would permit him to apply the funds from such a sale to the requirements of the other one-half of his property," Castro reported to the commissioners. Less than a month later, Phillips was in Albany, meeting with representatives of the Office of Parks, Recreation, and Historic Preservation (OPRHP), seeking, as Castro summarized it, "a State financial vehicle" that would allow Phillips to "avoid subscribing any land for development." Before the state could respond, the *Times Herald Record* confirmed that the Ulster County Planning Board had recommended to town of Rochester officials that rezoning of the property be denied. The proposed development "would be entirely out of character" with Minnewaska State Park, said board chairman Gifford Beal. He, too, urged the state to acquire the property.

In the spring of 1975, Castro and Phillips were talking again. On this occasion, Phillips proposed that the PIPC purchase 600 acres of his property, including a marginal downhill ski facility, for $1.8 million. Castro characterized this proposal as not "business-like," but he was determined to keep trying. In the meantime, the PIPC was completing transactions with various other landowners on the Shawangunk Ridge. At the May commission meeting, Castro confirmed that an additional 1,570 acres had been added to Minnewaska State Park, bringing the total holding to 8,644 acres.

The next item on the commission's agenda dampened this good news. Castro reported that, in an overheated moment of government restructuring, the New York legislature voted to consolidate all parkways in the state under the authority of the Department of Transportation. Most parkways near New York City had evolved into jam-packed commuter routes. Older design standards were proving inadequate to demand. The costs for maintenance and improvement were skyrocketing.

Still, the fact that the Palisades Interstate Parkway was swept along with the rest came as a surprise. Unlike the other New York parkways, the Palisades Parkway formed a historic connection between two states. It existed because of the foresight of early PIPC pioneers working in alliance with John D. Rockefeller Jr., and it was integral to the formation of the entire PIPC park and historic-site system. To summarily yank the parkway from the PIPC and turn it over to highway engineers who had little knowledge of its genesis and purpose posed a real threat to the integrity of the PIPC's operations.

Commissioner-emeritus Harriman was first in line to protest. He urged recently elected Governor Hugh Carey to veto the bill, arguing that "parkways are parks and not highways and should continue to be managed as such." Harriman's plea went unheeded. Although ownership remains with the PIPC, the thirty-mile New York section of the Palisades Parkway is maintained by the New York Department of Transportation (DOT). The fourteen-mile New Jersey section remains under the authority of the PIPC.

In June 1975, the *New York Times* carried a bold advertisement with a photograph of the Cliff House and Wildmere Hotels at Lake Minnewaska over the lead line "Equity Partner Sought." Following a detailed description of the property, the tag line declared, "We seek an equity partner of substance to co-venture the development of what must be considered one of the finest resort locations in the world." Interested parties were invited to contact Phillips and his son, Kenneth Jr. The search for venture capital was a race by the Phillips's against the financial clock. Time ran out in early August, as the *Daily Freeman* confirmed: "Foreclosure Action Is Filed Against Lake Minnewaska."

The accompanying article gave the details: $750,000 plus interest in default to the First National Bank of Highland; $134,173 in school taxes unpaid; a $35,000 federal tax lien. The "state financial vehicle" that Phillips sought through Castro to avoid the kind of headline appearing in the Kingston newspaper did not arrive. In response to a conversation with Castro, Phillips wrote, "Thank you for your telephone call. . . . Although your answer is negative, we want you to know that we appreciate your efforts in going to Albany to seek funds to acquire additional lands for Minnewaska State Park. . . . Your reply leaves us no choice but to pursue alternatives." Neither Phillips nor Castro knew what those "alternatives" might be.

Phillips and his son understandably wanted to avoid losing their livelihood and the property they cherished. If foreclosure somehow could be avoided by finding venture-capital copartners or funding from the state at a level that would satisfy the First National Bank, the Phillips family presence at Lake Minnewaska could continue in genteel style in the picturesque and historic setting. C. David Loeks, representing Mid-Hudson Pattern for Progress, an alliance of businesspeople, recommended that funds from a voter-approved 1972 Bond Act be used to acquire the Phillips land. Loeks reminded readers of the *Times Herald Record* that only $22 million of the $66 million bond fund had been spent, leaving an obvious source of readily available dollars for the purchase. The Adirondack Mountain Club, the Nature Conservancy, and other environmental groups echoed Loeks's plea.

These organizations found a careful listener in the person of Maurice D. Hinchey, representing the 101st New York Assembly District. Hinchey's legislative district covered a significant midportion of the Hudson River Valley. He was chairman of the New York Assembly's Environmental Conservation Committee and was known for paying close attention to the policies, presumptions, activities, and purse strings of the state's land-management agencies. Learning that Castro's initiative to find state funding for the Minnewaska purchase had been unsuccessful, Hinchey jumped in. On September 6, 1975, he toured Minnewaska State Park and the adjoining Phillips property with Castro and Orin Lehman, commissioner of the New York Office of Parks, Recreation, and Historic Preservation (OPRHP). Hinchey found obvious, ready partners in both men.

Within four days, Lehman, who was to gain a highly respected reputation as a defender of parks during his many years of service with the state, was in contact with Peter Goldmark, Governor Carey's budget director. Lehman wanted a $2 million commitment from Goldmark to buy all the Phillips property, using the Nature Conservancy as the agent to put together the deal. Castro concurred.

A new commissioner was appointed to the PIPC as the drive to acquire the Phillips property gathered momentum. She was Dr. Mamie Phipps Clark, a resident of Hastings-on-Hudson and the first African American to serve on the commission. Her background contrasted with that of the commissioners who had preceded her, but her professional skills and personal interests merged readily with one of the most basic purposes of the PIPC: to connect parks with inner-city children in need.

Born in Arkansas, Clark was a Magna cum Laude and Phi Beta Kappa graduate of Howard University. In 1944 she was awarded a Ph.D. in clinical psychology from Columbia University. On graduation from Columbia, Clark joined the staff of the Riverdale Home for Children, not far from

Wave Hill, the former estate of George Perkins Sr. The privately funded Riverdale agency specialized in providing protection and care for homeless black girls. Realizing the "critical need" to provide psychological counseling to Harlem's black children and their families, Clark and her husband established the Northside Center for Child Development, where she served as executive director for thirty-four years. Using the Northside Center as a catalyst, Clark successfully created the 110th Street Plaza Housing Development Corporation to provide low- and middle-income housing for six hundred families.

Clark found a kindred spirit among her commissioner colleagues. Mary Fisk, though light-years removed from Clark in childhood experiences and opportunities, had also chosen a path as an educator in the inner city. Without fanfare but with quiet determination, Fisk traveled from her home on the Harriman estate to Harlem, where she taught as a volunteer for many years. Fisk and Clark stood with inner-city children along a fault line between hope and despair. They devoted themselves to the talent and promise they found in the children they assisted. Both women saw the group camps in Harriman Park as meaningful extensions of their efforts in the city. On the commission, they made the camps their specialty, sharing in advocacy and friendship.

Ironically, a commissioner who had preceded them, a champion in his own time of the group camps, one of the great PIPC pioneers, became a victim of the hazy veils of history. The son of Dr. Edward L. Partridge offered a portrait of his father to the commissioners. Although records confirmed that Partridge served as a commissioner from 1913 to 1930, institutional memory was vague. In the brief discussion about accepting the portrait, there was no hint of Partridge's turn-of-the-century promotion for a national park in the Hudson Highlands or of the development of a children's camp on his own property at Cornwall-on-Hudson. Nor was there mention of his contact with E. H. Harriman, in which the conservation theme for the Highlands was rung like a bell and a crucial champion was enlisted in the cause. On due deliberation, the commissioners declined to accept the portrait, reasoning that acceptance "might set a precedent which would be difficult to control."

Partridge nonetheless would have cheered an October 1975 editorial in the *New York Times* that reflected growing public concern about the fate of the Phillips property. The *Times* confirmed that the Nature Conservancy was again standing in the wings, waiting for a signal to negotiate the purchase on behalf of the state. Speaking for the conservancy, Pat Noonan was quoted as saying that the conservancy was willing to move "as soon as the whistle is blown."

In an attempt to nail down a purchase price so that the state of New York and the PIPC knew precisely how much money had to be put on the table, a conservancy representative met "amiably" with Phillips and proposed $1.2 million for his remaining property. Phillips countered by claiming that he "had an offer in hand" exceeding $1.5 million, adding that the fair market value really was $3.5 million, the price he expected to receive. With prospective buyer and potential seller far apart on value, discussion shifted to the option of acquiring 2,000 acres of mostly undeveloped land, leaving the "core resort" and about 1,000 acres with Phillips.

Lehman rejected the $3.5 million price as well as the idea of a partial sale that would not include the centerpiece of the Phillips property, Lake Minnewaska. In the meantime, an ad appeared in the *Kruse Reports*, an Indiana-based investment journal, stating, "Lake Minnewaska, one of America's most beautiful resorts since 1879, now offered for sale. Owners will trade equity for stock. Mergers invited." Coincident with the ad, Phillips was back in contact with the local planning board, presenting a plan for a 120-unit trailer park near the location he proposed for the townhouse development. But the foreclosure auction block still loomed.

In April 1976, Castro alerted the conservancy's staff that "judgment will be returned today for the Judge's signature, and he will then appoint a referee for the sale. Once this is done, he can then proceed to prepare the Notice for Sale for publication. The publication will be made once a week for four weeks." The bank's attorney "looks for the property to be sold at the end of May or the beginning of June." The auction subsequently was scheduled for June 14 in Kingston, New York.

The date shifted to June 25, at which time a "stay" in the bankruptcy proceeding was granted by the court in response to a plea from the Phillips family attorney for more time to deal with delinquent tax and mortgage obligations. By then, the conservancy had a letter-of-intent signed by Lehman confirming that "the State of New York will make every effort to repurchase the property known as Lake Minnewaska from The Nature Conservancy within one year of its purchase."

In August 1976 Castro was in contact with Brad Northrup, the conservancy's Boston-based regional director. He advised Northrup that attorneys for the First National Bank of Highland were attempting to get the "stay lifted." Understandably, Phillips was doing everything in his power to convince the bankruptcy court that he had the ability to pay off his creditors. The judge continued a "stay" in the proceedings to provide Phillips even more time. While juggling trial preparations and creditor payments, Phillips maintained contact with the PIPC and Commissioner Lehman, suggesting that the state might wish to acquire the Lake Minnewaska property, then lease it back to the Phillips family to continue operation of the resort hotels.

Lehman's staff felt that "the odds are 70-30 in our favor of picking up the property at a foreclosure sale," and so communicated to Castro. Albert E. Caccese, counsel for the OPRHP, then advised Judge R. Lewis Townsend that, "although all parties made a bona fide attempt to arrive at an amicable and reasonable leaseback arrangement with the Phillips family, we are, unfortunately, unable to arrive at a meeting of the minds."

A bankruptcy trial date was set for December 8, 1976. In preparation, First National Bank attorneys filed an appraisal report with the court that estimated the fair market value for the entire Phillips property at $ 1.7 million. By this time, the Phillips family was claiming a value of $5.7 million based on estimates from its own "real estate experts." With the state and the PIPC contending that the purchase price would be no more than $1.5 million, the director of the Bureau of Outdoor Recreation, the agency created as a result of Laurance Rockefeller's work for President Eisenhower, confirmed that $500,000 was available from the federal government to assist with the purchase.

The trial date was rescheduled yet again. In March 1977, Lucille Phillips wrote to Castro, "It was so good to see you Saturday." She asked in the letter whether the PIPC might consider the idea of a "life tenancy" for the Phillips family at Lake Minnewaska. The notion was that the Phillips's would be paid for the property but could continue to live there. Phillips referred to other property owners who had made similar arrangements with the PIPC, thanks to "much kindly assistance from you and the Nature Conservancy." Castro put a margin note on the letter: "No."

Public interest in the acquisition took on the characteristics of a low-grade fever as weeks dragged into months, with the occasional burble of activity surfacing in the media to remind people that the property remained in jeopardy. In one of many letters received by the PIPC, Robert M. Watkins introduced himself as a "stockbroker, not an ecologist." He described Lake Minnewaska as a "fairy-land lake, the most beautiful I have ever seen. Its iridescent blue color and overall beauty are such they must be seen to be believed. Of particular significance is that it is pristine, completely untouched from the time two Victorian hotels were built on its cliffs shortly after the Civil War. To expose such a piece of property now to the wrong type of private or commercial development would represent nothing short of a rape of probably the most beautiful remaining spot in New York State, if not the entire northeast United States."

Staff members for Lehman and Castro kept up a steady exchange of information, confirming that "18 separate parcels" made up the Phillips family's holdings and that these parcels might finally come under the hammer of the auctioneer more than two years after the First National Bank of Highland initiated the bankruptcy action. David Strauss & Company, Auctioneers, announced in the *New York Times* that "2,600 acres of resort land" would be auctioned, including two hotels, a nine-hole golf course, ski area, twenty miles of riding trails, and tennis courts adjacent to a "crystal clear glacial lake." The auction date was set for noon on November 9, 1977, at Lake Minnewaska.

But before the auction could take place, Lehman's staff, reflecting lengthy behind-the-scenes dialogue with the PIPC, made a surprise offer of $1,050,000 for a partial purchase of the Phillips family property. On October 20, 1977, a press release from Lehman confirmed that the deal for a partial purchase had been struck with Kenneth and Lucille Phillips. Lehman praised the Phillips family "for conducting their resort operation with utmost concern for the preservation of the environment" and advised Judge Townsend on October 31, 1977, "We have reached agreement with the owner for the purchase of 1,300 acres of the Minnewaska property plus conservation easements on another 239 acres which include the lake."

The primary intent of the easement was to prevent the use of powerboats on the lake, but it also encompassed land on which a nine-hole golf course was located and restricted alterations to the natural landscape, including a strict limitation on tree cutting except to provide for scenic views. Any tree cutting required the PIPC's advance approval.

The total purchase price, including miscellaneous costs, came to $1,110,000. Using the Laurance Rockefeller–inspired Land and Water Conservation Fund, the Department of the Interior provided $550,000 of the total purchase price, placing it in a position to legally monitor the future use of the land, including the easement, to ensure that park purposes would not suffer "conversion" to other uses. Title to the land and easement were vested in the PIPC. With proceeds in hand from the sale and still owning more than 1,000 acres, the Phillips's were positioned to make temporary peace with their creditors, although the family's corporation remained in bankruptcy. An editorial in the Newburgh *Evening News* cheered, "Minnewaska Saved." Tucked into the text of the editorial was a telling comment that would come back to haunt the PIPC: "From a conservationist point of view, it would have been better if the state had acquired the entire mountaintop. But half a loaf is better than none." The half-loaf left out of the deal was to prove very hot to handle.

With the issue of Minnewaska seemingly resolved, Castro, the commissioners, and the PIPC's staff could focus more time on other matters of management. Good news came from Barnabas McHenry, vice president and chief counsel for the *Reader's Digest* Association. McHenry confirmed that Lila Acheson Wallace and DeWitt Wallace, founders of the *Digest,* were committing long-term financial support to the Tiorati Workshop for Environmental Learning in Harriman Park. Money in support of environmental education was difficult to find in the 1970s, but McHenry made a persuasive case to his employers. The Wallaces needed little convincing. Their charitable legacy in the New York metropolitan area and beyond would grow to huge levels, including the establishment of a trust fund that has allowed two organizations, Scenic Hudson and the Open Space Institute, to invest more than $100,000,000, and still counting, in land conservation in the Hudson River Valley.

Castro and the staff had to balance the good news of the Tiorati Workshop against the startling news that sixty-four historic manuscripts, some signed by George Washington, had disappeared from a secure storage area at the PIPC-managed Washington's Headquarters State Historic Site in Newburgh, New York. Suspecting the obvious, that someone on the staff with access to the storage area had stolen the manuscripts, the park police questioned every member of the on-site staff. Police Sergeant Daniel Shea was particularly persistent in the investigation. He tracked down leads that implicated a former part-time night watchman who had moved to Raleigh, North Carolina. With the arrest of the watchman, all sixty-four manuscripts were recovered and safely returned.

Even more startling than the theft was the disastrous fire, probably caused by vandals, that burned the Lake Minnewaska Cliff House Hotel to the ground in January 1978. The Phillips family had permanently closed the ninety-nine-year-old hotel, a tinderbox of dry wood, in 1972. With the closure, the sprinkler system was turned off and insurance coverage canceled. When the fire broke out in a second-story room, only two four-wheel-drive vehicles from distant volunteer fire departments were able to reach the scene over the steep, snow-covered access road. By the time the firefighters arrived, nothing was left of the hotel but a pile of glowing embers. The commission was more fortunate later that same year when a fire broke out in the kitchen of the Bear Mountain Inn. At the time, nine volunteer firefighters happened to be having dinner at the inn. They got up from their table and put out the fire.

About a year after the Cliff House fire, Phillips Jr. confirmed that the Marriott Corporation wished to build a new eight-story, 400-room "destination resort hotel" on the Phillips-owned site of the old Cliff House. According to Marriott, the hotel, anticipated to cost $31 million, would employ 400 workers and inject $16 million in tax revenue into local government coffers in the first ten years of operation. The resort would offer a separate sports building, restaurants, a disco, indoor and outdoor swimming pools, indoor tennis courts, an ice rink, an equestrian center, a skeet range, and five racquetball courts. To complement the hotel's operation, 500 condominium units were expected to be added to the Lake Minnewaska site.

Commissioner Malcolm Borg, among others, must have raised his eyebrows at the thought of the Marriott Corporation's arrival at Minnewaska. He and the other PIPC commissioners had been present in 1976 at the dedication of the new visitor center at the Fort Lee Historic Site in New Jersey. The commissioners had invested $3.5 million of the PIPC's funds in the visitor center in recognition of the nation's Bicentennial celebration, the largest amount spent on any project in New Jersey. One of the speakers at the dedication was Alfred T. Guido, New Jersey's director of Parks and Forestry, who reminded those in attendance that "we would be standing in the lobby of a high-rise hotel were it not for the Palisades Interstate Park Commission. Hotels are nice, but parks are better." Two decades earlier, Borg, as a young man, accompanied his commissioner father, Donald, to meetings with Willard Marriott during the "second battle" of Fort Lee. Now, the battle was shifting to Minnewaska.

Change was occurring within the PIPC as Marriott-Minnewaska came on the agenda. After serving almost forty years with the PIPC, Laurance Rockefeller advised Governor Carey that he would step down when his term expired in February 1979. Citing age as the reason, the sixty-eight-year-old commissioner recommended to the governor that a fourth-generation Rockefeller, Larry, his son, be appointed to take his place.

At his final PIPC meeting, Rockefeller said, "At a time like this, one has a lot of mixed emotions, a sense of achievement on the one hand, and gratitude to so many people for the things we have done together." Rockefeller was leaving behind a proud record. Even in the final few years of his service, he had directed almost $1.3 million in personal and Rockefeller family contributions to the PIPC. Laurance Rockefeller could claim specific credit for Rockland Lake and Tallman Mountain parks, for wresting Iona Island from the federal government, and for the Land and Water Conservation Fund that benefits the PIPC and other state and federal land stewards across the nation. Writing from California, PIPC commissioner-emeritus Horace Albright stated, "From afar, my old friend, I salute you. I rejoice in your achievements."

At the same PIPC meeting in which Laurance Rockefeller offered his farewell, Nash Castro confirmed that he was resigning after almost ten years as general manager of the PIPC to return to Washington, D.C. Castro had declined in 1974 the invitation of secretary of the interior Rogers C. B. Morton to accept the directorship of the National Park Service, but when a second invitation came to return to the nation's capital, he accepted the position of executive director of the White House Historical Association, the organization Castro had helped organize during the Kennedy administration in 1961. He was leaving with high praise from the commissioners for his adroit skills and graceful style.

A search committee was formed to recruit Castro's successor. The PIPC's deputy general manager, Donald B. Stewart, was charged with all responsibilities for day-to-day management pending the result of the search. Fortunately for the PIPC, the search progressed slowly. In the meantime, Castro discovered that the challenges and complexities of the PIPC far exceeded the demands of his new assignment in Washington. He reminisced, "I would arrive at the White House, get all my work done in the first two hours, and have nothing much to do for the rest of the day." To the relief of the commissioners, Castro asked to be reinstated. Seven months after he departed for Washington, Castro was back on the job at Bear Mountain. Despite additional job offers, three times to serve as New York park commissioner and once as commissioner of the New York Department of Environmental Conservation (DEC), Castro remained at the PIPC's helm until he retired in 1990.

During Castro's brief absence in Washington, D.C., Stewart responded to an article in the *Gannett Westchester Newspapers* entitled, "The View Today." In the article, the PIPC was taken to task for two unsightly radio towers that loomed over the New Jersey section of the Palisades Parkway, pointing out that a primary obligation of the PIPC was to protect the skyline from just such intrusions. Stewart could do little in response other than to explain that the largest of the two towers, a four-hundred-foot-tall behemoth with three one-hundred-foot-wide cross arms, had been constructed just off the PIPC's property in 1937 to transmit the world's first FM radio signals. The tower is so large that it appears as an official visual checkpoint on Federal Aviation Administration aeronautical charts for pilots flying in and out of Newark and Teterboro airports. AT&T had constructed a smaller tower, also just off the PIPC's property, as part of its long-distance telephone network.

Stewart could do nothing about the towers but did act when he was advised by a PIPC staff member that "someone has remembered us in his will." In the last will and testament of Albert E. Milliken, instructions were left that if his wife survived him the residuary of his estate, including land on the Shawangunk Ridge, would go to her. If she did not survive him, 165 acres adjoining lands at Minnewaska State Park would go to the PIPC. "In order to fully understand this material," stated PIPC staff member Ron Karner, "some background information concerning Mr. Milliken's untimely demise is necessary. It is alleged that Mr. Milliken murdered his wife and then committed suicide." A further confusing explanation followed: "Mr. Milliken had been married previously and had children by his first wife. It was these, his natural children, [to whom] he left his estate. Milliken's second wife's former husband filed a wrongful death action on behalf of his children, who had been left out of the will." By the time Castro was back at his desk, lawyers for the contending family members had untangled the issue, and the land came to the PIPC.

Within days of Castro's return, he received a letter from James J. Stapleton, chairman of the Ulster County, New York, Environmental Management Council, transmitting a report from Dr. Stephen J. Egemeier, chair of the council's subsidiary Land Use Committee. Egemeier had reviewed Marriott's development proposal for land still owned by the Phillips family at Lake Minnewaska and summarized his findings in the report. In his cover letter to Stapleton, Egemeier said, "I find it amusing that I'm accused in the July 7th *Poughkeepsie Journal* by Marriott's consultant of 'reflecting a typical environmental bias.' On television the other night, Marriott advertised its Essex House as on Central Park. The selling point was a desirable environment."

Egemeier then listed his concerns: Marriott had misrepresented the geological structure of the Shawangunk Ridge, referring to "granite" rock rather than the sedimentary conglomerate that formed the ridge. This was important, Egemeier claimed, because of the potential for groundwater pollution. The corporation claimed that no archaeological or historic sites existed in the area, contrary to findings of State University of New York archaeologists. Marriott denied any potential visual impact, although the hotel "would be visible all over the northern Wallkill Valley." Company officials claimed that the new hotel simply would replace the existing old hotels, but room capacity would increase from 333 to about 1,400, including the condominiums.

The most telling comments in the report involved water and sewage. Pointing out that Lake Minnewaska was a "sky lake," replenished seasonally by rainwater and snowmelt, Egemeier commented that the level of the lake would fluctuate over six feet in the wettest years, based on Marriott's projected water usage, and could not recover at all in drought years. He then described the sewage problem: "Presently, Minnewaska sewage is pumped from a septic tank and dumped on the top of a hill. From there it flows downhill, forming a pond and killing trees. This 'overland flow system' is the worst pollution mess and health hazard I have seen anywhere in the county. The sewage pond is black and foul-smelling with a green scum. Gasses bubble to the surface, which is covered with flies. Remnants of toilet paper litter the ground. Approximately 600 feet downhill from the pipe I took a water sample. It gave a coliform count of 1,500, an incredibly polluted sample."

During a heavily attended public forum sponsored by the Friends of the Shawangunks and Citizens to Save Minnewaska soon after the Egemeier report was released, Warren McKeon, a former New York Department of Environmental Conservation regional director, labeled the Marriott plan "environmentally unsound." According to an article in the *Times Herald Record*, "When one person from the audience suggested that the state take over the proposed 375-acre hotel site and convert it into public park land, spectators erupted in applause." Individuals began forming themselves into activist groups in opposition to the project. A few dissident voices began to take on the characteristics of a chorus.

The PIPC commissioners and Castro were looking north, working with Lehman, Caccese, and the OPRHP staff in Albany in the attempt to expand land conservation at Minnewaska, when, in the PIPC's backyard, they were caught off guard by a lengthy article in the September 9, 1979, *Sunday Record*, Commissioner Malcolm Borg's own newspaper: "At N.J.'s Edge, a Park Is Dying," declared the headline; "Management Fails to Halt Decline, Detect Abuse." Investigative reporters Philip Barbara and Bruce Locklin had spent weeks touring the PIPC's facilities in New Jersey, interviewing PIPC commissioners including their own boss, Malcolm Borg, talking with present and former employees, and mingling with park visitors. The result was a seven-page article highly critical of the PIPC.

To a large degree, the article was testimony to the value Borg placed on the independence of his reporters. As owner and publisher of the *Record*, Borg was obviously responsible for the economic vitality and growth of his newspaper, but he let the news flow, wherever legitimate reporting might lead, even on those rare occasions when he found himself criticized in his own paper. The Barbara-Locklin piece was a case in point. The thrust of the article was that a beautiful park had badly deteriorated, accompanied by severe loss of morale among the PIPC's employees, because senior managers, including the commissioners and Castro, were not paying attention. The commissioners were criticized for holding most of their meetings in a private room at Rockefeller Center in New York City, a tradition that had been followed for many years. Using the meeting location as one example,

Barbara and Locklin charged that the hierarchy within the PIPC was far too remote from the guts of the operation in New Jersey.

To drive home the point, vividly contrasting photographs accompanied the article. One photo highlighted a nicely remodeled bathroom in the PIPC-owned home of the park's assistant manager, a project accomplished in part with the PIPC's funds and the use of the park's maintenance force. Juxtaposed was another photo that showed a grungy, ill-maintained restroom at State Line Lookout, the premier tourist-stop on the New Jersey Palisades. Two other photos, similarly displayed, contrasted a lively beach scene in the Undercliff area of the park, vintage 1925, to a now-vacant beach littered with driftwood. A PIPC employee was quoted as saying, "I was born and raised here, I love the park so much it hurts to see it going to hell."

The *Record*'s article appeared at a time when the PIPC and other governmental agencies in New Jersey and New York were adjusting to new requirements for open public meetings, as mandated by both legislatures. The long-standing habit of holding many of the PIPC's meetings in a private room on the sixty-fourth floor of 30 Rockefeller Center was already being modified. The meetings were always conducted during lunch, served from the kitchens of the famous Rainbow Room restaurant, located a floor above. When the meetings were first opened as required by law, a few members of the public arrived with their brown-bag lunches. They sat around the wall of the room, observing the meeting and admiring the luncheon, while they munched on their sandwiches.

Barbara and Locklin underscored the need for the commissioners to hear from the public about park management's activities, good or bad. The few brown baggers who had the time, inclination, and determination to attend the Rockefeller Center meetings symbolized a much wider public interest, confirmed when eleven organizations, representing a combined membership of 76,450, petitioned the PIPC to open its meetings to wide and unrestricted comment. Among the organizations were the Environmental Defense Fund, the Hudson River Sloop *Clearwater*, the Adirondack Mountain Club, the Audubon Society, the Sierra Club, the Friends of the Earth, and the Rockland County Conservation Association.

Malcolm Borg and Larry Rockefeller, who had succeeded his father on the commission in October 1979, echoed this plea. Larry Rockefeller, a graduate of Columbia Law School, former VISTA volunteer in Harlem, and a senior member of the Natural Resources Defense Council, saw a clear need and benefit in regular PIPC dialogue with the public it serves. He supported Borg's view that the PIPC should schedule fewer meetings in the city and more at PIPC parks and historic sites. In a short time, the Rockefeller Center meetings ceased entirely.

Public input was helpful, but the root cause of the problem in New Jersey was money. The problem was traced back to 1976 when the PIPC tapped into $3.5 million of its funds to construct the Fort Lee Historic Site Visitor Center in celebration of the Bicentennial. At about the same time, the New Jersey legislature went into austerity mode and cut appropriations across the board for statewide park operations, including annual funds provided to the PIPC. Other than appropriations, PIPC revenues were generated primarily from sales at two gasoline stations near the southern terminus of the Palisades Parkway. The days of major donor access to abundant private funds were gone. Gas station revenue had been shifted from projects to paychecks to avoid staff layoffs. Money for the restoration, maintenance, and improvement of facilities was meager.

A split personality exists nationwide in park stewardship. Parks expand. Precious places are preserved. The public estate evolves to benefit future generations, attendant with public and media enthusiasm and political flourishes. Left in the wake of the creation of each new park and historic site are the professional managers struggling to make the case for continuing financial support from state legislatures or the U.S. Congress once the excitement is over. The PIPC's split personality was glaringly evident in the Barbara/ Locklin article. Paint was peeling in restrooms at State Line Lookout, and the Palisades beaches were neglected and cluttered while, at the same time, the PIPC was seeking millions to conserve natural beauty at Minnewaska. Nash Castro took a needed detour south to Trenton to make the case for more appropriated funds for the New Jersey operation. Appropriated funds upticked, and the maintenance backlog began to recede.

Phillips Jr. confirmed in a November 1979 news article that "we are in the process of liquidating numerous antiques and artifacts that have been stored at Wildmere since it was opened for business a century ago." Marriott planned a spring opening for its new resort. The Wildmere Hotel would be torn down to make way for condominiums. Headlines started to appear in regional newspapers every few days: "Test Polluting Stream, Opponents Say"; "Marriott, Ulster Leaders Laud Resort Plans"; "Minnewaska Sale Near, Marriott Says"; "Archeologist Urges Study of Minnewaska Hotel Site"; "Will Marriott Mar Lake Minnewaska?"; "Marriott Big Issue in New Paltz School Vote"; "Casinos Not Marriott's Issue"; "Marriott Studies Disputed."

The New York Department of Environmental Conservation, custodian of the state's environmental regulations and sister agency to the Office of Parks, Recreation, and Historic Preservation, appointed administrative law judge Robert S. Drew to preside at public hearings to gauge the accuracy and acceptability of data contained in plans prepared by Marriott as a requirement for obtaining necessary water and sewage permits for its development project. Announcement of the hearings drew in the principals who would vigorously debate the issue: Caccese, representing the OPRHP and PIPC; attorney Robert Kafin, representing Marriott; Phillip Gitlen, a former DEC attorney, representing the Friends of the Shawangunks and Citizens to Save Minnewaska.

Castro received a letter in June 1980 from a senior Marriott attorney stating, "In order to accomplish expansion of the golf course it will be necessary to modify certain portions of the Indenture (conservation easement) which limit development activities." Not knowing the exact meaning of the word "modify," Caccese asked that the DEC hearings be delayed pending more information from Marriott. Judge Drew rejected the request. A subsequent exchange during the hearing procedures contrasted the positions of the contending parties on the easement question:

Kafin for Marriott: "The golf course cannot be built without permission of the PIPC. We will go and ask them for permission. They can set restrictions, but until they do, it is very hard for us to give much detail."

Caccese for the State of New York: "We made a motion to adjourn at the pre-hearing conference on the basis that we don't have enough information here to determine what the impact is going to be. Now, if the impact is going to be very severe and harmful to the easement area, then we're not going to act at all. We're going to sit down. What we want to know is what is the proposal, not guess at this and guess at that. We have to know what sort of mitigation is necessary based on the environmental ills that will be caused by the project. We don't know that."

The hearings, expected to be concluded after a few days, extended into weeks, then into months. Opponents and proponents alike were astonished at a session held in August 1980 when Joseph C. Cataldo, a hydrologist with the Cooper Institute for the Advancement of Science and Art, took the stand. Cataldo claimed that projected water use by the Marriott resort would consume 80 percent of the water in thirty-two-acre Lake Minnewaska in one decade, lowering the surface by twenty-three feet. Kafin objected to accepting this testimony into the record, contending that Cataldo had used a "mishmash of tables" to arrive at his findings. Kafin was overruled. Nash Castro, in attendance at the hearings, labeled the Cataldo testimony "very dramatic."

At another moment in the hearings, Albert F. Smiley, grandson of the founder of the Mohonk Mountain House, found himself the target of a subpoena from Kafin. Trying to counter the apparent damage done to Marriott's proposal by Cataldo, Kafin demanded on a Friday morning that Smiley produce historic records on lake-level fluctuations at the two "sky lakes," Mohonk and Minnewaska, by 10:00 A.M. that day. Smiley appeared at the hearing as ordered.

"All right, Al, what do you have for me?" Kafin commanded.

"Nothing," replied Smiley.

"Nothing?" said Kafin. "See you in court, fella!"

On further questioning by attorneys and the judge, Kafin admitted that Marriott had never made a reasonable request of Smiley for the records, depending, instead, on information provided by the Phillips family. Water and the conservation easement held by the PIPC on a portion of the Phillips land were appearing more and more to be the blocking points to the Marriott venture. Kafin, struggling to prove that the impact on Lake Minnewaska would not be nearly as dire as that asserted by Cataldo, sought from Judge Drew, and won, an extended suspension of the hearings pending the development of more precise water data by engineering and hydrology consultants retained by Marriott.

Phillips Jr. reminded the public through a news reporter that he and his family retained all the prerogatives of private land ownership. "People from the Unification Church have been up there a number of times," he said. "I even met the Reverend Moon in person." Reacting to a claim by Assemblyman Hinchey that 55 percent of his constituents favored adding the Phillips holding to Minnewaska State Park, Phillips Jr. responded by saying that Hinchey "was playing to the rabble."

The hearings resumed in February 1981. In the interim, an editorial had appeared in the *New York Times* on December 20, 1980, that proclaimed, "A Peace Treaty on The Hudson," but the editorial was not focused on Marriott and Minnewaska. It marked the agreement, mediated by Russell Train, former administrator of the Environmental Protection Agency, to finally put to rest the Storm King Mountain controversy. Ross Sandler of the NRDC, Albert Butzel representing Scenic Hudson, and Charles Luce of Consolidated Edison had agreed to and accepted the details of the long-awaited truce.

Storm King officially was history, but Marriott was back at the DEC hearings with new data on water consumption. It had pumped 30 million gallons from Lake Minnewaska in January, mimicking the projected water usage by the proposed resort as a test of the capacity of the lake to recover. The lake level dropped only twenty-one inches, less than the hydrology experts had predicted. Marriott contended, too, that nearby test wells demonstrated "ample capacity," independent of the lake. Caccese and Castro continued to seek more details on the overall proposed development: how many condos? where? access? sanitation? In a letter to Phillips Jr. Castro said, "It is not business-like for us to treat this matter informally, and the information we have been provided thus far is precisely that, informal." Castro wanted plans that were not based on "guesswork, approximations, or vagueness."

After eight months, including lengthy delays, Judge Drew asked for closing written statements from the various parties who had participated in the hearings. In a Statement of Proposed Findings of Fact, Issues, and Recommendations submitted to the judge, Caccese contended, "Undisputed evidence presented at the hearing has established the fact that Marriott proposes a mountaintop complex with a footprint at least six times as large as the historic development on this site. At this scale, the project will have dramatic, adverse effect on scenic integrity as well as the water supply and the quality of trails, vegetation, traffic, noise, and air. We find that with its massive scale and adverse impact, the project presently proposed by Marriott does not respect the natural constraints of the land nor the public interest in surrounding parklands and affected resources. While approval of the present project would be contrary to our public trust, we believe that a project reduced in scale and more carefully sited is acceptable, given the historic presence of a resort hotel and its significance to the local economy."

In other words, state parks and the PIPC were choosing negotiation over outright opposition. Caccese recommended that an easement be granted by Marriott to the PIPC to ensure a more limited hotel and condominium development, with emphasis on protecting a ridge line on the east side of Lake Minnewaska, where Marriott wanted to build 300 condominiums. Marriott's representatives had not expected to have to negotiate. In anger, lawyers in the firm of Miller, Mannix, Lemery & Kafin submitted a brief in rebuttal, charging, "In an intemperate, near hysterical, unsigned, and undocumented paper entitled 'Proposed Findings of Fact, Issues and Recommendations,' the Office of Parks and Recreation and the Palisades Interstate Park Commission asked the Department (of Environmental Conservation) illegally and unconstitutionally to seize by confiscation the Cliff House site to prevent the Applicant from using it for its Project."

All contestants awaited the judge's opinion. Prompted by the promise of 450 new jobs and a $6,000,000 annual payroll in an economy with a 9.7 percent unemployment rate, county officials favored full Marriott development. Phillips Jr. reaffirmed that his family's bankrupt corporation had to sell. If the Marriott deal prevailed, the Phillips family stood to receive $1.85 million in cash, 1.5 percent of the sales price for each condominium unit sold, and 35 percent of hotel revenues for twenty-nine years, a compensation package that Phillips said would "bring us out about even." If the Marriott deal fell through, Phillips suggested that the family's alternative would be to auction off all land parcels that constituted its Lake Minnewaska holdings to the highest bidders. Advocates for the preservation of the site, led by Friends of the Shawangunks and Citizens to Save Minnewaska, remained staunch supporters of adding all the Phillips property to Minnewaska State Park.

Commissioner Robert Flacke of the New York Department of Environmental Conservation endorsed on June 2, 1981, a 129-page decision by administrative law judge Drew allowing for the construction of Marriott's 400-room hotel and fifty condominiums on Phillips property. Judge Drew had found in favor of Marriott, but restricted condominium development to fifty units unless and until Marriott proved that the "total supply of water from its permanent wells can provide sufficient water to meet the total yearly water demands for the entire project."

In the administrative finding, the judge granted Marriott a sewage permit to discharge sewage effluent onto the planned eighteen-hole golf course while, in concert, confirming that a decision about amending the conservation easement to actually allow for expansion of the existing nine-hole golf course rested squarely with the PIPC. According to Marriott, the eighteen-hole course was termed "essential to the success" of the entire venture, now elevated in estimated cost from $18 million to $78 million.

Kafin, representing Marriott, was pleased that DEC commissioner Flacke had "given his general approval" to the project. The Friends of the Shawangunks and Citizens to Save Minnewaska expressed confidence that Marriott would never be able to prove sufficient ground-water supplies, thus holding the scale of the proposed development below a level considered economically viable for Marriott.

With the Administrative Court ruling, the *Times Herald Record* confirmed that Marriott's opponents were shifting their focus to the PIPC. Paul Lowry, president of the Mid-Hudson chapter of the Sierra Club, charged that the PIPC had a cavalier attitude about the conservation easement. "The Commission," Lowry said, "thinks of the easement as something they can trade on, but our position is it should not be violated. It's an important tool to environmentalists. I know of no other easement that has been changed anywhere in the country."

Castro, speaking for the commissioners, confirmed that, in fact, the PIPC was anticipating use of the easement as a bargaining chip to win the corporation's guarantee of maximum protection of scenery and water, as well as assurances that the public would have the right of transit through the Marriott property.

But a key PIPC commissioner, Larry Rockefeller, was emerging as a proponent for not amending the easement to expand the golf course unless Marriott agreed to relocate its proposed condominiums to a less visible section of the Phillips property. Rockefeller suggested that Marriott expand easement protection to include 750 ridge-line acres surrounding Lake Minnewaska that Marriott had designated as the condo-development site. This, according to Kafin, would sink the project.

Edward Bednarz, director of Hotel Development for Marriott, labeled Larry Rockefeller's stance "hard-line," adding that Rockefeller was "making extreme demands we can't meet." Bednarz failed to say that all ten members of the commission unanimously adopted Rockefeller's "hardline" position.

Phillips Jr. labeled the commissioners' position "improper" and was quoted by the *Times Herald Record* as saying, "It's a hell of a thing. I knew there would be battles left, but I didn't expect this." In an editorial, the *Kingston Daily Freeman* chided the commissioners for being "outsiders [who] meet in New Jersey" and who "have the last word on a local issue."

The PIPC commissioners held a special meeting in June 1981 at Bear Mountain to hear from the public about Marriott's project. This prompted Joseph Penzato, a pavement contractor and president of the citizens' group Friends of Marriott, to say that "the Commission is behaving like a spoiled child crying: 'it's my baseball, so we'll play by my rules.'" Kafin added, "The PIPC had two years to get their point across; it's a little bit late to be redesigning the project now."

About two hundred people attended the meeting. A large majority spoke in favor of protecting the natural beauty of Lake Minnewaska and voiced strong opposition to any change in the terms of the conservation easement. At the conclusion of the meeting, a PIPC negotiating team was appointed to continue discussions with Marriott. The team included commissioners Larry Rockefeller and Jon F. Hanson, Castro, Caccese, and special counsel Arthur Savage.

Within weeks, the PIPC's and Marriott's negotiators reached agreement that the conservation easement would be amended to allow an expanded golf course on condition that condominiums would be set back fifty feet from the Lake Minnewaska ridgeline and would be limited to 150 in number. Marriott's opponents immediately expressed resistance to the compromise and vowed to fight on. Castro wrote to Orin Lehman, thanking the state park commissioner for his support "throughout the interminable proceedings," adding, "[I have] profound admiration for Al Caccese. I have not witnessed a more dedicated public servant than he."

The "interminable" proceedings were far from over. The Friends of the Shawangunks, Citizens to Save Minnewaska, and the Appalachian Mountain Club initiated a lawsuit in state court against the New York Department of Environmental Conservation, contending that the DEC had violated its own regulations when it approved a "conditional" water-taking permit for Marriott.

The Sierra Club and the Appalachian Mountain Club filed a lawsuit in federal court against the PIPC, claiming the PIPC should have received approval from the Department of the Interior and the New York and New Jersey legislatures before reaching agreement with Marriott to amend the conservation easement. Several months later, while these lawsuits were pending, a federal bankruptcy court judge directed the Phillips family to provide confirmation that Marriott was in fact prepared to spend $80 million to construct the new resort. Personal liability for the Phillipses stood at about $1.5 million; $820,000 in accumulated tax bills were going unpaid, and the Wildmere Hotel had lost $15,374 in its operations the previous year. More than six years had passed since bankruptcy proceedings were initiated against the Lake Minnewaska Mountain Houses company owned by the Phillips family. Creditors and tax collectors were understandably frustrated.

The bankruptcy court was not to receive a clear answer from Phillips and Marriott anytime soon. This was assured in May 1983 when the Appellate Division of the New York Supreme Court, Third Department, struck down the earlier decision by administrative law judge Drew, who had approved condominium construction on "condition" that sufficient water supply could be proven. The higher court said, prove water supply first and condos second, if enough water is found. Judge Drew confirmed that the appellate decision meant that "full-scale" environmental hearings would be required to resolve the water-supply question one way or the other.

Marriott's Bednarz expressed concern that his company "cannot justify additional expense to get all that's needed done to fully develop the property." Marriott claimed to have spent $1.5 million during four years of project planning and preparation, and although it still had a 400-unit hotel and fifty condominiums on the table, it did not want to spend additional money on drilling wells in the hope that enough water could be found for the additional 250 condominiums it sought. The project was becoming economically shaky for Marriott.

With the project in doubt, new players appeared. Marriott announced that the Investors Management Group of Baltimore and Tokyo-based AOKI Construction Company Limited were joining in the project. In a parallel financial race, a coalition of environmental groups was attempting to join in partnership with the Trust for Public Land (TPL) to purchase the remaining Phillips property. After speculation in the press about various "investors" newly arrived on the scene, none seemed to muster the dollars needed to make a serious bid for the property. With no "white knight" stepping forward to financially strengthen the Trust for Public Land, the judge in the bankruptcy proceeding approved the exclusive right of Marriott and its recently recruited investment allies to close the long-awaited transaction with the Phillips family within one year.

But before signatures were affixed to the deeds, the Sierra Club and Appalachian Mountain Club prevailed in court proceedings against the PIPC. The U.S. Court of Appeals for the Second Circuit handed down its ruling, finding in favor of the plaintiffs. The 239-acre conservation easement owned by the PIPC, but purchased with matching federal funds, could not be amended without federal approval. There would be no expanded golf course and adjusted condo numbers and siting unless representatives of the Department of the Interior agreed, and the DOI already had signaled opposition to an amended conservation easement. Ironically, by this time, most of the PIPC commissioners favored the court ruling. Commissioner Larry Rockefeller was especially persuasive in his advocacy for preservation of natural scenery versus what he viewed as a misguided development compromise.

A headline in the *New York Times* on March 13, 1985, confirmed, "Marriott Calls Off Plans to Revive Ulster Resort." Bednarz, speaking for Marriott, said that the company "could not conclude an acceptable agreement" with the Phillips family to purchase additional acreage the investors deemed necessary to continue the project. Even if an agreement had been reached on more acreage, Bednarz knew that a redesigned project and more water questions guaranteed a return to the hearing arena as required by the State Environmental Quality Review Act (SEQRA)—and probably a trip back to court.

In a comment months later to Diana Shaman of the *New York Times*, Bednarz confirmed that the federal court ruling on the conservation easement was "the straw that broke the camel's back." From a marketing perspective, Marriott had from the beginning placed strong emphasis on an eighteen-hole golf course, now out of the resort development equation. Still faced with bankruptcy, the Phillips family was left without a major resort investor.

Responding to obviously strong public support and following through on the purposes of their respective agencies, Caccese and Castro entered again into direct contact with Phillips Sr. On October 22, 23, and 24, 1985, Caccese and Castro spent twenty-six hours in negotiation with Phillips Sr. and his attorneys. During this marathon negotiation, Caccese accepted an invitation to stay overnight at the Phillips's home. Unprepared for the stay, Caccese was lent pajamas and underwear and was served breakfast in bed. When the negotiations concluded, Phillips told Caccese that he did not have to return the borrowed underwear.

On October 29, the state made an offer of $3,000,000 for the remaining 1,200-acre Phillips property. The offer was declined. With the property again under threat of the auction block, the state initiated eminent-domain proceedings against Lake Minnewaska Mountain Houses, Inc., in December 1985.

Almost symbolic of the last hope for massive new resort development at Lake Minnewaska, the uninsured Wildmere Hotel, closed and vacant since 1976, caught fire in June 1986 and, like its predecessor, the Cliff House, burned to the ground.

New York governor Mario Cuomo stood at a cliff-side location overlooking Lake Minnewaska a year and a half after the eminent-domain proceedings commenced to accept a deed to the Phillips property from Frank D. Boren, president of the Nature Conservancy. To avoid more months of the condemnation proceedings in which claim and counterclaim for the value of the Phillips family's holdings would be batted back and forth, the conservancy had stepped forward as promised, bringing along its substantial checkbook. Working with Caccese and Castro under an agreement that ensured eventual reimbursement to the conservancy from bond-act funds approved by New York voters, a final accord was reached with the Phillips's to buy their entire property for $6.75 million. For the Phillips family, its estimate of value had been pleasantly validated.

Governor Mario Cuomo referred to Minnewaska as an "environmental heirloom," reminding those in attendance that "at issue was the limited capacity of this beautiful and fragile land to absorb private development." Lehman and Castro confirmed that the newly acquired property would be opened to the public for hiking, picnicking, and scenic enjoyment within a matter of days. They credited the steadfast organizations and individuals that participated in the seventeen-year effort to bring the Phillips holding into Minnewaska State Park.

Mentioned were Assemblyman Maurice Hinchey, the Mary Flagler Cary Charitable Trust, the Nature Conservancy, the Catskill Center, the Open Space Institute, the Regional Plan Association, the Friends of the Shawangunks, the New York–New Jersey Trail Conference, Mohonk Preserve, Inc., the Citizens to Save Minnewaska, the Cragsmoor Association, the Trust for Public Land, the Appalachian Mountain Club, Scenic Hudson, the Sierra Club, David Sive, Alfred and Albert Smiley, and Bernard Brennan.

As part of the purchase agreement, seventy-six-year-old Kenneth Phillips Sr. and his wife, Lucille, were granted life-tenancy rights, allowing them to remain in their artistically designed home on the clifftop overlooking Lake Minnewaska. All the property acquired by Alfred Smiley more than a century earlier was now encompassed within 13,000-acre Minnewaska State Park. (In recent years, Minnewaska has doubled in size through acquisition efforts led by the Open Space Institute.) The PIPC, acting on behalf of the state of New York and the federal government, had gained permanent stewardship of one of the premier park ecosystems in the eastern United States.

17

‖‖‖

Sterling Forest

October 15, 1951.

Dear Mr. [John D.] Rockefeller [Jr.]:

The Sterling Lake area, a tract of 17,000 acres to the southwest of Bear Mountain-Harriman Parks, has been placed on the market by the Harriman interests (Sterling Iron and Railway Company). Once the real estate promoter takes hold, the tract will be gone forever. In addition to the need for this wild land for recreational use, the area must be preserved for water supply purposes. The purpose of this letter is to inquire whether you or one of your sons would be moved to look further into this subject. If the subject does interest you, even to the extent of considering a conditional gift in cooperation with certain New Jersey communities now so concerned with protection of their water supply, I should be glad to confer with any representative you might appoint.

Very truly yours,
Ridsdale Ellis, Chairman, New York-New Jersey Trail Conference

February 2, 1986

To the Editor:

There is an exciting possibility to preserve an important part of the Town of Warwick, and a large cost if we fail to act. Many residents take it for granted that Sterling Forest will always be accessible to the public. Sterling Forest is now for sale and there is no guarantee that any of its 19,995 acres will be protected for future enjoyment. We need help to plan an educational campaign on the watershed and recreational importance of the land. We hope that community support will ensure that our children can enjoy Warwick's beauty fully. We owe that to them.

JoAnn L. and Paul R. Dolan

"STERLING FOREST WAS nearly saved last year," wrote Bill McKibben in the November/December 1996 issue of *Audubon* magazine, "but by a bill that would have paid for the land's purchase by selling 55,000 acres of federal grassland in Oklahoma. Even the New York and New Jersey congressmen and women crazy for a deal wouldn't do that, nor would they tie their bill to one that would have opened Utah's Redrock Wilderness to commercial exploitation. The bill died. If one of the bills now before Congress doesn't pass soon, the developers may get to work."

Malcolm Borg's newspaper, the *Record*, referred to Sterling Forest as a "sleeping giant waiting for the future. For the land, nothing has changed yet. Twenty thousand acres, much of it pristine, barely 40 miles from New York City. Bare, black trees against the snowbound knolls, rags of ice fringing the steel-colored lakes, a pair of hawks flat against the wind, hungry deer stripping the low branches of cedars. Here also, curved into the dips and rises, are handfuls of houses and a half dozen sleek research laboratories."

"Sterling Forest is on the razor's edge," suggested the *Record*. "In an ideal world, New York and New Jersey should purchase the property from its current owner, Home Insurance Company." The previous owner, City Investing Corporation, owned the property for twenty-five years, hoping to develop a community "blended into the lovely woods that would attract scientists and statisticians who could walk to their jobs in laboratories and raise their families in idyllic settings." The City Investing concept gained only slight traction before falling to economic vagaries and the tough, uncompromising nature of the Sterling Forest landscape. The corporation liquidated in 1985, ceding its assets to the insurance company that had been its subsidiary. The vast forest property was on the razor's edge because Home Insurance wanted to get rid of it as quickly as possible. If events were allowed to "take their course, ticky-tacky sprawl that mars so much of the landscape will swallow Sterling Forest."

Sterling Forest was the last great, single-owner undeveloped tract within sight of New York City's skyscrapers. By the standards of the western United

States, where open-space acreage is commonly measured in multiples of hundreds of thousands of acres, 20,000 acres might not seem particularly significant, but to the residents of the nation's most densely populated metropolitan region, the intrinsic value of such a sweep of unbroken wildland is inestimable.

The high ridgelines, steep rocky slopes, sequestered valleys, lakes, ponds, streams, deep woodlands, wetlands, and marshes of Sterling Forest were so taken for granted that in the 1986 edition of the *Hagstrom Atlas* the area was depicted in the color of publicly owned parks. Hunters had roamed the forest for years. Anglers knew of the rich treasure-trove of native trout swimming in Sterling Lake, a sparkling natural water body that formed the centerpiece of the entire property. Those who ventured into the deep forest were rewarded with a step back in time, a step away from the rumbling metropolitan region onto woodland paths leading through quiet vales where the sounds of soft breezes easily join with bird songs, rustling leaves, and tumbling clear water, all gracefully in tune.

But Sterling Forest, once part of a vast Eastern Seaboard wilderness, was no longer pristine. The forest had not escaped the stone implements, axes, flintlocks, and zeal of Native Americans and early European explorers. Anyone roaming the soft forest pathways was walking through centuries of history. Evidence of prehistoric settlement dating back at least 12,000 years had been found on the shorelines of Sterling and Blue lakes. Native Americans logically would have chosen the lake habitats, abundant with fish and gracefully framed by woodlands plentiful with wild game, edible plants, and building materials, as suitable places to live.

The disruptions of colonial settlement left silent the Native American sites by the time Scotsman Cornelius Board entered the forest in 1736, bringing men carrying tools to uncover extensive iron-ore deposits precious to the struggling colonies of North America. For the next 150 years iron ore was pulled from mine shafts driven deep under the forest floor. Some shafts extended downward, then were driven laterally under the bottom of Sterling Lake. Mining transformed the forest into a colonial-era industrial site. The ore, pried with hand tools from dark tunnel walls, was muscled to the surface. Large stone furnaces, standing three to four stories tall and fueled by wood taken from surrounding hardwood forests as rapidly as woodsmen could swing their axes, were fired to melt the ore into heavy ingots for ox-team transportation to market.

In 1740 most of the forest came into the ownership of William Alexander, heir to the Scottish title "Earl of Stirling." Alexander's son, James, known to his colonial neighbors as Lord Stirling, and William and Abel

Noble, other claimants to portions of the forest, were swept up in the events of 1776. James Alexander's allegiance was to the struggling colonies, and he played a key role in making Gen. George Washington aware of the strategic value of the Sterling Forest iron resources. When war came to the Hudson River Valley, Washington sent contingents of Continental Army troops and militiamen to defend the logical routes that British troops might follow from New York City should the redcoats attempt to capture the iron mines and furnaces.

Washington's tactical defense of the forest was transformed into one of the boldest acts of the Revolutionary War when the iron masters at the forges were ordered to produce individual links of chain, each two feet long and weighing about 180 pounds. When connected, these links formed enormous chains that were floated on barges across the Hudson River narrows in a desperate attempt to stop British ships from maneuvering upriver to divide and conquer the rebellious colonies.

National crisis again affirmed the strategic importance of the iron mines in Sterling Forest during the Civil War. Peter Parrott, owner of the mines and forges on the eastern edge of Sterling Forest, designed and produced the "Parrott Rifle," a newly designed cannon with spiraled groves cut into the bore. The Parrott Rifle had more range and accuracy than the smooth-bore artillery pieces used by Confederate troops, giving federal artillery-men a huge tactical advantage.

Parrott was considered a hero of the Civil War, but first and foremost he was a businessman, and his business went sour. Wartime honors could not save the Parrott family business when abundant deposits of iron ore, more easily mined and less costly to transport, were discovered in Minnesota. Forced to sell their holdings, members of the Parrott family saw E. H. Harriman prevail in a land auction in 1890 when he bought 9,500 acres of their land for $52,500, forming the core of what would become Harriman's vast estate of more than thirty square miles.

Although Harriman claimed that the Parrott property was "more land than I really needed," he followed this acquisition in 1895 by gaining complete control through stock purchases of the Sterling Iron and Railway Company, a business entity that, like the Parrott family enterprise, no longer was competitive in the iron-ore market. The company owned more than 20,000 acres of Sterling Forest. Through control of the company and his earlier purchases, Harriman could look west and east from his Arden House mansion and see his land stretching to the horizons.

His heirs continued to own this vast acreage through succeeding decades until 1953, when the Sterling Iron and Railway portion of the holding was sold for $452,000 to City Investing. In 1951, when Ridsdale Ellis wrote to Rockefeller Jr., suggesting that Sterling Forest be conserved, the PIPC was purchasing land in the vicinity. It acquired 1,200 acres owned by the Tuxedo Park Association, thanks to contributions by W. Averell Harriman and George W. Perkins Jr. But two years later the PIPC's annual report stated, "Aside from land acquired for parkway purposes, the only land acquired during 1953 was a small parcel purchased with gift funds, comprising 0.22 acres." Opportunity missed. The PIPC was on the sideline when 20,000 acres on its parkland doorstep went to City Investing.

City Investing subsequently attempted various real-estate development maneuvers, none of which was more than marginally successful. One bright spot was the opening of Sterling Gardens in 1958, a 125-acre site planted with 1.5 million tulips. Princess Beatrix of the Netherlands presided at the opening ceremony.

Other forest parcels were sold piecemeal to companies and institutions interested in the isolated character of the property, a factor that was particularly appealing for research and data-storage purposes. Among the purchasers were IBM, International Paper, Union Carbide, International Nickel, Xicom, Wehran EnviroTech, and New York University. Union Carbide operated a small nuclear reactor in Sterling Forest for medical-research purposes (later operated by Cintichem). New York University maintained a primate center in association with research on infectious diseases.

For more than thirty years, the plea by Ridsdale Ellis to protect the remaining open land in Sterling Forest went unheeded until Paul and JoAnn Dolan, joined by a small corps of like-minded conservationists, took up the cause once again. Paul was an executive with ABC News, and JoAnn was executive director of the New York/New Jersey Trail Conference. The Dolans and their colleagues were prompted, in part, by an examination of forest development possibilities by Arthur T. Ross, professor of planning at Yale University. Ross reported to the Tri-State Transportation Committee that "Sterling Forest would become a major new community of 500,000 persons and will encompass a prime area of 100 square miles."

Development of the property, though, seemed star-crossed. City Investing, caught in the economic scandals and excesses of the 1980s, was flattened financially and forced to liquidate. By default, City Investing's subsidiary, Home Insurance of Hartford, Connecticut, found itself owner of the mortgaged property and placed it on the market for sale. The asking price was $50 million. The insurance company created a subsidiary of its own, the Sterling Forest Corporation (SFC), to sell or develop the property, whichever option promised the best and quickest financial result.

Paul Dolan, thinking the time might be right for the states of New York and New Jersey to step forward as willing buyers, surfaced the concept of a twelve-mile-long "green belt" extending from the Appalachian Trail on the northern edge of the forest to the New Jersey state line to the south. The idea was to protect watershed resources crucial to both states. Dolan formed the Greenwood Trust and began lobbying for the purchase of at least 9,000 acres of Sterling Forest Corporation holdings. He urged New Jersey government leaders to purchase an additional 2,000 acres of contiguous SFC land on their side of the border.

The New York/New Jersey Trail Conference endorsed Dolan's plan. James D. Rogers, planning director for Passaic County, New Jersey, also was quick to note Dolan's initiative and offered encouragement, confirming that the SFC holding was located in the headwaters of a reservoir system being developed by the state to provide drinking water for 2,000,000 people, 25 percent of New Jersey's population.

Dolan owned a log house on the western fringe of Sterling Forest at Greenwood Lake that he used as a weekend getaway. He and his future wife, JoAnn, hiked in the forest on their first date. At the time, she was a freelance writer developing educational materials for people with disabilities. Paul directed the One-to-One Fund, a charitable organization that advocated civil rights for people with disabilities and served as a watchdog against abuse of patients at mental and medical institutions. JoAnn subsequently became executive director for the New York/New Jersey Trail Conference, and Paul joined the world of television news, steadily advancing within ABC-TV.

The routine of using the log house as a pleasant weekend getaway was unhappily complicated for the Dolans in May 1985 when their newborn son, Jamie, was diagnosed with Down's Syndrome. For a time, Jamie's very survival was in doubt, but he fought through the early crisis in his life, allowing the Dolans to begin taking him on outings along their favorite woodland paths. "Even though he can't walk," Paul Dolan said to a reporter for the *Record*, "he responds to the winds and the birds and leaves, and he lights up. He lights up. He watches everything." The Dolans credit Jamie with reinforcing their activist campaign to save the forest from what appeared to be almost certain fragmentation at the hands of real estate developers.

Using their Greenwood Lake home as a rendezvous, the Dolans began organizing a series of walks into the forest for anyone willing to participate.

Among the handful of friends plotting a strategy to protect Sterling Forest was John Humbach, a law professor at Pace University. The small group, closely aligned with the New York/New Jersey Trail Conference, gained welcome encouragement from Hooper Brooks of the highly regarded Regional Plan Association, an organization respected for its reasoned analyses of metropolitan growth patterns.

In a matter of months, this alliance of friends and colleagues sparked to life enough media and political interest to prompt state of New Jersey and Passaic County officials to take the first major step toward preserving the last great open space in the New York/New Jersey metropolitan region. In September 1986, New Jersey governor Thomas Kean approved a $2,000,000 grant from the state's Green Acres Fund to Passaic County for the purpose of attempting to acquire the 2,074 acres of Home Group holdings on the New Jersey side of the border. The county's obligation was to match this amount.

Water was the motivating factor. On its side of the border, New Jersey was doing what it could to protect clean, abundant, free-flowing water if a deal could be struck with the Home Group. Governor Kean looked northward with the hope that the remaining 17,500 acres held in corporate ownership in New York would be similarly pursued.

But before New York could follow New Jersey's lead, the Sterling Forest FOR SALE sign was taken down. In March 1987, Robert L. Woodrum, vice president of Home Insurance, announced, "We are sitting on an appreciating asset." The message was clear. The owners of Sterling Forest were opting for a development scenario, encouraged by rumors of construction of a new $5 million interchange on the New York State Thruway that would funnel traffic directly into the forest.

Watching from New Jersey, U.S. congressman Robert G. Torricelli spoke for his constituents and many others by saying, "This is the last chance for the states of New York and New Jersey to plan for a greenbelt around the metropolitan area. It should have been done a century ago." Elaborating in a lengthy article in the *New York Times*, Torricelli labeled Sterling Forest an "irreplaceable asset for the entire metropolitan region that is almost without parallel."

The congressman referred to the "livability of old-world cities" in England and France, contrasting them to "a loss of open space" adjacent to city centers in the United States "that has reached crisis proportions." He argued that urban sprawl must be controlled and said, "There is no better place to start than Sterling Forest," adding that "the Palisades Interstate Park Commission stands ready to assist."

Torricelli's reference to the PIPC was no coincidence. His neighbor in Englewood, New Jersey, was Malcolm Borg, owner of the *Record* and a very active Palisades Interstate Park Commissioner. Borg, Torricelli, and their neighbors in New Jersey had only to look at their water taps to understand the implications that heavy development of the Sterling Forest wildlands would have for them.

James Dao, writing for the *Record*, captured the message when he reported, "Amateur geologist Jerome Wyckoff stalks the narrow road into Sterling Forest like a detective hot on a criminal's tail. About a mile up, after the road has turned to mud, he finds his first clue: sandstone. What the rock tells him is this: eons ago, a glacier ripped it from the New York ridge and deposited it in the granite hills of Ringwood. There, it was polished by centuries of rainwater rushing inexorably toward the Atlantic. To Wyckoff, this explains why Sterling Forest is so important. 'You can't fool the water,' Wycoff says, 'Sooner or later, it will find its way to the ocean.' He might have added, water and effluent."

Ella Filippone, also watching from the New Jersey side of the border, would put the matter more starkly. "Building sewer plants in the high headwaters of any stream and river system is ludicrous." Filippone, executive director of the Passaic River Coalition, a small nonprofit organization, was being pulled like so many others into a full-scale conservation battle. In Filippone, both allies and opponents would discover a splendid master of political combat.

The scale of the contest was suggested in 1988 when the corporate owners of Sterling Forest refused to attend a gathering called by the Regional Plan Association and the New Jersey Audubon Society to explore options for the future use of the forest. David A. Wilkinson, speaking on behalf of the Home Insurance Group, said, "The forest has immense economic value," and pegged the asset at $150 million, a threefold jump from the $50 million estimate expressed by corporate representatives only two years earlier.

Two years had passed since Governor Kean approved Green Acres funding for the acquisition of the Home Group property in Passaic County. Republican Richard DuHaine, an elected county official, reported that negotiations with the Home Group for a willing-seller purchase had failed. "They told us point-blank they were going to develop the property." The Home Group was threatening to begin logging the property as a first, environmentally blunt step toward development. On November 1, 1988, the Republican-controlled Passaic County Freeholders (elected officials) exercised eminent-domain authority and wrested title to 2,074 acres of land from the corporation. The immediate risk to northern New Jersey reservoirs was removed, but thousands of acres just north of the state line remained in development crosshairs.

In 1989, the Home Group hired Robert E. Thomson to assume duties as the new chairman of the Sterling Forest Corporation. On arrival in the Hudson River Valley from previous political and development activities in California, Thomson said, "If modern land use planning means anything, it means citizen involvement" and pledged as one of his first actions to meet with John Humbach, now president of the Dolan-inspired Sterling Forest Coalition. In an article in the *Times Herald Record*, Thomson's expressed willingness to meet with Humbach was juxtaposed with a quote from Michael Manley, president of SFC's parent company, the Home Group. Asked about working with those who sought to conserve the remaining 17,500 acres of corporate holdings in New York as open space, Manley said, "I don't know who they are. I don't know whether they've ever been on the property. If they have been, they've been trespassing."

Manley added, "I'd be very pleased to sell the whole thing for a reasonable price," which he pegged at $200 million, another leap of $50 million in Home Group's estimate of market value in only five months. Manley's statement was followed a few weeks later by the release of a traffic study, sponsored and paid for by the SFC, that reaffirmed the need for construction of a new interchange on the New York State Thruway to feed traffic directly into the property, a step guaranteed to enhance the value of the corporate holding at taxpayer expense, perhaps even to a level approaching Manley's claim of value.

Local businesspeople and representatives of the New York Orange County Chamber of Commerce communicated their support for the $5 million interchange project to Governor Mario Cuomo and New York comptroller Edward V. Regan, but the reaction in Albany was underwhelming. A spokesperson for the Thruway Authority, the agency that would be responsible for constructing the interchange, said, "The project remains on hold indefinitely."

At Bear Mountain, Nash Castro and his PIPC staff withheld public comment on the interchange idea pending detailed review of a Sterling Forest Corporation traffic study. One impact was clear: if constructed, the interchange would take a big bite out of Harriman State Park. The PIPC, therefore, held approval authority over the interchange proposal.

The prospect that the commissioners would agree to slice off a chunk of the park was somewhere on a scale of slim to none. Thomson was nonetheless planning to make a vigorous push for the interchange within the context of overall development potential for the Sterling Forest property.

Thomson retained the services of Sedway Cooke Associates, a respected but distant San Francisco-based planning firm. A principal with the firm, Thomas Cooke, said, "The land tells you what to do," and he began flying back and forth from West Coast to East to listen to Sterling Forest.

With SFC's development planning in high gear, the *Philadelphia Inquirer* carried a lengthy article on June 26, 1989, headlined, "Open Space Around Our Cities Is Shrinking." The article was written by Democratic U.S. congressman Peter H. Kostmayer, representative of Bucks and Montgomery counties, Pennsylvania. As chairman of the House of Representatives Interior Subcommittee on Oversight, Kostmayer had developed a deep personal concern for what he saw happening on the edges of cities across the nation.

Echoing Congressman Torricelli's similar message in the *Times*'s article, he wrote, "A single day's drive through our congested suburban landscape demonstrates that the traditional fragmented and strictly local approach to controlling development has failed. We need a new ethic in this country that acknowledges the value of preserving a national landscape."

Citing greenbelt protection around London and farmland conservation in France and the Netherlands, Kostmayer scheduled seven subcommittee hearings around the nation to examine in detail the issue of open-space protection near urban areas. First on his list was a hearing in Tuxedo, New York, intended to closely examine Sterling Forest, rumored to be the site of the largest residential and commercial real-estate development project on the drawing boards anywhere in the United States.

About 200 people attended the hearing on October 3, 1989. Congressional members Torricelli (D), Marge Roukema (R), and Benjamin A. Gilman (R) were in attendance. Gilman, the senior New York Republican member in the House of Representatives and chairman of the powerful House Foreign Relations Committee, represented the New York congressional district in which the 17,500-acre property owned by SFC was located. Roukema's New Jersey district included parts of Passaic County where SFC property had been taken by eminent domain. She was Gilman's immediate congressional neighbor to the south.

JoAnn Dolan was there, representing the New York/New Jersey Trail Conference, as was the Passaic River Coalition's Ella Filippone. So, too, was, John Humbach, there to speak for the Sterling Forest Coalition. The PIPC's Nash Castro was in attendance, as were representatives of the Nature Conservancy, Sierra Club, New Jersey Conservation Foundation, North Jersey District Water Supply Commission, New Jersey Audubon Society, Environmental Defense Fund, Scenic Hudson, Orange County

Chamber of Commerce, and representatives of various local communities and businesses.

Thomson was among the first to comment during the five-hour session. Questioned by Kostmayer about the rumored large-scale housing development, Thomson responded, "Do I envision it? No. Is it an option? I'm not ready to say." Those who spoke for the business community urged development of the forest "because we need ratables," which prompted Humbach to counter that for every $1.00 in taxes generated from residential developments, communities ended up paying $1.36 to provide for police, fire, educational, transportation, medical, and utility services.

Humbach repeated this point when quoted in a *New York Times* article, citing the Cornell University Cooperative Extension Service and the American Farmland Trust as his sources of statistical information. Another person at the hearing put the issue of a promised property-tax windfall in terms of the city a few miles away: "If property taxes were the financial answer for communities, New York City would never be in debt."

Filippone, continuing to express concern about the impact of immense development in the Sterling Forest watershed, said, "We cannot afford to have massive amounts of siltation, lead, chlorides, pesticides, and fertilizers pollute the rivers, streams, and creeks flowing from the forest." Congressman Gilman expressed the hope that well-managed development could coexist in a "pristine environment."

This is where I entered the debate. Several months earlier I had been invited by New York Department of Environmental Conservation (DEC) commissioner Thomas Jorling to become one of Jorling's deputies, and subsequently was appointed to the position by Governor Mario Cuomo. Jorling said he selected me based on my twenty-five years of experience with the National Park Service (NPS), including an assignment as superintendent of Yosemite National Park, as well as my involvement in the early 1970s helping Mrs. David Rockefeller establish the Maine Coast Heritage Trust.

By bringing me onto the DEC staff, Jorling rescued my professional career. In 1986, after almost seven years at Yosemite, I had been abruptly "reassigned" to a desk job at the NPS Regional Office in San Francisco after confirming that I had tape-recorded a conversation with NPS critic and property rights advocate Charles Cushman. Although the recording was legally acceptable, it had been made without Cushman's permission. This was poor judgment on my part, for which I later apologized to Cushman, but the media outfall swept me from a job I cherished to a desk that felt like an anchor.

On arrival in Albany, New York, in February 1989, one of my assigned responsibilities was to oversee DEC's portion of New York's 1986

Environmental Quality Bond Act, a $1.2 billion fund authorized by voters for expansion of forests, parks, and wildlife areas throughout the state of New York. The DEC was responsible for management of almost 800,000 acres of state forest land and 3,000,000 acres of lands within the Adirondack and Catskills parks. Of the fund amount, DEC was allotted $250 million to enhance these holdings and to improve public access to river, lake, and ocean shorelines. Emphasis was placed on large privately owned tracts whose corporate or individual owners might be willing to sell.

One of these properties was land owned by Orange and Rockland Utilities, Inc., in the serene Mongaup Valley a few miles west of Sterling Forest. Early in my DEC assignment, I worked with Rose Harvey of the Trust for Public Land to negotiate the purchase of 6,313 acres of Orange and Rockland holdings and to secure conservation easement protection on an additional 5,542 acres. The property includes superb bald eagle nesting sites in a richly diverse natural habitat shared by white-tailed deer, turkey, ruffed grouse, coyote, fox, porcupine, black bear, trout, and flocks of happy waterfowl. Thousands of acres similarly were being added to the DEC's statewide land inventory, resulting in a rapid decrease in Bond Act funds.

When I arrived at the Kostmayer hearing in October 1989 to testify on behalf of the state regarding Sterling Forest, only $35 million of the $250 million available to the DEC was uncommitted. We had developed A, B, and C lists within the DEC for the use of these remaining funds. The A list consisted of properties where appraisals were in hand and purchase was imminent. The B list included properties where strong willing-seller signals had been received. The C list included scores of other desirable properties whose owners apparently were disinterested in selling or had expressed outright opposition to state land-acquisition inquiries.

The remaining $35 million was insufficient even to cover the properties on the A and B lists. Sterling Forest was on the C list. I testified to this financial shortfall to the obvious disappointment of Kostmayer, Torricelli, Roukema, and the many citizen advocates for Sterling Forest protection who were in attendance. Kostmayer concluded the hearing by proposing further study by the U.S. Forest Service of the New York–New Jersey "Skylands" region.

Many people in the hearing room at Tuxedo that day would be drawn tightly together during the last decade of the twentieth century by the vexing question of what was to become of Sterling Forest. I had no idea that I would be joining them at a level of intensity that would personally challenge my endurance like nothing I had encountered professionally or could imagine.

Only two days after the hearing, another person appeared on the scene who would become the point man for the Sterling Forest Corporation. Louis Heimbach, a native of the southern New York region, farmer, and businessperson, had served for twelve years as the elected executive for Orange County, New York State's fastest growing county. Sterling Forest was within the county boundaries. According to the *Times Herald Record*, Heimbach was reputed to be "a promoter of development in the mid-Hudson Valley." Heimbach was recruited to assume responsibilities as president and chief operating officer of the Sterling Forest Corporation. Robert Thomson would continue in his role as SFC chairman.

Just a few days later, news spread rapidly through the Hudson River Valley environmental community that Nash Castro was choosing to retire after twenty years with the PIPC. In addition to the artful management he had provided for the PIPC, Castro was leaving behind a special contribution. He wrote the PIPC's *Second Century Plan*, a guiding document that celebrated the history of the commission and anticipated the demands and opportunities that it would face in the coming century.

Wrote Castro, "The first history of the Palisades Interstate Park System, published in 1929, makes the proud boast: 'The Palisades Interstate Park of New York and New Jersey is the most notable example in the United States of interstate cooperation for the conservation of outstanding scenic features and the promotion of outdoor recreation.' It was a bold claim 60 years ago, but in subsequent years the cooperation has continued unabated. . . . The park has . . . withstood relentless growth at its boundaries, encountered unprecedented demands, adapted to rapid changes, and still managed to maintain the essential integrity of 85,000 acres of open space."

The commissioners had held Castro's retirement intent in confidence while a search was underway for his successor. Commissioners J. Martin Cornell and Malcom Borg comprised the search committee, assisted by Castro himself. During my months with the New York Department of Environmental Conservation, I had occasionally spoken with Castro about various state land matters. I had known of him for years and respected him as a distinguished and nationally prominent administrator of parks and historic sites. Yet when Castro made a surprise telephone call inviting me to stand as a candidate for the Palisades Interstate Park executive director position, I at first declined. I had been in my appointed post with the DEC for slightly less than a year, admired the tenacity and abilities of DEC commissioner Tom Jorling and my many workplace colleagues, and was grateful for the excellent opportunity Jorling had given me to join his staff. But Castro's persuasive skills prevailed, my hesitation collapsed, and on a gray

December day, I was interviewed by Borg and Castro at the Bear Mountain Inn and subsequently by Cornell at his law office. Borg made a point during the interview of assuring me that, if I were selected, he would "be there" whenever I needed to reach him for advice about the history, hopes, challenges, and expectations of the PIPC. Borg's word proved to be his bond.

In mid-April 1990, I joined the PIPC staff as successor to Nash Castro, just two months after the Sterling Forest Corporation released a plan announcing its intention to build 14,500 residential units and 7.4 million square feet of commercial and light industrial space in a scattered pattern throughout the forest. Thomson described the plan as "compact siting of development using the model of a New England town center and European mountain villages."

Advocates for the preservation of Sterling Forest variously described the plan as "fracturing the forest," "Swiss cheese," and "a shotgun blast at the heart of the Highlands." Lee Wasserman of the Environmental Planning Lobby captured the thoughts of many when he asserted, "It is simply fantasy to expect that you could drop a city of 35,000 to 40,000 people into the last privately owned forested open space in the metropolitan region without forever destroying the magnificent ecological benefits it provides."

Despite the development plan, corporate ground shifted under Thomson and Heimbach because of the financial struggles of AmBase, the Florida-based parent holding company of Home Insurance Group and its subsidiary Sterling Forest Corporation. The net second-quarter loss for AmBase in 1990 was $106 million. This loss was caused in part by unpaid debt owed to AmBase by the bankruptcy of Drexel Burnham Lambert, whose bond-trading executive Michael Milken pleaded guilty in April of that year to various tax and securities violations. In the third quarter of 1990, AmBase was the second-worst performer on the New York Stock Exchange. Before year's end, SFC had become a subsidiary of a new ownership group. For $970 million, AmBase sold its holdings to a consortium of investors led by Trygg-Hansa AB, Sweden's second-largest insurance company. Other investors included London-based Vik Brothers, international insurance advisors, and New York City–based Donaldson Lufkin Jennrette.

John Humbach, watching the transaction on behalf of the Sterling Forest Coalition, said, "I'm not sure whether they will pursue development or liquidate Sterling Forest as an asset. I know they didn't acquire Sterling Forest with the view of donating the land as a national park." In response, David McDermott, speaking for the SFC, said, "There will be no change whatsoever." Development might proceed, or not. Robert Thomson kept repeating that "we will sell for a fair price."

Those seeking protection for the forest placed substantial hope in a 1990 Environmental Quality Bond Act, a $1.975 billion funding package proposed by Governor Mario Cuomo. The package included $800 million for land acquisition. The PIPC and New York state agencies were standing by, ready to seek bond-act funds for their highest-priority land-protection projects, with Sterling Forest sharing top billing. But New York voters, who had approved many previous environmental quality bond acts, this time failed to do so. By a 51 percent to 49 percent margin in light voter turnout in a November off-year election, the 1990 Environmental Quality Bond Act was rejected.

By May 1991, Gordon Bishop, writing for the *New Jersey Star-Ledger*, labeled the Sterling Forest situation a "last chance effort" and reported that Congressman Kostmayer had won bipartisan support from members of the New Jersey congressional delegation to push through $250,000 in funding for a U.S. Forest Service study of the 1.1-million-acre New York/New Jersey highlands region. Kostmayer found ready bipartisan support from Democrats Torricelli, Roe, Frank Lautenberg, and Bill Bradley and from Republicans Roukema, Dean Gallo, and Richard Zimmer. Congressman Ben Gilman (R-NY) thought the study unnecessary, given that SFC was still in the process of refining its development plan. For a moment in the congressional budget process, Gilman knocked the Kostmayer funding out of an appropriations bill, but he was subsequently convinced by his Republican colleagues from New Jersey of the importance of the study and relented.

Joseph A. Michaels, a seasoned veteran with the U.S. Forest Service, was tapped to lead the team. For the PIPC, I added park ranger Tim Sullivan to the team. Sullivan's selection surprised some members of the PIPC staff who assumed that someone with more planning experience would be better qualified for the task. Sullivan, though, proved more than worthy of the task. For forty years he had been serving on the PIPC staff and was the first person assigned to Minnewaska when that park was created. He could hold his own with any ranger in the nation. Sullivan was a self-taught naturalist, search-and-rescue expert, forest fire specialist, deeds and records authority, and master of the Hudson Highlands terrain. No trail in Harriman and Bear Mountain state parks has escaped Sullivan's attention, and he had expanded his firsthand knowledge far and wide into the highlands on the premise that a response to emergencies is much more effective if familiarity with the terrain already is in the hip pocket. Sullivan brought a wealth of local knowledge to the U.S. Forest Service deliberations. He did not have to guess about the watercourses, rocky slopes, valleys, and ridge lines of Sterling Forest; he had been there, on his feet.

While the Forest Service inquiry was underway, John Humbach continued to challenge the Sterling Forest Corporation development plan. In a May 1991 article published in the *Times Herald Record*, Humbach contended that the corporation's comprehensive plan "is a fine example of advanced urban planning. That is not, however, the issue. The question is whether, as a matter of rational, far-sighted planning, a special regional asset like Sterling Forest ought to be developed at all. Two major impacts of pursuing the comprehensive plan are immediately apparent: First, it will constrict and fragment a key link in the Delaware [River]-to-Hudson greenway corridor, permanently altering and degrading the forest's traditional high-grade wildlife habitat, recreational space and natural lands. Second, development in the watershed of the North Jersey District Water Supply Commission will replace a historically reliable source of clean, safe drinking water used by two million people with permanent dependency on government-mandated sewage treatment machinery."

Speaking to the Sterling Forest Corporation contention that almost 75 percent of the 17,500-acre property would be left as open space, Humbach responded, "More than 100 separate, discrete areas or chunks of development will be distributed within the forest. Extensive portions of the surface area shown in green are, in reality, nothing more than buffers interlaced among chunks of development. This network of green membranes would be quite considerable if lumped together. As a latticework, however, it no more retains the open, wild character of Sterling Forest than the wooded medians of the [New York] Thruway."

As U.S. Forest Service representatives learned in nine public hearings, hundreds of people agreed with Humbach. Speaker after speaker approached the microphone in packed meeting halls to echo the message about the potential loss of recreational open space, wildlife habitat, and watershed. Thomson and his SFC representatives countered with arguments about the sensitivity of their comprehensive plan, the economic benefits to local communities, and the risk to Sterling Forest if, in the absence of public funding for a "willing seller" transaction, the forest were simply thrown on the open market and subdivided in keeping with existing zoning ordinances.

With the demise of the 1990 Environmental Quality Bond Act, and with continuing signals that SFC development plans were proceeding, an alliance of environmental organizations and government agencies began to coalesce around the PIPC in an effort to find another avenue to protect the forest. In addition to John Humbach, Paul and JoAnn Dolan, and Ella Fillippone, the group included village of Tuxedo residents Helmut Nimke and Al Ewert, Scenic Hudson's executive director Klara Sauer, the Regional

Plan Association's Rob Pirani, Judy Noritake of Congressman Kostmeyer's staff, chief engineer Dean Noll of the North Jersey Water Supply Commission, the Nature Conservancy's Olivia Millard, Jennifer Melville of the Appalachian Mountain Club, Environmental Defense Fund attorney Jim Tripp, and John Gebhards, the sole staff member of the Sterling Forest Coalition.

Within the PIPC, strength of purpose to rescue Sterling Forest was solid among the commissioners. The commission's president in 1991, Martin Cornell, a highly regarded trial attorney from New City, New York, was ready to lend his steady hand, optimistic outlook, and keen legal counsel to any PIPC action that might protect all or part of the forest. Barnabas McHenry, appointed to the commission in 1987 and serving as its vice president, signaled that he was ready to go anywhere, anytime, talk with anyone, organize any event, communicate in every way possible, and work day and night, if necessary, to find a way to save the forest. Malcolm Borg was in the treasurer's slot and placed Sterling Forest at the pinnacle of his personal agenda.

Larry Rockefeller, who had worked successfully to protect vast wildlands in Alaska and provided key leadership to preserve Minnewaska, brought this same strong interest to the Sterling Forest test. Mary Fisk and Anne Cabot, representing the Harriman and Perkins families, were ready to add to the legacies of their parents and grandparents. Nash Castro had stepped away from day-to-day commission activities, but not far. He, too, was eager to assist.

Sadly, a key person would be absent from the fight. On January 31, 1992, David P. McCoy, assistant director of the PIPC, adept landscape architect and devoted professional, died from complications of melanoma. The gentle, artistic, and unpretentious McCoy had provided a steady hand and guidance during the first months of my tenure with the PIPC.

Kenneth Krieser, a park veteran who had transferred to the PIPC in 1980 from his position as assistant director for the Thousand Islands Park Region on the St. Lawrence River, succeeded McCoy. During his college years, Krieser was enrolled in a Forestry major at the University of Massachusetts, Amherst. In his sophomore year a new major called Park Management was offered. Out of curiosity, Krieser attended a guest lecture on the subject by Harold Dyer, formerly a director of Maine State Parks and currently a regional director in the New York State Park system. Krieser changed his major and, on graduation, was offered his first park job by Dyer.

The Vietnam War interrupted. Krieser was drafted. After ten months of intensive training, he was commissioned an infantry officer and was a combat rifle platoon leader. He credits the army with the management style he brought to the PIPC, described by him as "showing respect, listening, mentoring, communicating clear expectations, delegating authority while ensuring accountability, and understanding all parts of the organization." Krieser said, "I never thought of myself as a visionary, but, rather, as a decent mechanic who can plan, sell, and coordinate project completion as part of a team." Regarding his chosen profession, he added, "Working in a park system celebrates nature, history, and the outdoors for the public good. No two days ever are the same."

That certainly was true of the Sterling Forest contest. An overarching problem was that Robert Thomson would "sell for a fair price," pegging the value at $150 million, but the PIPC and its collaborators had exactly zero dollars with which to negotiate a purchase. Somehow, funds would have to be cobbled together so that a serious purchase offer could be made before the SFC pulled the trigger on its development shotgun.

Jennifer Melville, watching from her Appalachian Mountain Club office in Boston, was concerned about different financial numbers. In various presentations, representatives of SFC were suggesting that local communities would receive a $35.5 million tax windfall from the proposed development. In alliance with the PIPC, Melville retained the services of Vermont-based Ad Hoc Associates to examine the tax-windfall promise. The Associates reported that school and municipal costs of almost $34 million were not included in the corporation's numbers, nor were the increased costs of treating the effluent that would begin to flow down from the forest into New Jersey's Monksville and Wanaque reservoirs. Water company engineer Noll estimated these additional sewage treatment costs at $150 million over a twenty-year period.

A *Wall Street Journal* article appearing in September 1991 seemed to confirm these findings. Under the headline "Boom County, Bust Budget," David Bergman, writing about a study of DuPage County, west of Chicago, said, "Curiously, the belief that development is lucrative may be one of the reasons it isn't. Across the country, developers have been able to thwart growth management movements by arguing that development is the only way to check rises in property taxes. If the association between taxation and development suggested by the DuPage study is confirmed, a powerful new argument for managing growth may be emerging." Bergman described how a hoped-for tax windfall in economically strong DuPage County never materialized. As job growth and development increased in the county, so did property taxes. He found that in the postindustrial United States, the pattern was to seek less expensive land on the periphery of metropolitan areas for development, thus adding to municipal costs for everyone. He could have been describing Sterling Forest.

SFC had already implied that it would fade away once environmental approvals for its vast project were in hand. Thomson put the matter in simple terms: "We're not planning to build anything higher than a curb," he said. The corporation's business strategy was to rough in the road and utility systems, then sell off the approved development sites to other builders, pocket the profits, and exit as swiftly as possible.

To no one's surprise, the Forest Service found, on release of its 1991 draft "New York–New Jersey Highlands Regional Study," that the region offered potentially huge outdoor recreational benefits to 22 million nearby residents, 20 percent of the nation's population. These findings reflected similar conclusions reached in three parallel studies: "Communities of Place: Interim State Development and Redevelopment Plan," by the state of New Jersey; "Skylands Greenway—A Plan for Action," by the New Jersey governor's Greenway Task Force; and "Conserving Open Space in New York State," by the New York Department of Environmental Conservation.

Congressman Kostmayer, now chairing the House Subcommittee on Energy and the Environment, used the Forest Service study as a springboard to ask for a $25 million federal appropriation to help acquire Sterling Forest, contending that the forest was "a pearl in a necklace of linear greenways . . . stretching through the northern tier of states." Congressman

Sterling Forest and Lake (PIPC Archives)

Gilman promptly made known his opposition to the appropriation, arguing in favor of SFC's development plan and pointedly stressing that Kostmayer had no business getting involved in a "local matter" in Gilman's congressional district. From his neighboring congressional district in New Jersey, Congressman Torricelli favored the appropriation and firmly maintained that all 17,500 acres of the corporate holding should be conserved. In this contest, Gilman prevailed; a House appropriations subcommittee denied Kostmayer's request.

Despite this setback, the persistent Kostmayer scheduled a June 1992 hearing before his subcommittee to further debate the fate of Sterling Forest. After hearing from Thomson about the comprehensive plan, the unimpressed Kostmayer responded, "You can't preserve a forest intact as an ecological unit if you break it up. . . . It won't be a forest anymore. . . . [You can] call it Sterling Acres, Sterling Vistas, or Sterling Meadows, but don't call it a forest because it won't be a forest."

JoAnn Dolan and Ella Filippone were at the hearing to remind subcommittee members of the strategic environmental importance of the forest ecosystem. Representing the PIPC, Barnabas McHenry confirmed that the commission, the New York Office of Parks, Recreation and Historic Preservation (OPRHP), and corporate representatives were in contact, trying to find some means of ensuring environmental protection for at least the most environmentally sensitive portions of the forest, prompting Congressman Gilman to say, "If both the PIPC and the SFC can come to some reasonable agreement, I promise to do all I can to acquire the funding, as questionable as it may be, for such an acquisition."

Later, in response to comments by Torricelli that appropriations should be "pursued doggedly," Gilman said a push for a major slice of federal funding was "pie in the sky" thinking. Gilman had reason for pessimism; taxes had been cut, defense spending was up, and the nation was in a severe economic slump. Congress was in no mood to add more red ink to an already crimson financial ledger.

Outside the hearing room, Dolan, Filippone, McHenry, and I were at work with Kostmayer, Torricelli, Roukema, and New Jersey Republican congressman Dean Gallo, a member of the House Appropriations Committee. On a parallel track, Judy Noritake, Kostmayer's staff assistant, was networking with her colleagues among the staff corps on Capitol Hill. These efforts resulted in a small financial victory when the House of Representatives approved a $5-million appropriation earmarked to acquire Sterling Forest lands. But in the face of the entrenched budget-cutting mood in the 103rd Congress, a House-Senate Conference Committee cut the amount to $3 million, far from the number demanded by Robert Thomson and Lou Heimbach for serious negotiation. Still, the Kostmayer initiative represented a glimmer of hope that the drive to protect Sterling Forest warranted national attention.

While this effort was underway in Washington, Richard Curley, a land-acquisition specialist with the New York Office of Parks, Recreation, and Historic Preservation, obtained real-estate appraisals of SFC's land holdings, finding that the value was far below the $150–$200 million value claimed by Thomson. The appraisal pegged per-acre value of about $3,500, a total value in the range of $60 million.

With this figure in hand, OPRHP commissioner Orin Lehman encouraged discussion with Thomson to determine whether a partial purchase of corporate holdings might be possible. The idea was that if enough land could be identified to create a park of reasonable size, this specificity might prompt federal, state, and private sources to produce the necessary acquisition funds. This cart-before-the-horse approach was ridiculed by the Thomson group. Lou Heimbach kept challenging, "Where is the money?" The partial-purchase idea was dropped. The ball rolled back into the PIPC court.

With encouragement from the PIPC commissioners and the New York/New Jersey Trail Conference Board of Directors, JoAnn Dolan and I decided to reach for the collective brass ring by coordinating a meeting of representatives of a host of environmental organizations purposely to share information and expertise, critique strategy, and assign tasks, all with the goal of saving Sterling Forest. This would be asking a great deal. Environmental organizations obviously have their own agendas and priorities, and often compete for funding and media attention. The Sterling Forest rescue attempt promised to stretch forward for years. We were suggesting that the immense potential of laser-like attention on Sterling Forest be shared across a united environmental front. Fortunately, Samuel F. Pryor III, chairman of the board of the Appalachian Mountain Club and senior partner at Davis Polk & Wardwell, a prestigious New York City law firm, agreed with us. He offered meeting space. Larry Rockefeller agreed to sign the invitations.

Mixed with the stream of employees arriving at Davis Polk & Wardwell in midtown New York City in early 1992 was a group representing twenty-six distinct environmental organizations ranging from the well-known Nature Conservancy to the little-known Orange County Land Trust, all headed for a conference room reserved by Pryor. The conference room was large, handsomely decorated with lightly toned wood panels, a sweeping conference table to match, built-in kitchenette, plush chairs, and almost magical acoustics. A person standing at one end of the long room could talk in normal tones and be heard easily by those at the other end. Hidden in the ceiling were speakers for the telephone system. Anyone calling in could be heard in godlike style by those in the room, the voice drifting down from above. The attendees, more accustomed to folding tables, modest rooms, and paper cups, settled right in. There would be no bylaws for this group, no officers, no minutes to document discussions. The participants were coming together as a brain trust to fight for Sterling Forest. As expressed by John Humbach, the group would have four general purposes:

1. Identify sources of public and private funding for the purpose of acquiring Sterling Forest.
2. Monitor and participate in the environmental impact statement process.
3. Educate and inform the public.
4. Cooperate with one another.

Along with coats and hats, the members of the group instinctively left their egos and turf concerns outside the conference room. From the first moment an unstated rule applied: no prima donnas. At the suggestion of Nash Castro, who, though retired, was in attendance, the group chose to name itself the Public-Private Partnership to Save Sterling Forest (PPP). In Washington, a small corps of congressional staff members would mirror and reinforce the PPP initiative, encouraged to become involved by Judy Noritake and, beginning in 1993, expertly cajoled by Chris Arthur, staff aide to newly elected Congressman Maurice Hinchey.

Extending a compliment to the PIPC after the first meeting, JoAnn Dolan wrote, "It would have been fun to be around in the Major Welch years, but I imagine that this is the first time since the Welch era that a state park agency has acted so purposefully in a partnership with the public. Your leadership and steadfastness, from the very beginning, have given great credibility to the Sterling Forest project. You have been the energizing force for many groups to work in harmony. A partnership like this can be a very dynamic process, but rarely happens."

The gathering of the PPP came at a time of transition in Washington. With the election of Bill Clinton and despite Democratic gains across the nation, Congressman Kostmayer, the first elected official to gain funds for the preservation of Sterling Forest, was defeated in his home district in Pennsylvania.

President Clinton appointed Bruce Babbitt, former governor of Arizona, as Secretary of the Interior. All eyes in the PPP turned toward Babbitt, anxious to convince the new secretary that the federal government should legitimately join in a financial partnership to purchase Sterling Forest.

Elizabeth Gordon, a bright intern recruited by PIPC commissioner Barnabas McHenry, prepared a paper entitled, "Why Does California Get So Much?" The Gordon paper confirmed that over a recent five-year period, California had received $248 million in federal Land & Water Conservation Fund appropriations for various park stewardship projects while New York received only $9 million. Put another way as calculated by Gordon, per capita spending from the fund for Californians was $8.33, for New Yorkers, $0.03.

New York was not the only eastern state being shortchanged. Gordon found a letter to a congressman that said, "I am writing to ask for your help in support of an amendment to the Land & Water Conservation Fund Act. The proposed amendment would reestablish the formula by which annual L&WCF appropriations are divided between federal and state governments." The letter was signed by the former governor of Arkansas, Bill Clinton.

With the arrival of Bruce Babbitt in Washington, expectations ran high that the Public-Private Partnership to Save Sterling Forest might gain sympathy and support for the rescue effort, especially when George Frampton agreed to leave his post as executive director of the Wilderness Society to assume responsibilities under Babbitt as assistant secretary in the Department of Interior. Frampton's portfolio would include oversight of L&WCF expenditures. The PPP anticipated that Frampton, widely respected within the environmental community, could be quickly convinced to join in the initiative to protect rare open space in the crowded northeastern United States.

I was especially optimistic. In 1986, Frampton and I had participated in the formation of a not-for-profit corporation designed to win the concession contract for Yosemite National Park with the intent of reinvesting profits in park stewardship. Instead, the contract was awarded to a for-profit company, but this personal association and Frampton's impressive environmental record led me to anticipate that the new assistant secretary would be quick to see the pressing need for federal matching funds to acquire Sterling Forest.

Frampton, though, proved to be a tough sell. In a meeting in 1993, Frampton told me that money in the coveted L&WCF should be used almost exclusively for federal projects. Frampton recognized and appreciated that the fund, as originally envisioned, would be shared about fifty-fifty between the federal and state governments but explained that not enough money was being appropriated by Congress for even the most urgent federal projects of highest national importance. He asserted that there simply was not enough federal money to help with state-level projects like Sterling Forest.

On the shoreline of Sterling Lake, a different dialogue was taking place. Discussion continued in May 1993 between a New York State Parks/PIPC negotiating team and the SFC staff. The partial-purchase idea was replaced by an effort to pinpoint the maximum amount of acreage the corporation might be willing to sell, and at what price. A sluggish economy was incentive for continuing the dialogue. Speculation in the *New York Times* was that SFC might rather have money in hand than a grand development plan on the table.

The discussion was supposed to be confidential. For understandable business reasons, the SFC did not want to signal publicly its sales strategy or values associated therewith. The SFC was considering a sale of 13,000 acres for between $30 and $40 million, leaving the corporation with 4,500 acres for development of 6,000 homes and 4 million square feet of commercial space. As part of any deal, Thomson wanted airtight assurance that the hoped-for interchange on the New York State Thruway would be constructed. But almost before the negotiators were out the door, and more to the surprise of PIPC commissioner Malcolm Borg than anyone, his newspaper, the *Record*, broke the story. Someone had leaked. The details were out in public, and Thomson was furious.

Still, the leaked news that some level of an acquisition might be possible encouraged congressmen Gilman, Torricelli, and Hinchey to introduce House Resolution 2741, which would appropriate $35 million to the PIPC for Sterling Forest acquisition purposes. Gilman, who had earlier applauded the full Sterling Forest development package, expressed support for a compromise that would create a park while leaving significant acreage for a scaled-down development.

New Jersey's senator Bill Bradley, working closely with Torricelli, introduced a companion bill, S-1683. The rationale was that, separate and apart from the Land and Water Conservation Fund, the PIPC, as an interstate agency functioning under a 1937 compact approved by Congress and the president of the United States, could receive federal appropriations. The interstate structure of the PIPC positioned it to receive funds from Congress, channeled through the Department of the Interior, that neither New York nor New Jersey could receive directly. Congresswoman Marge Roukema was alert to the same possibility. She introduced her own version of an appropriations bill, asking for $25 million to be matched by New York and New Jersey, making clear that her bill was not intended as a substitute for the Bradley-Gilman-Torricelli-Hinchey legislation, but an addition to it.

The PIPC was hoping that Thomson might express active support for the funding initiatives being taken in Washington if, in fact, the corporation wanted payment now, not later. But Thomson's lobbyist, Leon Billings, dashed that hope. Billings was well regarded as the author of the 1979 Clean Air Act while serving on the staff of Senator Edmund Muskie, but, responding to the *Times Herald Record* in June 1993, Billings said, "I would not have taken the SFC as a client if I didn't think that their plans were environmentally defensible." Then, assuring the *Times Herald*'s reporter, "I have absolute admiration for wilderness," Billings added, "We're not talking Glacier National Park here; this isn't even farmland in Virginia; this is an area that was at one time industrialized. . . . There's significant urbanization, power lines, roads."

JoAnn Dolan did not agree. Writing directly to Billings, she said, "In this most crowded urban area in the United States, where there is no virgin land, people are holding on to the shreds of their sense of place. We hope that you will not write off the urban areas because you can only relate to preserving things on a grand scale . . . like Glacier National Park, which is so far away from the maddening crowds." Echoing Dolan's sentiment, I reminded the *Star-Ledger* that "the PIPC accommodates 8 million park visitors a year, compared with Yellowstone's 2 million."

Week by week, and despite lack of acquisition funding, the Sterling Forest conservation message was gaining traction. On a campaign swing with New Jersey's governor Jim Florio in September 1993, Secretary of the Interior Bruce Babbitt stood at a scenic vista on the Palisades overlooking the Hudson River and expressed his general support for the initiative to acquire Sterling Forest. Florio was in a tough campaign with Christine Todd Whitman, and he was well known to Babbitt through their earlier contacts at the National Governors' Conferences. Even a casual reading of public sentiment confirmed that Sterling Forest was a good campaign issue in northern New Jersey. Florio wanted Babbitt standing by his side on the issue, and Babbitt was happy to oblige.

Keen observers at the event, although delighted by Babbitt's expressed interest in saving the forest, noticed that he did not mention any amount or source of federal funding to help with the acquisition. But the very fact that a member of the cabinet was talking about Sterling Forest provided hope that the Clinton administration was about to join the bipartisan coalition in Congress to push forward the $35 million appropriations bill. The PPP needed a funding breakthrough.

The breakthrough came, but not in Washington, D.C. During a winter ice storm, Malcolm Borg and I drove down to Trenton to meet with Governor Florio and his staff to discuss Sterling Forest strategy. Scott A. Weiner, counsel to the governor, had been searching for funding sources for Sterling Forest. His exploration uncovered an almost forgotten Water Conservation Bond Act approved by New Jersey voters in 1969 that still held $16 million in nonobligated spending authority. Funds from the bond act could be made available for Sterling Forest acquisition through routine financial procedures administered by the New Jersey Department of Environmental Protection. With the governor's approval, Weiner advised Borg and me that at least $10 million of the amount could be so used, if New York State and the federal government came aboard to help prove to the SFC that serious acquisition money was on the table.

This welcome news proved to be a parting message of encouragement to the PIPC from Governor Florio. He did not prevail in the November election. The incoming Whitman administration would have to be convinced of the wisdom of using the Water Conservation Bond money as proposed. Thanks to Weiner's detective work, though, Florio was handing a relatively uncomplicated choice to Governor Christine Whitman. With financial authority already provided by statewide vote, even if that vote was dated by twenty-four years, the new governor had a clear opportunity to provide acquisition funding for Sterling Forest if she chose to do so.

The timing of the New Jersey funding discovery seemed good. Before the end of 1993, the *Wall Street Journal* was reporting that Trygg-Hansa, the parent company of SFC, was recapitalizing. After three years of struggle to staunch the flow of red ink from the Home Insurance Group, Trygg-Hansa took the insurance company public, hoping to raise $350 million in a stock offering. The result was disappointing. Only $127 million was raised, forcing Trygg-Hansa to increase debt for Home Insurance from $100 to $280 million. The debt-to-equity ratio stood at 70 percent. In addition, Robert Thomson confirmed that the annual red-ink carrying costs for Sterling Forest were in the range of $2 million. The bad corporate financial numbers raised the expectation that the corporation's interest in selling Sterling Forest would be elevated from maybe to please!

In early 1994, Klara Sauer of Scenic Hudson, an active PPP participant, convinced twenty-five of her environmental colleagues to sign a letter to Governor Mario Cuomo urging that Sterling Forest be given extra special attention. "A powerful team of public and private, state and federal, interests has emerged in a collaborative effort to protect the precious natural landscape," Sauer wrote. She had reason for optimism in contacting the governor. In addition to promising funding initiatives in Washington and New Jersey, the PIPC and New York/New Jersey Trail Conference were raising funds to ensure that experts could be retained to comment during the public review of SFC's still pending environmental impact statement (EIS) that, if approved, would open the pathway to maximum housing and commercial development. The PIPC raised $165,000 for this purpose. Principal contributors were Malcolm Borg, George Perkins Jr., the James H. Ottaway Jr. Revocable Trust, the Gladys and Roland Harriman Foundation, the National Fish and Wildlife Foundation, the George W. Perkins Memorial Foundation, the Geraldine R. Dodge Foundation, the Mosaic Fund, and the Victoria Foundation.

In addition, trustees of a Lila Acheson and DeWitt Wallace charitable fund specializing in Hudson River Valley conservation activities signaled that they would be willing to make a $5 million grant jointly to Scenic Hudson and the Open Space Institute for Sterling Forest acquisition purposes. PIPC commissioner Barnabas McHenry certainly must have smiled at this commitment. As general counsel at *Reader's Digest* for many years, he worked closely with Mr. and Mrs. Wallace to establish several charitable funds to benefit the arts and environmental values in New York City and the Hudson River Valley. Scenic Hudson and the Open Space Institute were two of the many organizations to benefit substantially.

While funding possibilities began to glimmer, Ella Filippone of the New Jersey Passaic River Coalition was performing some political magic of her own. She had noticed that the PIPC owned about 78,000 acres of land in New York and only 2,400 acres in New Jersey. Wondering why, Filippone learned that since 1900 the PIPC had been legislatively restricted in its New Jersey activities to the "highest elevations of the Palisades," a narrow north-south strip running parallel to the Hudson River.

Working with her friend Phyllis Elston, a staff assistant for New Jersey Assembly Speaker Chuck Haytaian, Filippone drafted legislative language that would expand the PIPC's land-protection authority into the 1-million-acre New Jersey Highlands Region. Her motivation was that New Jersey should be positioned to take full advantage of the interstate structure of the PIPC, especially to carry forward the momentum in watershed protection that might be achieved at Sterling Forest. With Elston's help, Filippone convinced Haytaian and cosponsor Maureen Ogden, chairwoman of the assembly's Environmental Committee, to introduce the legislation.

Filippone and I testified before Ogden's committee, which then acted unanimously to endorse the PIPC bill. The influence of Haytaian and Ogden promised favorable action by the assembly, but Filippone knew that a possible stumbling block existed in the person of Robert Littell, chairman of the Budget Oversight Committee in the New Jersey Senate. Littell, a fiscally conservative Republican reputed to be unenthusiastic about most environmental initiatives, had to be convinced if the PIPC legislation were to prevail.

Filippone wasted no time. She and I went straight to Littell with a presentation about the potential financial and environmental benefits of expanding PIPC capabilities into the Highlands. Littell made one demand. He did not want his county, Sussex, included in the legislation. Otherwise, he gave it his consent. In late April 1994, the assembly unanimously approved the bill that allows the PIPC to function in Bergen, Passaic, Morris, Somerset, Hunterdon, and Warren counties, New Jersey, a highlands swath of hundreds of thousands of acres extending southwesterly from the New Jersey–New York border to the Delaware River.

Senate endorsement and the governor's signature followed. After almost a century, New Jersey invited the PIPC to step far beyond the Palisades, bringing along its traditions for acquiring and managing parklands and historic sites. Filippone could easily translate the word "park" to watershed. For her, the distinction was small; parks are generally good sponges, soaking up and filtering rainfall essential to all life-forms.

In Washington, events were not progressing so smoothly. Senator Bill Bradley scheduled a hearing in May 1994 to garner support for the $35 million federal appropriations legislation. Torricelli, Gilman, Hinchey, and Roukema were there to record their support. Governor Mario Cuomo sent a letter to every member of the New York and New Jersey congressional delegations, expressing his backing for the federal initiative and pledging assertive follow-up by New York. Robert Thomson reconfirmed SFC's willing-seller posture. The Public-Private Partnership was strongly represented at the hearing. The mood and expectation in the hearing room were very positive, until Marie Rust took the microphone.

Rust was the Northeast regional director for the National Park Service (NPS). Appearing in uniform, she was carrying testimony vetted through the Department of the Interior and the Office of Management and Budget, the keeper of budget strategies for the White House. To the stunned audience, Rust's testimony slammed the collective effort to save Sterling Forest. She testified that any federal funding directed toward the acquisition of Sterling Forest lands "ignored" important NPS priorities. "With a current backlog in land acquisition of some $1.1 billion," she said, "we should not divert what limited funds are available to projects outside the system."

Adding that no federal interest would be served by protecting Sterling Forest, despite the fact that a brand-new park would be attached to the NPS-owned Appalachian Trail, Rust claimed that "a dangerous precedent" would be set if Congress appropriated money for Sterling Forest. She did offer a limp carrot by suggesting that the NPS would be willing to "study" Sterling Forest for a year and come back to Congress with recommendations.

Everyone in the room knew that Secretary of the Interior Bruce Babbitt stood on the Palisades with then-Governor Florio only a few months earlier and pledged support for the Sterling Forest protection initiative. Roger Kennedy, director of the NPS and a fan of the Hudson River Valley, offered similar strong assurances in an earlier meeting with McHenry and me. Questioned a day after Rust's testimony by one of his own reporters, Malcolm Borg was diplomatic: "It is our belief that if the federal government takes a leadership role in this, private funding will follow." Borg's frustration, as expressed privately to me, was much less diplomatic.

Congressman Torricelli was also quoted: "We are at a loss to understand how the Park Service could fail to see the federal interest in preserving this tract." Feeling heat from members of the New York and New Jersey congressional delegations, a spokesperson for Babbitt offered the disingenuous remark that the secretary's staff was trying to "track down" what happened between "last fall," when Babbitt expressed support for Sterling Forest, and the NPS's expressed opposition, now flapping like a red cape.

Congressman Hinchey dispatched his staff aide, Chris Arthur, to the Interior Department to investigate. Governor Mario Cuomo personally telephoned Babbitt. Steam from Congressman Torricelli clearly was evident. The *Record* carried an editorial declaring that New Jersey got a kick in the teeth from the National Park Service. One pundit said Babbitt's words were "chiseled in Jell-O." Referring to the gross imbalance between L&WCF money spent in the West versus the East, a *Rockland Journal News* editorial said, "Maybe the National Park Service is so blinded by the beauty of California redwoods . . . that it has written off the equally spectacular eastern part of the nation." Nash Castro and I were in shock. Between us, we had years of proud service with the NPS and viewed with deep regret a funding conflict that seemed to us to have no rational basis. Sterling Forest would be a superb addition to the nation's system of parks, federal, state, and local alike, and would be a huge asset to the NPS-administered Appalachian Trail. The PPP took on a war-room mentality.

Flak was still flying and telephone lines burning the following week, when the commission's president, Martin Cornell, appeared on behalf of the PIPC before the House Subcommittee on National Parks, Forests, and

Public Lands. To the general relief of Cornell, Dolan, Filippone, Tripp, and PPP colleagues, word circulated in advance of the hearing that the NPS had changed its position. In place of Rust, who had been delegated the unpleasant task of delivering the earlier message, Mike Finley, an NPS assistant director, explained the reversal.

Finley testified that the NPS would support a $17.5 million appropriation for Sterling Forest, but with strings attached. The NPS would make its own deal with SFC, he suggested, by buying land adjacent to the Appalachian Trail. He added that the Interior Department reserved the right to pay no more than 25 percent of the total purchase price for Sterling Forest lands. Assuming a total price of $55 million, the NPS would be obligated for just $13.5 million.

Although the PIPC and its allies were encouraged that a major level of federal financial support now seemed possible, the attached strings were unthinkable. The three towns that claimed portions of Sterling Forest for property-tax purposes had swung strongly in the direction of the PIPC's acquisition initiative, in large part because property and school taxes would continue to be paid on the acquired land but very few municipal services would be required in return.

In the long-ago days of Perkins Sr., foresighted political decisions had been made to keep almost all the PIPC's lands in New York on the local tax rolls rather than move them to tax-exempt status, as was the custom with parklands. Over decades, this strategy greatly aided the expansion of the park and historic-site system. The rationale was that the payment of property and school taxes on open space was less costly to taxpayers in the long run than the escalating tax demands required to indefinitely cover the costs of an aging infrastructure left in the wake of development.

In 1993, the New York General Fund paid more than $6.5 million in taxes to towns in which PIPC holdings were located. In the town of Tuxedo, the PIPC's lands provided the largest single source of tax revenue. By contrast, any land at Sterling Forest that the NPS might acquire under the strings-attached scenario would be promptly removed from the tax rolls. The specter of the NPS vying head-to-head with the PIPC for Sterling Forest acreage would also position the SFC to play one buyer off against the other.

To snip these strings, more trips to the nation's capital, telephone calls, faxes, letters, and entreaties were required of the PIPC and PPP. Congressman Hinchey's staff assistant, Chris Arthur, was spending more time at the Department of the Interior than he was in his office on Capitol Hill. His boss personally consulted with Secretary Babbitt about the problem.

Torricelli, Roukema, Bradley, and Lautenberg pitched in. The mood of the New York and New Jersey congressional delegations was that no new and restrictive federal rules should suddenly be invented for Sterling Forest when no such restrictions had been draped on other Land & Water Conservation Fund or related federal-state projects.

In the meantime, Filippone was continuing to work with New Jersey state senator Robert Littell, even though many of her environmental allies took a dim view of her willingness to collaborate with someone they considered to be a political enemy. She shrugged off these criticisms, judging that if she could not work effectively with the chairman of the New Jersey Senate Budget Committee, then any hope for funding from New Jersey, to be spent in New York, would be dead on arrival.

To the amazement of many of Filippone's friends, Littell's Budget Committee unanimously endorsed the use of $10 million in New Jersey bond funds for Sterling Forest acquisition purposes across the state line in New York, providing political strength for Governor Whitman. So persuasive was Littell's influence in the state budget process that the full Senate followed suit, voting in July 1994 to support a New Jersey fund transfer of $10 million to the PIPC. The vote was 37 in favor, 1 opposed. New Jersey was putting its money where its clean water was and ringing a signal bell for New York and the U.S. Congress that would be hard to ignore.

Robert Thomson would not be around to see where the New Jersey lead would take the acquisition initiative. Abruptly and without fanfare in mid-August 1994, he left his position as president and C.E.O. of the Sterling Forest Corporation to "pursue other interests." Lou Heimbach immediately took Thomson's place. Only three weeks later, the SFC released its environmental impact statement, all thirty-one volumes and 4,000 pages.

Imposing though the EIS was, and supposedly filled with objective data, it delivered one clear and simple message: the real-estate development project was in full gear and racing forward. In the section of the EIS that required a listing of "alternatives," should the entire development package not be approved, no mention was made of any possible public acquisition of the property. As far as the corporation was concerned, the only legitimate alternatives were development or slightly less development.

The "preferred alternative," around which all the data spun, affirmed that 13,000 housing units and 8 million square feet of commercial space could be constructed in Sterling Forest with only slight, inconsequential environmental impact. The commercial space, alone, would be equivalent in size to 240 football fields.

The PPP was scrambling to keep up. Against the millions claimed to have been spent by SFC on the statement, the PPP and the PIPC were turning to universities, various environmental organizations, pro-bono law-yers, and a few paid consultants to try to gather well-documented and sci-entifically defensible facts on hydrology, water supply, ecological fragmentation, endangered species, wetlands, transportation, scenic qual-ity, and fiscal impact.

At Bear Mountain, Jack Focht, director of the Trailside Museums and Wildlife Center, had been watching the contest unfold. Knowing that SFC was diligently guarding its land against trespassers, especially those who might be too curious about some of the conclusions reached in the EIS, Focht gathered his Volunteer League of Naturalists, a loosely arranged group of academicians, businesspeople, retirees, docents, and students who shared a common love of natural history.

Focht sent them for weekend walks along the section of the Appalachian Trail that passes through Sterling Forest. Mostly confining themselves to the narrow trail corridor, the naturalists invested a few days of volunteer time compiling a list of 369 different species of plants, birds, insects, fish, mushrooms, mammals, reptiles, and amphibians. The list included the rare American chestnut tree, dainty Queen Anne's lace, wild licorice, red-throated and common loons, wood ducks, bald eagles, wild turkeys, Amer-ican woodcocks, great horned owls, ruby-throated hummingbirds, and belted kingfishers. Twenty-two species of warblers and fifteen species of butterflies were found. Beavers were at work, and white-tailed deer, red fox, and various shapes and sizes of turtles were abundant. The naturalists found signs of black bears roaming the forest.

From high elevation points, Focht's troopers could study pigeon-worthy skyscrapers on the distant horizon in New York City while their feet were planted in a veritable wilderness. Forest fragmentation, the shotgun-blast that SFC was aiming at this intact and amazingly fruitful ecosystem, was predicted to leave only a relic of what the volunteer naturalists found all around them.

Despite Focht's initiative, the effort in Washington to gain approval of an appropriation for the forest was beginning to run up against the head-winds of a midterm election. By a 16-to-4 vote, the Senate Energy and Natural Resources Committee endorsed a $17.5 million appropriation for Sterling Forest in August 1994, adding the legislation to a truckload of bills trundling toward final action, but in jeopardy because senators and mem-bers of the House of Representatives were anxious to adjourn as soon as possible to attend to campaign needs.

Months of intense effort by the PPP had been narrowed to the last few weeks before the 103rd Congress would adjourn in anticipation of the November 1994 election. In October, I was glued to CSPAN, watching as senators passed to and fro in front of the unblinking television camera, conferring in small groups, asking for procedural votes, and claiming parliamentary privileges. At one point, Senator Bill Bradley could be seen chatting with Senator Richard Shelby, Republican from Alabama, who was objecting to legislation about a historic site in Pennsylvania that was packaged with Sterling Forest and other public-lands bills. Was Bradley trying to convince Shelby to let the Sterling Forest legislation go forward on its own? Only they knew.

No question remained, though, when Senate minority leader Robert Dole, acting at Shelby's request, formally objected to the legislative package. Based on the clubby rules of the Senate, objection by the minority leader killed the package. Moments later, the 103rd Congress adjourned so that members could hit the campaign trail. The November 1994 election is referred to as the "Republican Revolution," the first time in forty years that Republicans, led by Newt Gingrich, took majority control of the House of Representatives. The Senate, too, was firmly in Republican hands. The 104th Congress would formally take control in January 1996, and the PPP would have to start all over again in its quest for federal funds. Shrugging off the missed opportunity, Lou Heimbach of the Sterling Forest Corporation was quoted in the *Times Herald Record* as saying, "It's a major setback, sure, but we didn't really have anything before."

For those seeking to preserve Sterling Forest, the November election brought more change. In New York, Governor Mario Cuomo, seeking a fifth term, lost to State Senator George Pataki. About three months prior to the election, Orin Lehman, after years of service, had voluntarily retired with great honor from his post as New York's park commissioner. In Lehman's place, Governor Cuomo appointed philanthropist Joan Davidson, who picked up right where Lehman had been, by expressing immediate and strong support for the Sterling Forest initiative. But, in turn, Davidson was replaced by Bernadette Castro (no relation to Nash Castro) when George Pataki prevailed in the 1994 election. Thus, the way of political transitions.

Governor Pataki, Bernadette Castro, and a whole raft of new people in Albany and Washington, D.C., would have to be convinced of the increasingly imperative need to save Sterling Forest. The exception to this new challenge was in New Jersey. Governor Christine Todd Whitman and other state officials watched the election results from the sidelines. Except for the congressional races, the state of New Jersey operated on a different election-year calendar.

A few ripples in the political pond did occur in the New Jersey congressional races. Republican Rodney Frelinghuysen, who, until the Gingrich sweep, served in the state assembly as chairman of the Appropriations Committee, won against his Democratic opponent and joined the 1994 "freshman" class in Congress. In his position as the appropriations chair, Frelinghuysen heard testimony from Filippone and me and strongly supported the need to protect Sterling Forest. Just as had his counterpart, Robert Littell, on the senate side of the New Jersey legislature, opened the funding door for Sterling Forest, Frelinghuysen's assembly committee also had approved the $10 million commitment, and he was taking this message to Washington, D.C.

For the PPP, Frelinghuysen's move to Congress was good news, spiced even more by the discovery that he had been honored at the national level with a coveted appointment to the House of Representatives Appropriations Committee. Frelinghuysen more than landed on his feet in Washington. D.C. He was right where the purse strings were controlled. Joining Frelinghuysen in the 1994 freshman congressional class was another assembly colleague, Bill Martini, who rode the Republican wave to success in a normally Democratic district in New Jersey.

Soon after the election, on November 22, 1994, Governor Whitman, addressing an audience at Drew University, delivered an undeniably clear message about her position on Sterling Forest: "Most of you know that I recently took positive action on legislation that will bring us very close to preserving Sterling Forest. I am committing upward of $10 million toward this effort. I am counting on the State of New York and the Federal Government to commit their fair share of funds to complete this project. I pledge to you this evening that I will work with our new Congress and with Governor-elect Pataki to make this vision a reality as soon as possible. Sterling Forest naturally cleanses the water supply of some two million New Jerseyans, and it does this for a song. If we lose the forest to development, expensive treatment plants will take its place. That would be crazy! Once again, the right environmental choice is also the best economic policy!"

The North Jersey District Water Supply Commission's chair, Robert Rubino, confirmed the price that Governor Whitman had in mind if Sterling Forest were not protected. Rubino, on a track parallel to the PIPC as an activist in the battle for Sterling Forest, had made trips to and from Trenton and Washington, D.C., to testify for acquisition funding. Speaking to the financial numbers, he reported that if the forest were to be developed as proposed, chemical-treatment costs for drinking water would leap at full build-out from $3.1 million to a projected $77 million per year, calculations for inflation not included.

On a sunny day, I drove down to attend a ceremony near one of the New Jersey reservoirs where a large mockup of a $10 million check was on display for media purposes. On the drive back to Bear Mountain, I carried a little envelope with the real check tucked inside.

In early 1995 Governor Whitman wrote to New York's governor, "Dear George: As you know, last month I signed legislation essential to the preservation of Sterling Forest. With confidence that you share my strong interest in protecting the natural resources of the Highlands region and safeguarding . . . vital aquifers, I urge you to champion passage of similar legislation in your state. Your support of this important interstate effort can ensure its success."

Governor Pataki's response came quickly: "Thank you for your letter expressing your strong support for the preservation of Sterling Forest. As an avid outdoorsman, I am personally familiar with the property and I feel, as you do, that it must be protected." Pataki meant what he said. He was a son of the Hudson River Valley, having grown up on a farm near Peekskill, New York. He and his family had been frequent visitors over many years to Bear Mountain and Harriman Parks and had hiked along the Appalachian Trail in Sterling Forest. Pataki closed his letter to Governor Whitman by pledging his enthusiasm for "cooperative ventures between our two great States, beginning with Sterling Forest."

Whitman also acknowledged a letter to her from Larry Rockefeller: "I appreciate the support and strong interest of the Public-Private Partnership to Save Sterling Forest. . . . As you know, the next steps are for New York and Congress to pass similar legislation."

In a March 1995 letter to Pat Noonan, who had worked as a Nature Conservancy staff member to help preserve Minnewaska and who now was president of the Conservation Fund, I expressed my own view that something special might be happening in the partnership effort to save the forest: "Perhaps most importantly, the effort to acquire Sterling Forest may be taking on aspects which could apply in a positive manner to major land conservation initiatives in the 1990s. The informal, but increasingly effective 'partnership' which has materialized in pursuit of the acquisition goal suggests how federal-state-private alliances can capture the attention of Congress. . . . The convergence of two states, the Federal government, and the private sector in pursuit of a common land acquisition goal is not necessarily unique, except that in this case it is happening in the crowded northeast, and it is happening in the '90s."

New York parks commissioner Bernadette Castro, rapidly learning about the parks/historic-site portfolio handed to her by Governor Pataki, attended a spring meeting of the PPP held in New York City in one of the Davis Polk Wardwell conference rooms. She came away impressed, sensing, as she put it, that Sterling Forest might become a worthy environmental prize to be claimed during the first four-year term of the Pataki administration and would be a mark of personal success for the governor.

Her impression was reinforced during a follow-up visit to Bear Mountain. I had arranged for an aerial view of Sterling Forest via helicopter. Only after a safe return landing on the great lawn at Bear Mountain did Mrs. Castro confirm that she had just completed her first helicopter flight. She did like the bird's-eye view of the vast forest, and said, "If I have to rent an apartment in Washington to lobby this thing through, I will."

While political transitions were underway in New York and Washington, D.C., the Zurich Insurance Group replaced Trygg-Hansa as owner of Home Holdings and the Sterling Forest Corporation. This newest corporate owner promptly announced a $30 million line of credit to initiate proposed development of the forest, but media reports continued to characterize Home Holdings as a financial millstone, reporting $264 million in losses in the first quarter of 1995. Any hope that this red ink might tempt the Zurich group to unload Sterling Forest at a bargain price proved to be wishful thinking. When the New York Department of Environment Conservation announced that hearings would be held on SFC's Draft Generic environmental impact statement, Lou Heimbach welcomed the news, reaffirming that public dialogue was a necessary prelude to full-scale development.

During the spring of 1995, the DEC held four public hearings in communities adjacent to Sterling Forest. James Ahearn, writing for the *Record*, described the first hearing: "When the hearing got underway, eight of the two dozen witnesses were from the Town of Tuxedo. The town supervisor, Joseph Ribando, was the only witness who was not unreservedly opposed to the plan. The hearing was in a gymnasium, hung with red-and-white banners celebrating championship seasons in girls' soccer and boys' basketball. The microphone did not work. Until a replacement was found, witnesses had to stand with their backs to the hearing officials and bellow. The audience included many New Jerseyans, concerned about downstream effects of development. There were lots of green and white lapel buttons saying, 'Keep Sterling Forest Green.' The room, packed with 200 people, became warm. Big ceiling fans were turned on, creating a tremendous racket. The fans were turned off. Doors to the outside were opened to admit fresh air, revealing an orange-and-white dog chained to an exit stairway. The dog began lunging at his chain and barking ferociously, drowning out the witnesses. It was, all in all, a Tuxedo, not Tuxedo Park, event."

This first hearing lasted almost until midnight. People were still waiting their turn when the exhausted hearing officer called for adjournment. The succeeding three hearings mirrored the first.

Lou Heimbach was among the 200 people packed onto the gymnasium floor, there to listen, not to speak, as would be his style at all four hearings. John Gebhards of Sterling Forest Resources, a spare man with a graceful presence and granite resolve, would also attend all four hearings, and he did speak. Gebhards deserved much of the credit for the packed hearing rooms. Sterling Forest Resources was the local grassroots organization that, with a budget even more spare than its leader, served as the counterweight to the corporation's public-relations campaign. Gebhards's low-budget newsletter, combined with his willingness to speak in any forum, visit schools, talk with neighbors, debate with Heimbach, lead hikes, gather facts, refine strategy, respond to the media, reason with elected officials, and urge informed citizens to become activists won him great admiration and respect within the Tuxedo community and among his PPP peers. Many of the speakers at the hearings were exceedingly well informed because Gebhards made certain that they were.

By numbers alone, the public voice proclaimed overwhelming opposition to the development of Sterling Forest. Mary Yrizarry, a local resident and Gebhards's friend, said, "We're being asked to buy a pig in a poke. . . . We may be left with the remnants of a grandiose scheme that will become burdensome to the entire region." The PIPC argued tax benefits, flatly contending that acquisition of the forest for park purposes would be tax-positive to the local communities, while "Swiss cheese" development would clearly be tax-negative. Describing the hidden costs of development, the PIPC reported that a transportation consultant had pegged the need for road improvements alone at $41 million, not including SFC's sought-after interchange on the New York Thruway, which the PIPC firmly opposed.

Harvey Stoneburner, a mathematical economist at Hunter College, challenged the conclusions reached by SFC in its environmental impact statement as being based on a "selective approach to data." Labeling corporate assumptions about the type of lifestyle to be created at Sterling Forest as "utopia," another speaker said, "People who will inhabit the forest will be among the most unique human beings on the planet: They will have fewer children, fewer cars, ride more buses and create less garbage than anyone yet known."

David Startzell, representing the Appalachian Trail Conference, pointed out that SFC's statement suggested that a way to mitigate the impact of development near the Appalachian Trail was to "move the trail." Another local resident, Geoffrey Welch, a professional musician and self-taught water expert, criticized the corporation for what he judged to be its cavalier approach to water issues and defended his own carefully reasoned hypothesis about the true impact on water quality should the development go forward.

Dr. Richard Lathrop, a Rutgers University associate professor, used computer modeling to examine SFC's claim that the thousands of residences and acres of commercial space it intended to introduce into the forest ecosystem still would leave 75 percent of the land as "open space." Lathrop found that 86 percent of the property would be "severely or very severely" impacted. He did discover that ten acres of the 17,500-acre property could be safely placed in the "slightly impacted" category. Martin Lavenhart, using simple logic, remarked, "We have a forest. No assumptions are necessary to keep it a forest." Speakers ranged in age from eight years to ninety. Hundreds of pages of written testimony were stacking up on the hearing table, all requiring detailed review by the New York Department of Environmental Conservation staff.

The Environmental Defense Fund's Jim Tripp questioned in detail the legal sufficiency of the Environmental Impact Statement. Tripp, Larry Rockefeller, Barnabas McHenry, Sam Pryor, JoAnn Dolan, Ella Filippone, Timothy Dillingham, and the rest of the PPP knew that the road ahead would likely lead straight to court if the hearings failed to produce major corrections in what they judged to be faulty EIS data, assumptions, and conclusions.

Heimbach, listening throughout, characterized the hearings as "part of the process" and suggested that sound and reasonable answers to all the concerns brought forward at the hearings, emotional and factual, could be found in the EIS and the corporation's frequently reiterated intent not to damage the forest ecosystem. In this stance, Heimbach was distinctly in the minority. Later in the year, when Tuxedo supervisor Joseph Ribando and two town board members, considered to be overly sympathetic to SFC's development scheme, stood for reelection, they were voted out of office.

Another "part of the process" continued to play out in Washington. In July 1995, Senator Bill Bradley telephoned Malcolm Borg with good news: the Senate had authorized an appropriation of $17.5 million for Sterling Forest acquisition purposes! After months of maneuvers, steps and missteps, and optimism dashed by disappointments, the powerful and legislatively skilled New Jersey senator, joined by his equally adept colleague, Senator Frank Lautenberg, had scored mightily on behalf of forest conservation.

The authorization bill was the first step of a two-step process. Authorization is the green light that provides legitimate hope that appropriations— money in the bank—will follow. The authorization bill placed the U.S. Senate squarely on record in favor of the necessary appropriation. This achievement caused expectations to run high on the House side of Capitol Hill when a hearing was held before the Subcommittee on National Parks, Forests and Public Lands to consider a similar authorization bill.

Congresswoman Marge Roukema and her staff had been working hard behind the scenes to convince her Republican colleagues that protecting Sterling Forest was in the national interest. In this effort, she was joined by Gilman, Frelinghuysen, Martini, Sue Kelly, and other Republican members of the New Jersey and New York congressional delegations. Maurice Hinchey of New York and Robert Torricelli of New Jersey were carrying the flag for the House Democrats. Chairing the hearing that day was James Hansen, R-Utah. Other participating subcommittee members included Wes Cooley, R-Oregon, Wayne Allard, R-Colorado, Richard Pombo, R-California, and Maurice Hinchey, D-New York.

Allen Freemyer, counsel to the Republicans on the subcommittee, set the tone by referring to Sterling Forest as "radioactive." Bernadette Castro, testifying for the first time before Congress, tried to make the case for federal participation in the forest preservation initiative. Pombo, a rancher from the San Joaquin Valley in California and first-term member elected to Congress in 1994, made an obvious point of chatting with a staff aid while Castro testified.

Speaking on behalf of the PIPC, I attempted to explain that the commission functioned under a 1937 mandate approved by Congress and signed by the president, but fared no better. The western congressmen made no secret of their distaste for what they deemed to be a misguided attempt to use federal money to buy a "state park." Reminded by Hinchey that the initiative to save Sterling Forest enjoyed strong bipartisan support within the New York and New Jersey delegations, the western members were unimpressed. "Just local pressure," one said.

James Hansen, enjoying his new post as subcommittee chair, was a well-known advocate for mining, grazing, hunting, trapping, logging, and off-road motor-vehicle access to public lands, and equally suspicious of environmental motives, disdainful of wilderness protection, and resistant to expansion of the public land base. "I can hardly wait for the full committee to get this one," he said. His meaning was clear. The Committee on Resources, to which Hansen referred, was chaired by Don Young, R-Alaska, a former taxidermist and kindred spirit to the subcommittee chair.

The fate of federal funding for Sterling Forest was likened to a "high-stakes poker game." Passage of the authorization bill in the Senate put pressure on the House to follow suit. Senator Bradley made known that if the House did not act favorably on Sterling Forest, he would bottle up fifty or so bills being sought by Hansen and Young that required Senate endorsement.

Marge Roukema, the senior member of the New Jersey Republican delegation, decided to work around Hansen's subcommittee by appealing directly to Speaker Newt Gingrich. In a campaign swing to the Hudson River Valley several months earlier to aid New York Republican Sue Kelly in her successful bid for election to Congress, Gingrich had applauded the idea of protecting Sterling Forest. None other than the intrepid Ella Filippone had counseled him about the forest preservation effort. While in Washington on other business, Filippone had impulsively decided to swing by Gingrich's office to try to talk with a staff person about Sterling Forest. Telephoning me afterward, she said, "I went into the office, asked if I could see Gingrich or a staff person, and the next thing I knew, I was talking with Gingrich."

Now, weeks after the Speaker's campaign visit with Sue Kelly, Congresswoman Roukema wanted to remind him of his stated support for the Sterling Forest conservation initiative. She did so by sending him a letter, signed by thirty-eight members of the New Jersey and New York congressional delegations, urging that funds be appropriated to purchase the forest.

But, then, to Roukema's utter surprise, her fellow New Jersey colleague, Congressmen Bill Martini, and his 1994 freshman-class colleague from California, Richard Pombo, put an unanticipated bid on the political poker table. They proposed that the federal government auction off 56,000 acres of national grasslands in Oklahoma and use the proceeds to help buy Sterling Forest. Added to this ploy was that a 326-acre Washita National Battlefield Site would be established in Oklahoma to commemorate the deaths in 1868 of 150 Cheyenne, surprised while sleeping and killed by cavalry troops under the command of Lt. Col. George Armstrong Custer.

Martini and Pombo thought they had found a creative trade-off that would respond to a conservation need in the northeastern United States while lowering the inventory of federal land in the West. The Cheyenne and Arapaho tribes in Oklahoma did not agree. They had a stake in the grasslands based on tradition, spiritual meaning, and personal enjoyment. Neither did the Oklahoma Wildlife Federation, joining in this instance in an unusual alliance with the National Audubon Society. From different perspectives, the two organizations saw risk in losing access to and use of 5 percent of the public land in the state, known particularly as excellent pronghorn antelope and quail habitat. Oklahoma hunters, backed by the New Jersey State Federation of Sportsmen's Clubs, saw the proposed grassland auction as a "land grab" by ranchers.

Roukema believed that she was making progress with Gingrich, but the Oklahoma component produced a sudden complication that threatened to bring momentum to a halt. At the Department of the Interior, Secretary

Babbitt left no doubt about his view: "Throwing public land on the auction block to generate revenue for other conservation projects would set a very bad precedent," he contended. Major environmental organizations quickly concurred. But bringing the heavy weight of his influence to bear, Gingrich stepped forward to express support for the Pombo-Martini maneuver, claiming that it offered to pay for the Sterling Forest purchase "in a way that fits in with getting a balanced budget."

Despite the political obstacle course in Washington, D.C., enough thrust was being developed to save Sterling Forest that Lou Heimbach was convinced to begin serious negotiation of a possible purchase. There still was no answer to the logical question, "How much acreage at what price?" Speculation about purchase price was running all over the chart. To reel in federal, New York, and additional private money, a precise answer was required.

Heimbach made known that any purchase discussions must be draped in strictest confidentiality and that the PIPC, especially, was not welcome at the table. But two very adept organizations were allowed at the negotiation table: the Trust for Public Land (TPL), represented by Rose Harvey, and the Open Space Institute (OSI), represented by Kim Elliman.

Harvey and Elliman recruited Steve Horowitz, a talented attorney with the law firm of Cleary, Gottlieb, Steen & Hamilton, to provide what proved to be thousands of dollars of pro-bono counsel. TPL and OSI lawyers Phyllis Nudelman and Robert Anderberg were part of the exclusive team, as was OSI's Katie Roberts, the niece of former PIPC commissioner Frederick Osborn. According to Elliman, "grinding" was the best word to describe the process. On occasion, a negotiating team staff member or lawyer would check with the PIPC on details related to park-management needs, but otherwise the shades were tightly drawn.

While negotiations proceeded in secrecy, the Sterling Forest-for-Oklahoma-grasslands deal in the U.S. Congress fell of its own controversial weight, only to be replaced by an even more questionable tactic. In early February 1996, I arranged for a series of meetings in Washington, D.C., for Bernadette Castro. Barnabas McHenry and I rode the train yet again from New York City to the nation's capital to rendezvous at the office of Utah's James Hansen with Mrs. Castro and Mary Ann Fish, widow of Congressman Hamilton Fish Jr. and a member of Governor Pataki's congressional liaison staff.

After exchanging pleasantries with Hansen and Allen Freemyer, counsel to Hansen's Subcommittee on National Parks, Forest, and Public Lands, an entirely new legislative strategy for Sterling Forest was introduced. Hansen proposed to open millions of acres of federally owned wilderness in southern

Utah to mining, dams, and off-road vehicles. He asserted that the only chance for Sterling Forest rested with support by the New York and New Jersey congressional delegations for the Utah maneuver. Hansen was talking about the famous Redrock country of southern Utah that includes Zion, Bryce Canyon, Capitol Reef, Arches, and Canyonlands National Parks among its many scenic assets, a spectacularly arid landscape full of vibrantly weather-sculpted geologic forms, once the hiding place of Butch Cassidy and the Sundance Kid, and spiritually precious to Native Americans.

Federal lands outside the national park boundaries in southern Utah were largely under the management of the Bureau of Land Management and were known to include vast coal deposits. These lands were protected in part by wilderness designations, and there was speculation that President Bill Clinton intended to strengthen this protection even more by proclaiming national monument status for the vast heart of the region. Hansen wanted the opposite: a region of millions of acres available to industry, especially coal mining, and saw Sterling Forest as a bargaining chip to get his way.

No one in the room yearned for Sterling Forest preservation more than Barnabas McHenry. He had devoted countless volunteer hours to the bid for environmental common sense in the Hudson River Valley and its surrounding highlands, always had been available at a moment's notice when needed, was carrying around a personal calendar filled with appointments and meetings about Sterling Forest, was sending out letters to anyone and everyone who might be nudged toward the preservation camp, and was striding forward with unshakeable perseverance among the growing legion of advocates for forest protection.

He and I made such pests of ourselves on Capitol Hill that congressional staff members had started calling us "Sterling and Forest." McHenry, though, reacted to Hansen's ploy like someone shot from a cannon. He was on his feet, staring furiously at Hansen. "We want this very badly, but we will never sacrifice Utah wilderness to save Sterling Forest!" Hansen sat back in his chair, gave a dagger stare to McHenry, then rose, claiming scheduled business elsewhere, and left the room. His staff aide, Allen Freemyer, was left in charge of the meeting, but not much of a meeting was left. Freemyer urged us to "think about" the need to bring western members of Congress into the Sterling Forest mix.

Hansen's Utah card was on the table. He wanted mining and off-road access to 4,000,000 acres in the Redrock region. When this news surfaced in the media, Secretary Babbitt said he would recommend a veto if any such legislation reached President Clinton's desk. Hansen's Republican colleague, Congressman Ben Gilman, took an equally dim view. Maurice Hinchey, serving in the minority on Hansen's subcommittee, joined aggressively with western-based environmental organizations to head off the Utah wilderness grab and keep Sterling Forest from becoming entangled in a legislative package that would certainly be vetoed.

The Hansen card was in play in March 1996 when Speaker Newt Gingrich agreed to return to the Hudson River Valley to host a fundraising reception in Fishkill, New York, for his colleague, Congresswoman Sue Kelly. The Speaker was then scheduled to travel by van to the Monksville Reservoir in New Jersey to make an announcement about Sterling Forest. Congresswoman Marge Roukema had ensured that the Monksville visit was added to the itinerary.

On learning of the Speaker's plans, I took a chance and telephoned Rob Hood, a Gingrich staffer who was frequently in the chain of Sterling Forest communications, and suggested that New York parks commissioner Bernadette Castro be given a seat in the van for the hour-and-a-half drive from Fishkill to the Monksville Reservoir. Hood agreed. On the day of the fundraiser, the PIPC's Ken Krieser spotted the location of the van and alerted Castro. Krieser and Castro maneuvered through the crowd to the van as the event concluded. Gingrich and his staff were still involved with photographers and well-wishers. Castro hesitated to enter the van, not knowing with certainty whether the Speaker even knew of Hood's agreement that she could be a passenger, but the opportunity was just too good to pass up. Kreiser urged, "Get in the van, get in the van!" When Gingrich and his staff arrived, the van doors closed and the passengers, Mrs. Castro included, were on their way to New Jersey.

A reporter for the *Record* wrote, "Waiting for Newt Gingrich, we heard the northwest wind whistling. After it discovered our mass huddled in the slushy parking lot and invaded our puny jackets, leaving our blood curdling, the wind could be heard merrily whistling at a job well done. An hour late, Gingrich devanned and strode to the little rubber mat laid down to protect North Jersey's tundra from the depredations of politicians' feet. 'We are going to save Sterling Forest,' a beaming Gingrich announced, warming the crowd."

The Speaker of the House of Representatives assured the gathering that a bill "is going to pass this year" and would not be "held hostage" in Congress. Castro, standing with a group close behind the Speaker, was also beaming. The van ride had provided her a much better forum than the one experienced in Congressman Hansen's office. Governor Christine Todd Whitman, Congresswoman Roukema, and Congressmen Gilman, Frelinghuysen, and

Martini were there to add their enthusiastic endorsements to the Speaker's message.

Even so, the Sterling Forest roller coaster continued. Senator Bill Bradley mounted a three-day filibuster against an "omnibus" package of parkland legislation sent to the Senate by the House of Representatives that included more than fifty mostly noncontroversial bills, but also embraced the Sterling Forest-Utah "hostage" link sought by Hansen and supported by Gingrich, despite the Speaker's comments at the Monksville Reservoir. When it became obvious that sixty-one votes could not be mustered among the senators to break Bradley's filibuster, Majority Leader Bob Dole withdrew the entire package.

Over in the House of Representatives, Hansen's Republican-controlled subcommittee regrouped. It inserted a provision in the much-sought-after $17.5 million Sterling Forest authorization requiring the forest to be legislatively declared a wilderness. This legislative maneuver was approved on a 10-to-7 party-line vote. "Senator Bradley stands up there and says how great wilderness is," Allen Freemyer was quoted as saying. "Well, here is an opportunity to create some in his backyard." If written into law, the wilderness designation would require existing roads, power lines, and buildings to be taken out of Sterling Forest. One subcommittee staffer gleefully crowed that local residents who commuted on the existing roads through Sterling Forest would have to go "on foot or horseback." Roukema termed the amendment "absolutely absurd." Hinchey's staff member Chris Arthur kept saying, "I want to get this thing off my desk," and dug even deeper by networking with other congressional staff colleagues to gain approval of a clean appropriations bill.

With flak flying in Congress, Kim Elliman, Rose Harvey, Steve Horowitz, and their negotiating team emerged from the shadows of confidentiality in May 1996 after nine months of on-again, off-again sessions to announce that they had initialed a purchase agreement with the Sterling Forest Corporation. The corporation would retain 2,220 acres of its holdings. For a purchase price of $55 million, it would sell 15,280 acres and eventually donate another 525 acres when the transaction was completed.

The corporation would retain the right to draw millions of gallons of water from Sterling and Blue Lakes for development purposes, but drawdown limits were imposed on both lakes, two feet at Sterling Lake, one and one-half feet at Blue Lake. Logging contracts already in place would be allowed to run their course, even if title was transferred before the contracts expired. Lou Heimbach signaled that he anticipated higher-density development on the SFC's retained 2,220-acres, and he put safeguards in the agreement to protect the corporation's right to the higher densities. A $5 million down payment was required.

Finally, purchase price and acreage were out there for everyone to see, and the clock was ticking on an escrow agreement and final closing. Heimbach wanted the closing within a year—two years maximum. The Lila Acheson and DeWitt Wallace Fund would be the source of the $5 million down payment, thanks especially to the influence of Barnabas McHenry. New Jersey's $10 million was in the bank. But that was it. The Washington, D.C., roller coaster had yet to land on anything resembling firm financial ground. New York had not authorized any funds for Sterling Forest. The negotiators came out of the room with an agreement, but $40 million short in needed funds.

Even though Governor George Pataki and the New York legislature had yet to authorize funding for acquisition of Sterling Forest, the governor's resolve to match or exceed New Jersey's $10 million was assumed. Pataki frequently expressed his admiration for the conservation legacy of Theodore Roosevelt and seemed to be judging the need to protect Sterling Forest based on his own strongly evident Roosevelt-style environmental convictions. High on Pataki's first-term priority list was a $1.75 billion Clean Water/Clean Air Bond Act he intended to place before New York voters for approval. He signaled that money for Sterling Forest would come from the bond act, if approved, or from other state sources if necessary.

In Washington, the once ridiculed venture to protect this "ragged collection of trees somewhere in the northeast" had taken on such a high profile that it was beginning to attract unpopular "riders" like the "Rangelands Management Act," an attempt to open public lands in western states to more intensive cattle grazing on the assumption that momentum on Sterling Forest in the omnibus parks bill, carrying endorsements by President Clinton and Speaker Gingrich, had become a politically attractive vehicle on which to try to hitch a ride. Fortunately, the rangeland maneuver fell by the wayside, as had the Oklahoma grasslands and Utah Redrocks gambits. (On September 18, 1996, President Clinton established by proclamation the 1.8-million-acre Grand Staircase-Escalante National Monument in southern Utah. In 2017, President Trump ordered reduction in size of this national monument by about 1 million acres.)

Three key sections were driving forward the omnibus package of park-related legislation. One was Sterling Forest. The second was proposed transfer of the Presidio in San Francisco to National Park Service stewardship. The third was a proposed federal-private land exchange in Utah in preparation for the 2002 Winter Olympics.

On July 7, 1996, the House of Representatives approved the $17.5 million Sterling Forest authorization bill. The House also voted favorably on a $9 million appropriation for the acquisition, with the stated intent of adding the remaining $8.5 million in 1997. Congressman Bill Martini must have breathed a sigh of relief. He had timed his election to Congress perfectly, joining the 1994 sweep that put Republicans in majority control of the House of Representatives, but Martini knew that he was in for a tough reelection bid in a district centered on Paterson, New Jersey, where registered Democrats outnumbered Republicans. Delivering on Sterling Forest, for which Martini could claim some credit, was a feather in his cap. Gingrich wanted all of the Class of '94 Republicans reelected. Martini's vulnerability in a swing district was part of the reason the Speaker opened the door for Sterling Forest. Sterling Forest helped Martini, but not enough; he lost in the November election.

On the Senate side, adjournment was in the air, and Senator Bill Bradley had announced his intention to retire. Senators were exercising all sorts of parliamentary privileges, slowing legislation, blocking legislation, speeding legislation, ignoring legislation, and hooking on amendments to the most popular bills. Somewhere in this flux, surfacing from time to time, was the omnibus park package that included Sterling Forest and a long list of other park and public lands projects that would benefit forty-one of the fifty states.

Most threatening to the bill was a provision to extend logging for fifteen years in the Tongass National Forest, Alaska. The Alaska provision prompted a veto threat from the White House. Senator Bradley spent his last days in Congress brokering a two-year logging deal on the Tongass that Senator Frank Murkowski of Alaska found acceptable. The veto threat was removed. During this final endgame, Bradley used Senate rules to "hold" forty-seven bills on his desk that other senators wanted passed to ensure that Sterling Forest was not again maneuvered to the sidelines. In the predawn hours on October 3, 1996, the very last piece of legislation passed by the 104th Congress, supported by the very last vote cast by Senator Bradley, was the omnibus parks bill, including the $17.5 million authorization for Sterling Forest.

The Sterling Forest acquisition purse, once containing moths, had increased to $32.5 million, counting $10 million from New Jersey, $5 million from the Lila Acheson and DeWitt Wallace Fund, and $17.5 million from the federal government. In the November 1996 election, New York voters approved Governor Pataki's Clean Air/Clean Water Bond Act, providing a source for the money he had pledged for Sterling Forest. Governor

Pataki approved $16 million of bond act funds for Sterling Forest, bringing the purse to a substantial $48.5 million. A $6.5 million gap remained to reach the negotiated acquisition price.

The New Jersey–based Victoria Foundation, already on record with a timely contribution to the PIPC in response to the earlier environmental impact statement challenge, responded again, this time with a $1 million contribution. In addition, the PIPC, the New York–New Jersey Trail Conference, and Sterling Forest Resources were regularly receiving contributions from the general public. Among the PIPC donations were $1,034.74 from students and faculty at the E. G. Hewitt High School, Ringwood, New Jersey, and $158.00 in lunch money from students at the Dwight D. Eisenhower Middle School, Wyckoff, New Jersey.

Harvard student Jamie Fitzgerald organized an "Ocean-to-Ocean" cross-country bicycle trek with the intended goal of raising $12,000 in per-mile-pledges for the Sterling Forest purchase. Only days after graduating from Harvard, Jamie, along with his sister, Shannon, classmate Alan Ferency, a Cornell University junior, and Patrick Farley, a neighborhood friend from the village of Palisades, New York, set out from Seattle, Washington. Their destination was the summit of Bear Mountain, 2,875 miles distant. When they arrived two months later, lean and muscled, sunburned, happy, and successful, there was a sense among the welcoming crowd that the Sterling Forest rescue would succeed.

Private contributions, small and large, closed the gap to $5 million. Kim Elliman and Rose Harvey reached out to the Doris Duke Charitable Foundation. The foundation, with assets of $1.25 billion, ranked among the top ten in the nation. Tobacco and hydroelectric heiress Doris Duke, who died in 1993, specified in her will that, among various charitable purposes, she supported protection of wildlife and ecology. Settlement of her will and establishment of the foundation was contentious and time-consuming. Three years after her death, foundation trustees were just finding their way in clarity of purpose and priorities, and no environmental grants had yet been made.

The trustees retained the services of the Conservation Fund's Pat Noonan for a due diligence review of the Sterling Forest funding request. Noonan dispatched a staff colleague to the PIPC New Jersey office to spend most of a day reviewing the details of the acquisition effort with Elliman and me. In December 1997, with the Conservation Fund findings in hand, Joan Spero, president of the Doris Duke Charitable Foundation, announced a $5 million grant for Sterling Forest. In response, the PIPC commissioners agreed that a 1,000-acre portion of the forest would be designated the Doris Duke Nature Sanctuary and that this section of the forest would be permanently closed to hunting. Lou Heimbach had asked, "Where is the money?," and, finally, the answer was in hand.

The devil-in-the-details of the $55 million acquisition fell into the capable hands of New York's assistant attorney general Henry DeCotis. DeCotis sat at an intersection within New York state government through which all land transactions must pass. He supervised a real-property staff of twenty-three attorneys, fourteen title-search specialists, and thirty-two additional support staff. Working with attorney Megan Levine and real-property specialist Steve Lewis representing the New York Office of Parks, Recreation, and Historic Preservation, Jim Economides, a lawyer with the New York Department of Environmental Conservation, and Donald King, chief of the Land Acquisition Field Office for the National Park Service, DeCotis labored straight through the December holidays and into the new year to pull together the pile of real-estate closing documents.

On February 5, 1998, I sat quietly with various lawyers for several hours signing documents in the Sterling Forest Corporation office. The mood was businesslike and subdued, with papers being shuffled from one signatory to the next. With all signatures in place, notarized, and deeds ready to be filed, handshakes ended the session. I walked alone from the building feeling an immense sense of calm, thinking of the years, close calls, and good people that produced this supreme conservation moment. The SFC still owned over 2,000 acres in Sterling Forest, and that was a problem, but the development "shotgun blast" that had so threatened the last, great open space in the New York/New Jersey metropolitan region had narrowed by about 90 percent. Six days later, on February 11, governors Pataki and Whitman formally announced the acquisition at a press conference at the Bear Mountain Inn.

The PIPC was owner of a new 15,805-acre park, "the largest to have been created in the northeastern United States in the last half-century," as reported in the media. The new park promised unending benefit for warblers and wanderers, historians and ecologists, children and grandparents, hikers and poets, frogs and flowers, people of the region, people from afar, wildlife inhabitants of all kinds—promise fulfilled in a gentle, timeless, intact forest ecosystem.

Determined effort continued to wrap up dangling threads, the acreage still owned by the SFC and a few parcels owned by others. In 1999, Rose Harvey of the Trust for Public Land and Kim Elliman of the Open Space Institute, ever capable, succeeded in arranging for the acquisition of 659 acres of Sterling Forest land from New York University. In 2000, a negotiated

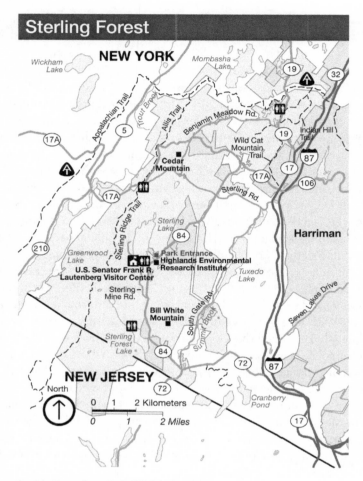

Sterling Forest State Park (PIPC Archives)

acquisition of 1,065 acres for $7.89 million from the SFC was achieved by Harvey and Elliman with financial support from New York, New Jersey, the federal Forest Legacy Program, the Lila Acheson and DeWitt Wallace Fund, the Trust for Public Land, and the PIPC.

In 2003, the Frank R. Lautenberg Visitor Center opened in Sterling Forest. Senator Lautenberg, who had been among the key advocates in the U.S. Congress for federal funding to acquire Sterling Forest, made a personal $1.75 million donation to the PIPC from construction of the visitor center.

In 2006, the "last piece of the puzzle," a 525-acre parcel referred to as the "hole in the doughnut" on which the SFC proposed to construct 100 luxury homes and an eighteen-hole golf course, was purchased for $13.5 million with New York environmental protection funds. The Trust for Public Land handled this transaction with strong support from Governor Pataki. A celebrated new park now stood on the doorstep of the metropolitan house, merged with other splendid PIPC holdings more than a century in the making.

Sterling Forest is a living monument to the conservation ethic. There were many moments in the Sterling Forest saga when just giving up might have been the rational choice. Every person involved was busy with other responsibilities, but there was the forest, and it was in dire need of rescue. The Public-Private Partnership to Save Sterling Forest had a clear goal, even if that goal was framed against seemingly hopeless odds. The PPP forged ahead, improvising, gathering allies, and finding a kind of collective chemistry that overcame those odds. Political and charitable leaders responded in a manner that their predecessors would admire and that their successors should appreciate.

18

⣿

Honor and Electronics

"The General, ever desirous to cherish virtuous ambition in his sol-diers, as well as to foster and encourage every species of Military merit, directs whenever any singularly meritorious action is per-formed, the author of it shall be permitted to wear on his facings, over his left breast, the figure of a heart in purple cloth or silk edged with narrow lace or binding." By general orders, General George Washington.

"By order of the President of the United States, the Purple Heart, established by General George Washington at Newburgh, August 7, 1782, during the War of the Revolution is hereby revived out of respect to his memory and military achievements." By order, Douglas MacArthur, General, Chief of Staff of the Secretary of War.

STANDING IN RANK in 1932 at the historic New Windsor Cantonment (encampment) near Newburgh, New York, were more than 100 World War I Army veterans, many carrying wounds of combat, there to be awarded the Purple Heart Medal for meritorious service. On orders issued by Gen-eral MacArthur in recognition of the bicentennial of the birth of George Washington, these veterans were symbolically connecting over a span of 150 years with Sergeant Elijah Churchill of the 2nd Continental Dragoons, Sergeant William Brown of the 5th Connecticut Line Infantry, and Ser-geant Daniel Bissell, 2nd Connecticut Continental Line Infantry, the first three recipients of the Purple Heart.

Churchill and Brown were summoned in 1782 from the Continental Army encampment at New Windsor to Washington's Headquarters on the Hudson River in nearby Newburgh, New York, to be presented with the small cloth hearts. Bissell would later receive his award at a separate ceremony. There was special meaning in these ceremonies. For the first time, soldiers below the rank of officer were being recognized for military merit. By designing and awarding the Purple Heart to enlisted men, General Washington was signaling his respect for the citizen-soldiers who fought gallantly in the Revolutionary War and now were holding the British in check, awaiting an end to hostilities.

But waiting through months of inaction in an encampment was hard. Soldiers were thinking of home and family, of fields in need of plowing or of crafts and commerce to be pursued. General Washington was thinking of morale, and of holding his restive army of about 7,000 men together in 1782–83 while a treaty to end the conflict was being negotiated in Paris. Downriver, British forces in control of New York City also waited. If negotiations failed, combat most likely would resume.

Standing at attention, twenty-six-year-old Sergeant Elijah Churchill must have been proud to receive the first of the Purple Hearts personally awarded by General Washington. He had distinguished himself for gallantry,

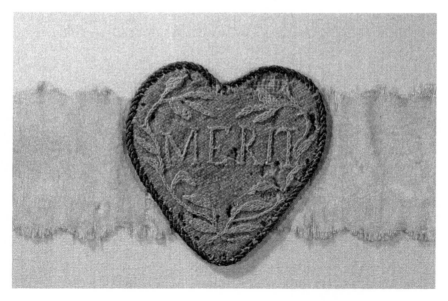

Badge of Military Merit awarded to Sergeant Elijah Churchill by General George Washington (PIPC Archives)

successfully and bravely leading his men in attacks against British strong-holds, confirmed by the scars of combat wounds on his body. Unknown to him at the time, while standing at attention in front of the legendary gen-eral, his cloth heart would be passed down through history, eventually to be donated by his descendants to the Washington's Headquarters Historic Site, where it was cherished and on public display.

After the Revolutionary War, the Purple Heart award had fallen into disuse until it was revived 150 years later by the historically astute General Douglas MacArthur in recognition of General George Washington's con-cept of "meritorious military service." Then, in 1942, President Franklin D. Roosevelt took the medal to new symbolic heights by issuing an executive order affirming that the Purple Heart would be awarded to members of all branches of military service who were wounded, died in combat, or later died of combat wounds. Subsequently, the Purple Heart was made retro-active to include veterans of World War I, providing distinguished recogni-tion for those who truly place their lives on the line for their country. More than 1.8 million Purple Hearts have been so awarded—and counting.

Elijah Churchill had enlisted in the Continental Army in his hometown of Enfield, Connecticut. Residents of the town were aware of this gallant native son and in the 1980s began an effort to establish a Purple Heart museum in Enfield to commemorate all recipients of the award. In 1995, the initiative underway in Enfield caught the attention of Patric Morrison, a resident of Newburgh, New York, who wrote a letter to the editor to his local *Sentinel* newspaper suggesting that a much more appropriate location for a Purple Heart museum would be in the Newburg area. The historic importance of Washington's Headquarters/New Windsor Cantonment is unquestioned. In 1850, the state of New York acquired the Hasbrouck farmhouse and property, commandeered by General Washington as his headquarters in the closing months of the Revolutionary War, to establish the first publicly owned historic site in the nation. When Morrison's letter arrived at the *Sentinel*, the headquarters and cantonment were under the stewardship of the Palisades Interstate Park Commission.

Everett Smith, publisher of the *Sentinel*, was quick to see the logic in Morrison's proposal. He became a strong journalistic advocate for a Purple Heart facility and registry to be located where enlisted men and noncom-missioned officers notably held firm under the command of their respected general until the Treaty of Paris was signed on September 3, 1783.

Morrison's reasoning also found timely favor with the Palisades Interstate Park Commission. The commission's deputy director, Kenneth Krieser, had been awarded a Purple Heart for his service in Vietnam and was a lifetime member of the Military Order of the Purple Heart, a congressionally chartered charitable organization that provides support services to all veterans and their families.

Prompted by Jane Townsend, the PIPC's onsite manager at the New Windsor Cantonment, Krieser and I had been discussing the idea of adding a wing to the cantonment's existing museum to tell a broader story of the Purple Heart. The idea was only visionary at that stage but had gone far enough that a name for the project was chosen, the Purple Heart Hall of Honor. Spurred on by the *Sentinel* article and Jane Townsend's enthusiasm, a worthy race with Enfield, Connecticut, began.

The Enfield, Connecticut, advocates had acquired a fourteen-acre site for the proposed museum and obtained an architectural rendering for an elegant facility estimated to cost $20 million, but, even after years of effort, fundraising had stalled. By comparison, the PIPC had the extraordinary ground at New Windsor so strongly associated with the Purple Heart and the precious two-inch bit of violet cloth shaped in the form of a heart that had been awarded by the hand of General George Washington to Sergeant Elijah Churchill.

For the PIPC, an additional advantage existed in the person of New York state senator William J. Larkin Jr., a decorated military veteran who represented the political district in which the New Windsor Cantonment is located. Larkin had enlisted in the army in 1944 at age sixteen, served in World War II and the Korean War, and advanced from the rank of private to lieutenant colonel before retiring from military service. He was known for his support for veterans and as a fiscal conservative holding tight to New York State purse strings.

Krieser and I invited Senator Larkin to the PIPC administrative office at Bear Mountain to discuss the New Windsor project. The senator was noncommittal but listened carefully. Strength was brought to the New Windsor initiative when Senator Larkin joined with Everett Smith of the *Sentinel*, Lieutenant General (ret.) James D. Hughes, and veterans advocate Joe Farina to form the Genesis Group with the specific purpose of achieving the Hall of Honor goal.

For both projects, New Windsor and Enfield, funding was the challenge, and the Enfield attempt was not going well. Months after the Bear Mountain meeting an article appeared in the *Hartford Courant* entitled "Purple Heart Museum Has Two Suitors." Reporter Anne Hamilton wrote, "Retired General, Colin Powell, former Chairman of the Joint Chiefs of Staff, and General Barry McCaffery, who is President Clinton's drug czar, have endorsed the proposal. And the Military Order of the Purple Heart, a group of 30,000 recipients of the medal awarded to those wounded in battle, has also backed the efforts in New Windsor."

Endorsement by the Military Order of the Purple Heart resulted from attendance at the organization's annual conference by Ken Krieser. At the gathering in Norfolk, Virginia, Krieser presented the PIPC plan. In addition to winning support for the plan, a significant adjustment to the name of the facility was achieved. Exercising their congressional mandate, the Military Order members affirmed the name *National* Purple Heart Hall of Honor.

The *Courant* article continued, "New Windsor's rapid preparations pose big problems for the group of dedicated men and women in Enfield, Conn., that has struggled more than 10 years to fulfill their vision of a $20 million museum dedicated to the Purple Heart. They've had many barriers to contend with, including a slow trickle of donations and the anger of a group of disillusioned Purple Heart recipients who last week asked the museum officers to resign and demanded half the seats on the board of directors. So far, all the organizers have to show for their decade of work is a set of architect's plans and a flagpole that stands in a field off I-91 in Enfield. Even the flagpole has caused problems because the Purple Heart veterans are angry that the flag flies at night without illumination. Meanwhile, just a year after being spurred by a letter to the editor of a local newspaper that suggested New Windsor was a more appropriate home for the Purple Heart museum, plans are moving forward apace."

I was quoted in the article: "This is the first award given to enlisted men in the Continental Army. It is very appropriate to have the hall of honor located where the soldiers were encamped. New Windsor is a destination. Assuming we succeed in letting people far and wide know about the Purple Heart Hall of Honor, many people who go to West Point will deliberately make the choice to extend their visit." (The United States Military Academy at West Point is twelve miles from New Windsor.)

This was among my last media comments about the PIPC, its purposes, expectations, and priorities. After four decades of public service, including ten years with the PIPC, the time had come in December 1999 for me to retire. The executive director baton was gladly passed to Carol Ash, who brought strong environmental credentials gained with the Nature Conservancy, the New York Port Authority, and the New York Department of Environmental Conservation. Any management transition at the PIPC brings with it a long list of active projects, and always the need to secure funding. For Carol Ash the list included the nascent National Purple Heart Hall of Honor project, securing the final pieces of the Sterling Forest puzzle, a host of maintenance needs including resurfacing of the Palisades Parkway, and bringing the PIPC into the world of high-tech publications, online services for park visitors and patrons, and the typical demands of overseeing a large and diverse staff.

The PIPC functions seasonally, depending on a core staff of permanent employees during the cold months of the year and then morphing into an employment center for hundreds of college students and other part-time workers who greet the public at nature centers, beaches, swimming pools, golf courses, on hiking trails, in the campgrounds and picnic sites, group camps, museums, and at cultural sites. Safety for thousands of daily visitors is always a concern. The chiefs of two separate police departments reported to Carol Ash, one in New Jersey, one in New York. In New Jersey, the officers are responsible for patrol of the parkway as well as the adjacent parklands. In New York, the State Highway Patrol handles the parkway while PIPC officers provide protection in an intricate web of interior road systems and facilities that accommodate millions of visitors.

Maintenance of such a complex and widespread park and historic site structure is key to overall success. Buildings, utility systems, machinery, trails and road systems, groomed beaches and fairways, interpretive exhibits, refuse disposal, signage, electronics, fire prevention, and restoration of historic structures and artifacts are essential to quality public services.

The transition was smooth, the baton passed without a misstep. Ash benefited from the deep experience and dedication of the PIPC staff, those professionals who meet the day-to-day challenges of maintaining such an intricate park and historic site system, assuring that maintenance needs are dealt with, security is provided, educational and recreational opportunities are widely accessible, natural and historic resources are carefully preserved, and, above all, that millions of visitors can enjoy the landscape and cultural treasures protected for them.

One of the projects that would quickly capture Ash's attention was buried under foliage and all but forgotten. The twin forts at Bear Mountain where militiamen fought fiercely against British, Hessian, and Loyalist troops in October 1777 had receded into afterthought. Just across the road from the Bear Mountain Inn, Fort Clinton had a subdued interpretive profile as part of the Trailside Museum, but the fourteen-acre Fort Montgomery site, less than a mile away on the other side of the Popolopen Creek gorge, had disappeared under a hardwood forest canopy. Led by archeologist Jack Mead, the PIPC had conducted extensive investigation of the site in the 1970s, but otherwise the old fort remained concealed by Mother Nature. Only a few people, primarily keen students of the Revolutionary

War or less worthy adventurers just looking for artifacts, ventured through the tangled underbrush to the site.

Residents of the small hamlet of Fort Montgomery were aware of the historic ground in their community and had formed the Fort Montgomery Battle Site Association in 1997 for the purpose of working with the PIPC and the state of New York to bring the fort out of the wilderness, so to speak. It was here that the famous chain was floated across the Hudson River narrows and where, on demand by the British to surrender, Lieutenant Colonel William S. Livingston of the Continental Army, speaking for his vastly outnumbered troops, demanded that the British, instead, surrender.

Ken Krieser lived in the hamlet of Fort Montgomery and had many friends there. He and Carol Ash worked closely with the local Battle Site Association to advance the need for restoration of the old fort. Ash found a willing listener in the person of Governor George Pataki. The governor approved a $1 million budget allocation in 2001 to allow for site restoration. The next year, on the 225th anniversary of the battle, Governor Pataki dedicated the artfully restored site and a pedestrian suspension bridge spanning the creek gorge that, together, provide the public with a superlative learning opportunity. In addition to the history lesson, visitors are rewarded with a captivating view of river scenery, looking right down the Hudson River narrows. With a bit of effort, they can imagine British frigates sailing right at them.

Carol Ash also assured that momentum developed for the Purple Heart project would continue. Ken Krieser was right there to assist. Jane Townsend, the onsite manager at the New Windsor Cantonment, coordinated plans for expansion of the museum and design of exhibits. The Newburgh-based William and Elaine Kaplan Family Private Foundation signaled its willingness to help fund the project. And there was the Genesis Group, including Senator Larkin, steadfast from the earliest moments of the Hall of Honor idea.

In the early stages of the New Windsor project, an inquiry was made by the PIPC to the Department of Defense to obtain a roster of Purple Heart recipients. The response was underwhelming; no official roster existed. Clearly, part of the task would be to begin the assembly of a registry of the estimated 1.8 million recipients by inviting active and inactive military personnel, their families, or associates to provide information on Purple Heart awards. This became and continues to be an important project goal.

Superb historic significance, strong leadership from the New York governor, firm involvement by credentialed private citizens and charitable donors, endorsement by the Military Order of the Purple Heart, and the in-place management capability of the PIPC were a persuasive combination. As the slow machinery of securing state funding ground forward in New York, the Enfield initiative faded. Time passed, but the combination stayed strong.

On Veteran's Day in November 2006, Governor Pataki opened the ceremony to dedicate the National Purple Heart Hall of Honor. The idea of adding just a small wing to the existing New Windsor Cantonment museum had evolved into a stately, full-sized structure. Assembled at the New Windsor Cantonment were men and women representing all branches of military service and all regions of the nation. Families and friends were in attendance. Generations were represented by gray hair and creaky bones, the bloom of youth, and the ease of middle age. Blackhawk helicopters flew overhead, a fife and drum corps marched, and the United States Military Academy Band performed.

Senator Bill Larkin spoke for all: "This is an historic moment, a truly important event in our nation's history. After many years of planning, we now have a very special place to honor the lives of every single person throughout the nation's history who has ever received the Purple Heart Medal. Those who have received this award know exactly the price they paid to preserve our nation's freedom, and so do their loved ones. The [National] Purple Heart Hall of Honor will honor those who suffered greatly; those who still suffer with the scars of battle; and those we lost. It will be a place where family members can feel the pride and greatness that we all feel for the brave individuals who so valiantly defended our country, and a place where we can all reflect on the meaning of their sacrifice. The Hall of Honor will preserve the stories of these ordinary men and women and the extraordinary things they have done in the name of freedom."

State of New York funds of about $4 million were complemented by donations from the Kaplan Foundation and the Military Order of the Purple Heart. A registry of Purple Heart recipients already included more than 12,000 names and would increase rapidly. In the years since the dedication, thousands of visitors have seen in one of the display cases a small cloth heart with the stitched word, "Merit," the Purple Heart awarded to Sergeant Churchill in 1782.

For Carol Ash, Ken Krieser, Jane Townsend, and the PIPC staff the day of the dedication was one of quiet pride, a confirmation of confidence in public stewardship stretching back for more than a century, and, in the same moment, a salute to historic treasurers recognized and preserved for future generations. (In November 2019, New York governor Andrew Cuomo announced a $17 million expansion of the National Purple Heart Hall of Honor.)

Preservation on another scale and for a different reason was glimmering at the Awosting Reserve, a 2,500-acre property abutting Minnewaska State Park. John Atwater Bradley had purchased the property in the 1950s and was its dedicated guardian. He allowed limited public access to his private land holding, but always with a careful eye to assure no abuse of natural features. Though of smaller scale, the Awosting Reserve was otherwise an identical twin to Minnewaska, preserving waterfalls and cliff faces, a deep forest, abundant wildlife, and miles of trails and carriage paths. Bradley lived there part-time in a small home tucked artfully next to a tumbling mountain brook.

He and Barnabas McHenry roamed the same social circuit in New York City, meeting each other at various gatherings and sharing deep admiration for the charms of the Hudson River Valley. On more than one occasion, McHenry visited the Awosting Reserve to gently suggest to Bradley that such an outstanding natural property might one day be merged with Minnewaska State Park. In these visits there was a hint of like-minded camaraderie. Bradley knew well the years'-long record of McHenry's many conservation achievements, including preservation of Boscobel, an early nineteenth-century neoclassical mansion on the shoreline of the Hudson River near Garrison, New York, and of his dedicated service with the PIPC, the Open Space Institute, and the Hudson River Valley Greenway. But Bradley was suspicious of public land stewards and those working for non-profit conservation organizations who he claimed were too casual about "spending other people's money." He was strongly in the camp of private land stewards, so much so that he had a local reputation among neighbors for being "lord of the manor" and a "fierce guardian," as so reported by Alan Snel for the *Times-Herald Record*.

After almost a half-century of ownership, the financial weight of main-taining such a large property caused Bradley to turn, cautiously, to a devel-opment partner who proposed construction of 349 luxury homes and a 296-acre golf course on the property. This was a startling departure from all the years of low-impact use Bradley had made of his property. On hear-ing the news, McHenry and PIPC executive director Carol Ash soon were again visiting the Awosting Reserve, hoping to persuade Bradley to choose a course that would add his land to Minnewaska State Park rather than burden it with so much speculative development.

McHenry and Ash were backed by Kim Elliman, the executive director of the nonprofit Open Space Institute (OSI) who had brought such significant negotiating skill to the rescue of Sterling Forest. Beginning in the 1970s, the OSI had focused strong effort on preservation of the forty-mile-long Shawangunk Ridge, a northern extension of the Appalachian Mountains. Abutting Bradley's property to the south was the OSI's 5,000-acre Sam's Point holding (later to be merged with the state park). Elliman and his colleagues, including McHenry, were determined to do all they could to shepherd Bradley's precious and as yet untrampled real estate into the public preservation camp.

Their chance came in 2005. Bradley and his development partners had fallen into legal dispute, so much so that a judge ordered sale of the property at auction to settle financial claims and counterclaims. In November 2005, the property came under the hammer of an auctioneer. One of the bidders was famed actor Robert De Niro, a friend and neighbor of Bradley. De Niro was representative of a small group of wealthy second-home owners who opted for the rugged outdoor charm of the Shawangunk environs rather than more trendy places like the Hamptons, Newport, or the Maine Coast. Bradley, himself, entered a bid, trying in a last-ditch effort to reclaim rights to cherished property that had drifted from his control.

The winning bid was $17 million, submitted by the Open Space Institute and the Trust for Public Land. The state of New York would reimburse OSI and TPL, and an immensely important addition to Minnewaska State Park was achieved. Governor George Pataki, in his final term in office, considered acquisition of the Awosting Reserve to be a strong plus near the top of his long list of environmental and historic preservation accomplishments.

Carol Ash could breathe a sigh of relief. Protecting the Awosting Reserve had been a close call. It was achieved on her watch, thanks to the type of public/private partnership that is a hallmark of the PIPC. In this instance, the partnership team included McHenry and Ash for the PIPC, the OSI's ever-intrepid Kim Elliman, equally resolute Rose Harvey for the Trust for Public Land, and Governor Pataki and park commissioner Bernadette Castro for the state of New York. Ash would soon take this success to Albany. In 2006, with a change in political administrations, Ash was selected to succeed Bernadette Castro as commissioner of the New York Office of Parks, Recreation, and Historic Preservation. She would hand the PIPC portfolio to her successor, now including a new, nationally prominent museum at New Windsor, a rehabilitated Fort Montgomery, and exceedingly strategic land acquisitions at Sterling Forest and Minnewaska.

Jim Hall, a veteran of New Jersey state government and, for ten years, superintendent of the New Jersey section of the PIPC, was chosen to take Ash's place. He was taking on forward-looking management responsibilities but would be unexpectedly confronted with a severe challenge that would take the PIPC almost back to square one.

On June 30, 2011, Hall sent a letter to the secretary of the Englewood Cliffs Board of Adjustment expressing concern about the proposed construction of a building adjacent to PIPC parklands in New Jersey. This was not just any building. On a twenty-seven-acre site only feet from PIPC holdings and the Palisades Parkway, the giant Korean firm LG Electronics was proposing the construction of a 490,000-square-foot North American corporate headquarters. LG had contacted officials in the borough of Englewood Cliffs asking for a variance to zoning restrictions that limited the height of buildings to thirty-five feet. The building proposed by LG would soar upward for 143 feet, providing picture-window views of the Hudson River and the nearby New York City skyline. The twenty-seven-acre site was just the right place to present a bold, exclusive architectural presence standing tall on the Palisades that could be seen for miles. The spectacular building would be a daily reminder for millions of urban residents of the corporation's motto, "Living Good."

In his letter, Hall pointed out that "the Palisades Interstate Parkway and associated park land are listed on the National Register of Historic Places; the Natural Landmarks and as a state Scenic Byway." He urged that a "visual impact analysis" be required of LG for consideration by borough officials before a decision was made on the requested zoning variance.

L. G. Electronics, North American corporate headquarters, Palisades (Artist's Conception, Scenic Hudson)

The officials might also have been reminded of the moral commitment made by the PIPC to John D. Rockefeller Jr. in the 1930s when Rockefeller donated about thirteen miles of property "to preserve the land lying along the top of the Palisades from any use inconsistent with your ownership and protection of the Palisades themselves." According to a *New York Times* article at the time, this commitment included protection against "disfiguring structures."

In a follow-up letter in October 2011, Hall seemed to nudge open the door for PIPC acquiescence to the project by stating, "We believe the visual impacts resulting from the height of the project can potentially be screened and/or mitigated to a significant extent." Landscape screening might soften the visual impact of the base of a very large building, but the LG corporate office building would soar high above ground level, bursting into the sky in solitary splendor with no structural competitors nearby to complicate the privilege. Especially when the Palisades scene was viewed from the New York side of the Hudson River, the LG building would stand elegantly alone in a bucolic landscape or, as many would quickly conclude, stick out like a sore thumb. The risk, too, was that a claim of corporate visual privilege would not stand long in a commercially competitive world, especially in the high-stakes New York/New Jersey marketplace. If LG could blot the Palisades landscape, others might soon follow.

For the borough of Englewood Cliffs, the LG proposal was a very big deal; a prestigious $300 million corporate project; the promise of at least 1,000 construction jobs; a big boost in property tax revenue; the ripple effect that LG employees would bring to local businesses and the housing market. A thirty-five-foot height ordinance had been in place for decades to protect the natural scene, but the zoning board saw an economic opportunity too good to pass by.

At a series of public hearings hosted by the board, a few dissenting voices were heard. Kevin Tremble, a member of the PIPC's Citizen's Advisory Council, asked of LG planner Joseph Burgess whether a multistory building looming in full view above the Palisades tree line was consistent "with a state and regional interstate resource that has been protected for 100 years." Burgess replied that "planning is a balance of competing interests." When Tremble referred to an artist rendering of what the proposed building would look like when observed from the New York side of the Hudson River, Burgess responded that the visual impact would not be "significant" in any way.

Attorney Daniel Chazin, also a member of the PIPC Advisory Council and a member of the New York/New Jersey Trail Conference, pointed out that the Palisades were protected, in part, to "preserve the view of the cliffs from the river and from New York."

Jim Hall said at one of the hearings that the PIPC had "concerns about the height of the building with its visual impact on the park and the scenic corridor, as well as its national register listing." He was referring to approvals by the U.S. Congress of National Historic Landmark designation for the Palisades in 1965 and of its designation in 1983 as a National Natural Landmark.

Soon after the hearings, the Englewood Cliffs Zoning Board approved the zoning variance requested by LG Electronics. Despite the hearing record, the board referenced twenty-one factual findings, including how the building would be screened from nearby properties in the borough and how natural landscaping would benefit portions of the twenty-seven-acre parcel. There was no mention in the findings of the Palisades or of potential visual impact beyond the borough limits.

Not everyone in the local community agreed with the zoning board findings. Carol Jacoby and Marcia Davis, residents of Englewood Cliffs, took the bold step of legally challenging the variance. They contended that the zoning variance would have detrimental visual impact well beyond the borough limits and that this expansive impact should have been considered by the zoning board. But LG had the green light, and steps were put in motion to begin construction. In response to the legal challenge from Jacoby and Davis, LG spokesman John I. Taylor said, "We understand the position of a handful of folks who are opposed to this, but the project enjoys significant support among those in New Jersey who understand the economic impact." In August 2013, Judge Alexander Carver III upheld the decision of the zoning board. LG began clearing the construction site.

Environmental beauty is held in high favor by a very large majority of the nation's residents; it's usually a quiet value expressed by personal outings to parklands, forest preserves, ocean shorelines, and wildlife sanctuaries, or simply by appreciating the view. Taylor's calculation that only a "handful of folks" were concerned about a construction invasion of the Palisades scene was far off the mark.

One of those "folks" was Larry Rockefeller. He had served as a PIPC commissioner for twenty-seven years, from 1979, when he took a leadership role in acquisition of key properties at Minnewaska State Park, to 2006 when the final pieces of the Sterling Forest puzzle largely were put in place. Rockefeller is very aware of the legacy left by his great-grandfather,

grandfather, and father to protect natural scenery in the Hudson River Valley. He is heir to that family tradition, but he also has an agenda of his own. He is a graduate of Columbia Law School, worked as a lawyer, and now is a trustee of the Natural Resources Defense Council, a cofounder of the nationally active League of Conservation Voters, and president of the American Conservation Association. In his early days as an environmental lawyer, Larry Rockefeller participated in the Alaska Lands initiative that, in 1980, with the stroke of a pen by President Jimmy Carter, doubled the size of the National Park Service.

A signal was sent to Larry Rockefeller by the PIPC staff that the New Jersey Palisades was in jeopardy. The PIPC seemingly had no authority to overturn a zoning decision beyond its jurisdiction, despite the pending negative impact on its property, and had taken no action to challenge the LG project beyond the cautionary statements submitted by Jim Hall. Larry Rockefeller was not so constrained. Carol Jacoby and Marcia Davis must have been amazed and pleasantly surprised when the New Jersey State Federation of Women's Clubs, Scenic Hudson, the New York/New Jersey Trail Conference, the New Jersey Conservation Foundation, the Natural Resources Defense Council, and Margo Moss and Jakob Franke intervened in the legal action against LG. The "Protect the Palisades" campaign was underway.

Soon, four former New Jersey governors (two Republicans and two Democrats) expressed their concern for protection of the scenic beauty of the Palisades. They were joined by six mayors of neighboring Englewood Cliffs communities and by the editorial boards of the *Bergen Record*, *Newark Star-Ledger*, *New York Daily News*, and *New York Times*. New York governor Andrew Cuomo, represented by none other than Rose Harvey, who had succeeded Carol Ash as commissioner of the New York Office of Parks, Recreation, and Historic Preservation, sent a letter to LG asking for reconsideration of the project. New York senator Charles Schumer did the same. Kristina Heister, chief of Natural Resources for the National Park Service, said that the LG project "threatens the integrity of the scene in a startling and major way." Three former PIPC executive directors, Nash Castro, Carol Ash, and I, added our voices in support of opposition to the LG plan.

A billboard on the West Side Highway in New York City showed graphically how the LG building would impact the Palisades. Larry Rockefeller reached out personally to LG executives, expressing the collective message that no one was opposed to a beautiful corporate headquarters in Englewood Cliffs, just please spread the building out among trees rather than push it into the sky.

The initial LG reaction was predictable: too much had been invested in building design and site preparation; construction jobs were at stake; the project was underway; if LG was blocked by litigation it might have to move elsewhere. But there was a glimmer of a silver lining. LG has a strong environmental reputation, specializing in high-quality, energy-efficient products. The chief operating officer of the corporation, Seog-Won Park, was well known in Korea for his interest in wild birds and natural beauty. This interest was reflected in the building design proposed for Englewood Cliffs. The building was expected to meet the highest worldwide standards of Leadership in Energy/Environmental Design (LEED).

In February 2014, the PIPC commissioners gathered at the Fort Lee Historic Site in New Jersey for a regular meeting. Usually, such meetings attract only a few members of the public, those who are particularly interested in all the minute details of managing scores of parks and historic sites. This meeting, though, would be different. About 400 members of the public were in attendance, there to urge the PIPC to unequivocally state its position on the LG project. They were not disappointed.

By unanimous vote, commissioners Philip White, David Mortimer, Barnabas McHenry, Sam Pryor, Keith Cornell, and Lloyd Tulp resolved that "the Palisades Interstate Park Commission finds that the height of the proposed LG USA Headquarters building above the tree line is not in accordance with our stewardship mission and the public trust to preserve the scenic beauty of the Palisades and would create a precedent inconsistent with our mission and again urge the management of LG USA, Inc., to lower the proposed height of the building to below the tree line in order to preserve the scenic beauty of the Palisades."

To assure no misunderstanding, the commissioners also requested LG USA "to refrain from implying, in correspondence or its website, that the Palisades Interstate Park Commission does not have concerns with the proposed 143-foot building height."

Larry Rockefeller and his colleagues continued with what he referred to as "quiet time" discussions with LG representatives. And because of public pressure, the zoning ground started to shift in Englewood Cliffs. In August 2014, the town council restored the general thirty-five-foot building height restriction, but the problem was that the zoning variance granted to LG remained in place. In that same year, and because of the LG proposal, the nonprofit World Monuments Fund included the Palisades on its annual list of endangered cultural sites.

Months of on-again, off-again "quiet time" discussion led to a cordial respect among the contending parties. LG concluded that it did not want to invest millions in a structure that might stand conspicuously as a symbol of environmental insensitivity, and the Rockefeller group accepted that a redesigned structure might still peek a bit above the Palisades tree line.

In the meantime, the legal wheels continued in motion. In October 2015, an appeals court ruling overturned the lower court finding that the zoning variance had been legitimate. In a reversal of the lower court decision, the appeals court confirmed that the proposed multistory corporate headquarters would "dramatically affect the view of the historic Palisades cliffs," which the court described as a "national treasure," and that impact on surrounding neighborhoods "means all reasonable visual vantage points."

In February 2017, William Cho, chief operating officer of LG Electronics North America, stood near Larry Rockefeller. Along with twelve other people, they were smiling for the photographer, fancy shovels in hand, ground broken. Political and labor representatives were present. So, too, were representatives of the New Jersey State Federation of Women's Clubs, Scenic Hudson, the New York/New Jersey Trail Conference, and the New Jersey Conservation Foundation.

Larry Rockefeller said, "This truly is a win-win resolution. I would like to publicly commend LG for its willingness to listen and work with us in seeking a mutually agreeable solution. The new design, which results in an outstanding new headquarters for LG, will help preserve the Palisades as a treasured natural landmark."

A precedent was set, not of disruption of nature's handiwork, but of a cooperative effort to find that win-win resolution. Future building initiatives in towns adjacent to the New Jersey Palisades will feel the influence. The LG corporate building is elegant, a LEED-Platinum structure sixty-nine feet in height, spread horizontally through the greenery of the twenty-seven-acre site. More than 1,000 employees work there, benefitting from the beauty of their surroundings. Wetlands on the site are preserved, 1,500 additional trees have been planted, wild birds are abundant. From the New York side of the Hudson River, only a person with a powerful telescope might see even a hint of the building hidden in such pleasant green space.

When that "slip of an Irish girl," Alice Haggerty, pushed a plunger in 1898 to dynamite the Indian Head, "one of the most widely known and splendid pieces of scenery in North America," she symbolically set in motion an experiment in life's quality in the Hudson River Valley that informs twenty-first-century stewards of the land.

There is a sense of optimism in the fact that representatives of the New Jersey State Federation of Women's Clubs helped celebrate the LG compromise more than a century after their predecessors sounded the alarm to save the Palisades. The lesson is clear: nurture our home planet and be continually ready to defend its health.

The PIPC now is responsible for stewardship of thousands of acres of parkland and historic treasures that honor the past, invite current generations to cherish and enjoy life, and stand as symbols of respect for those who will inherit the products of our civilization. To visit, soak in the views, share space with wild flora and fauna, explore, discover, and learn, gather lifetime memories, hike the trails or just ease along, the people's park beckons.

APPENDIXES

Appendix A

Palisades Interstate Park Commission Parks and Historic Sites

Bear Mountain State Park, N.Y.
Blauvelt State Park, N.Y.
Bristol Beach State Park, N.Y.
Fort Lee Historic Park, N.J.
Goosepond Mountain State Park, N.Y.
Greenbrook Nature Sanctuary, N.J. (nonprofit managed)
Harriman State Park, N.Y.
Haverstraw Beach State Park, N.Y.
Highland Lakes State Park, N.Y.
High Tor State Park, N.Y.
Hook Mountain State Park, N.Y.
Knox's Headquarters State Historic Site, N.Y.
Lake Superior State Park, N.Y.
Minnewaska State Park, N.Y.
National Purple Heart Hall of Honor, N.Y.
New Windsor State Historic Site, N.Y.
Nyack Beach State Park, N.Y.
Palisades Park, N.J.
Palisades Interstate Parkway, N.J./N.Y.
Rockland Lake State Park, N.Y.
Schunnemunk Mountain State Park, N.Y.

Senate House State Historic Site, N.Y.
Stony Point Battlefield Site, N.Y.
Sterling Forest State Park, N.Y.
Tallman Mountain State Park, N.Y.
Washington's Headquarters State Historic Site, N.Y.

Appendix B

The Public-Private Partnership to Save Sterling Forest

Bob Binnewies—Palisades Interstate Park Commission
Malcolm Borg—Palisades Interstate Park Commission
Nash Castro—Palisades Interstate Park Commission (retired)
Charles Clausen—Natural Resources Defense Council
Martin Cornell—Palisades Interstate Park Commission
Tim Dillingham—New Jersey Sierra Club
JoAnn Dolan—New York–New Jersey Trail Conference
Kim Elliman—Open Space Institute
Ella Filippone—Passaic River Coalition
Wilma Frey—New Jersey Conservation Fund
John Gebhards—Sterling Forest Resources
Rose Harvey—Trust for Public Land
John Humbach—Sterling Forest Coalition
Andrew Lawrence—New York Sierra Club
Barnabas McHenry—Palisades Interstate Park Commission and Hudson
    River Valley Greenway Council
Jennifer Melville—Appalachian Mountain Club
Olivia Millard—The Nature Conservancy
David Miller—National Audubon Society
Louis V. Mills—Orange County Land Trust
John Myers—New York–New Jersey Trail Conference
Bill Neil—New Jersey Audubon Society
Dean Noll—North Jersey District Water Supply Commission

Jerry Notte—North Jersey District Water Supply Commission
Rob Pirani—Regional Plan Association
Samuel F. Pryor III—Appalachian Mountain Club and Palisades Interstate Park Commission
Larry Rockefeller—Natural Resources Defense Council and Palisades Interstate Park Commission
Steven Rosenberg—Scenic Hudson
David Sampson—Hudson River Valley Greenway Council
Klara Sauer—Scenic Hudson
David Startzel—Appalachian Trail Conference
Jim Tripp—Environmental Defense Fund
Richard White-Smith—New York Parks & Conservation Association
Neil Woodworth—Adirondack Mountain Club
Neil Zimmerman—New York–New Jersey Trail Conference

ACKNOWLEDGMENTS

MIDGE BINNEWIES VOLUNTEERED for two years to research the story of the Palisades Interstate Park Commission. She was joined in this task by commission historian Sue Smith. Together, they rediscovered thousands of pages of long-forgotten documents and news articles, many of which were found in dank, mice-infested storage.

Commissioners Barnabas McHenry and Malcolm Borg assured that the book project would be launched and gave many hours of their time, experience, and wit to keep the writing wheels in motion.

Palisades Interstate Park Commission staff members Ken Krieser, Mary Thomas, Jack Focht, and Kathryn Brown provided timely counsel. Their colleague, Elizabeth van Houten, was responsive to the need for administrative detective work. For this second edition, Sue Smith offered important advice on graphics and title. Matt Shook, Director of Development and Special Projects for the commission, has been key in selecting images that illustrate this story. Attorney Daniel Chazin volunteered to read carefully for accuracy. Aldene Fredenburg provided superb copy-editing services.

Saverio Procario agreed that *Palisades* should be published as the first offering in the Fordham University Press series on the Hudson Valley. Fredric Nachbaur, the current Director of Fordham University Press, found that the message should resonate and gave the signal to proceed with this second edition.

I salute these associates and all who nourish quality of life.

SOURCES

NOTE: EITHER COPIES or originals of all sources listed are located in the Palisades Interstate Park archives.

The author thanks the Rockefeller Historical Archives, located in Tarrytown, New York, for granting permission to copy and use much relevant information.

Chapter 1: Boss Blaster

"Indian Head Destroyed." *New York Times* (March 5, 1898).

Mack, Arthur C. *The Palisades of the Hudson*. 1909; reissued by Walking News, 1982.

"Palisades Interstate Park 1900–1929: A History of Its Origin and Development." PIPC: 1929.

Chapter 2: The Commission

Books

Bradley, Stanley W. *Crossroads of the Hudson: The Story of Alpine, New Jersey*. Alpine Bicentennial Committee.

Ford, John. *The Life and Public Services: Andrew Haskell Green*. DD, Page, 1913.

Garraty, John A. *Right-Hand Man: The Life of George W. Perkins*. Harper and Bros., 1957.

Howat, John K. *Hudson River and Its Painters*. American Legacy Press, 1983.

Humphrey, J. A. *Englewood: Its Annals and Reminiscences*. J. S. Ogilvie, 1899.

Lattimer, John K., M.D., Sc.D. *This Was Early Englewood*. Englewood Historical Society, date unknown.

O'Brien, Raymond J. *American Sublime: Landscape and Scenery of the Lower Hudson Valley*. New York: Columbia University Press, 1981.

Rybczynski, Witold. *A Clearing in the Distance*. New York: Scribner, 1999.

Serrao, John. *The Wild Palisades of the Hudson*. Plind, 1986.

Pamphlets and Periodicals

"A Century of Service." Englewood Woman's Club, 1995.
"A Century of Challenge." New Jersey State Federation of Women's Clubs, 1995.
"The Palisades Interstate Park 1900–1929: A History of Its Origin and Development." PIPC, 1929.
"Preserving the Palisades." *American Monthly Review of Reviews* (date unknown).
Brown, Edward F. "Perkins: Park Builder." PIPC Archives, 1919.
Hopkins, Franklin W. "Preservation of the Palisades." *American Scenic and Historic Preservation Society,* vol. 2, nos. 3 and 4 (December 1930).
Martin, Neil S. "Saving the Palisades." *Gannett Westchester Newspaper Sunday Magazine* (January 7, 1979).
Strouse, Jean. "Annals of Finance: The Brilliant Bailout." *New Yorker* (November 23, 1998).

Newspapers

"To Save the Palisades." *New York Times* (November 25, 1894).
"A Light Legislative Day." *New York Times* (February 15, 1895).
"For the Protection of the Palisades." *New York Times* (July 13, 1895).
"The Palisades of the Hudson." *New York Times* (July 29, 1895).
"To Save the Palisades." *New York Times* (August 27, 1895).
"Save the Palisades from Ruin." *New York Times* (September 29, 1895).
"Park, Not Military Post." *New York Times* (October 7, 1895).
"Preserve the Palisades." *New York Times* (October 8, 1895).
"New Plan for Palisades." *New York Times* (October 9, 1895).
"To Save the Palisades." *New York Times* (September 23, 1897).
"Question of the Palisades." *New York Times* (October 29, 1897).
"To Save the Palisades: Recommendations to Include Them in a Public Park." *New York Times* (January 18, 1900).
"For a Palisades Park." *New York Times* (January 20, 1900).
"Outlook for the Palisades." *New York Times* (January 14, 1900).
"And 'Fishermen's Village' of Undercliff." *Sunday Star Ledger* (March 31, 1963).

Correspondence

Files: Greenbrook Nature Sanctuary, Palisades Interstate Park.
From William Welch, March 13, 1929.
"35th Anniversary of PIPC." Raymond Torrey press release, June 12, 1935. PIPC Archives.

Chapter 3: Upriver

Books

Fosdick, Raymond B. *John D. Rockefeller, Jr.: A Portrait.* Harper and Bros., 1956.
Muir, John. *Edward Henry Harriman.* Doubleday, Page, 1912.

Reports

"Palisades Interstate Park Commission Financial History 1900–1927." PIPC.
"6th Annual Report." American Scenic and Historic Preservation Society, 1901.

Pamphlets and Periodicals

Journal: *American Scenic and Historic Preservation Society*, vol. 2, nos. 3 and 4 (December 1930).
"The Preservation of the Highlands of the Hudson First Publicly Advocated by Edward Lasell Partridge, M.D." *The Outlook* (November 1907).

Correspondence

PIPC Minutes and Correspondence, 1900–1907.
John D. Rockefeller to Governor O'Dell, March 18, 31, 1902.
George W. Perkins to Governor O'Dell, March 31, 1902.
James P. McQuade to John D. Rockefeller Jr., March 14, 21, 1902.
Starr J. Murphy to Timothy L. Woodruff, March 10, 1906.
Starr J. Murphy to John D. Rockefeller Jr., March 22, 1906.
From Mrs. Barrie Tait Collins, January 6, 1998.

Newspapers

"Palisades League Formed." *New York Times* (May 4, 1900).
"Palisades Plans in Danger." *New York Times* (January 27, 1901).
"Palisades Bill Signed." *New York Times* (March 23, 1901).
"The Palisades Park." *New York Times* (April 26, 1901).
"White in a Dilemma." *Observer* (May 5, 1901).
"Was Mr. Morgan's Gift." *New York Times* (May 15, 1901).
"Palisades Interstate Park: A Landscape Engineer Employed to Study and Preserve the Rocks." *New York Times* (October 23, 1901).
"Abram S. Hewitt Dead." *New York Times* (January 19, 1903).
"The Majestic Park on the Palisades." *Nyack Evening Star* (August 1, 1903).
"To Save the Palisades." *New York Times* (January 14, 1906).
"William B. Dana." *New York Times* (October 11, 1910).
"Dr. E. L. Partridge Dies at 77 Years." *New York Times* (May 8, 1930).
"Abram DeRonde Dies in South, 88." *New York Times* (February 24, 1937).
"J. DuPratt White, Former Park Head Dead." *Journal-News* (July 14, 1939).
"Governor in Tribute to White and Stagg." *New York Times* (July 1, 1939).

Chapter 4: Harriman

Books

Abramson, Rudy. *Spanning the Century: The Life of W. Averell Harriman 1891–1986*. William Morrow, 1992.
Myles, William J. *Harriman Trails: A Guide and History*. New York–New Jersey Trail Conference, 1992.

Reports

13th Annual Report, *American Scenic and Historic Preservation Society*, 1908.
PIPC Minutes, 1908–1910.
8th Annual Report, PIPC, March 9, 1908.
10th Annual Report, PIPC, January 1, 1909.
"Report of the Work of the Commissioners of the Palisades Interstate Park Made by
    George W. Perkins, President of the New York State Commission on the Occasion of
    the Dedication of the Park at Alpine, New Jersey, September 27, 1909."
"Highlands of the Hudson Forest Reservation." 15th Annual Report, New York State
    Forest, Fish and Game Commissioner, 1909.
11th Annual Report, PIPC, January 31, 1911.
17th Annual Report, American Scenic and Historic Preservation Society, March 28, 1912.
"Dedication of Palisades Interstate Park." The Hudson-Fulton Celebration.

Pamphlets and Periodicals

Hopkins, Franklin W. "Preservation of the Palisades." *Scenic and Historic America* (June
    1930).
"Dr. Edward Lasell Partridge Passes." *Scenic and Historic America* (June 1930).
"To Improve the Harriman Forest." PIPC Archives, 1899.

Correspondence

Of J. DuPratt White, 1908, 1909.
Of Leonard Smith, 1909.
George W. Perkins to General Woodford, June 22, 1909.
From William Welch, October 20, 1936.

Newspapers

"Blasting Away of Bear Mountain." *City and Country* (February 11, 1908).
"The Patriotic Effort to Save River Scenery." *Rockland County Messenger* (March 23, 1967).
"Palisades Park Opened." *New York Times* (September 28, 1909).
"60-Mile River Park Insured to the State." *New York Times* (January 6, 1910).
"The Proposed Park in the Highlands." *City and Country* (January 22, 1910).
"Killed in Senate." *City and Country* (May 28, 1910).
"Palisades Park for the People." *City and Country* (May 28, 1910).
"The Voters Say Yes to Park Bond Issue." *New York Times* (October 1, 1910).
"Harriman Park Passes to the State." *New York Times* (October 30, 1910).

Chapter 5: Legend and War

Books

Bedell, Cornelia A. *Now and Then and Long Ago in Rockland County*. The Historical
    Society of Rockland County, 1968.
Dunwell, Frances F. *The Hudson River Highlands*. New York: Columbia University Press,
    1991.
Leiby, Adrian C. *The Revolutionary War in the Hackensack Valley*. New Brunswick, N.J.:
    Rutgers University Press, 1962.

Reports

Carr, William H., and Richard J. Koke. "Twin Forts of the Popolopen: Forts Clinton and Montgomery." PIPC Archives, July 1937.

Chapter 6: Welch

Books

Garraty, John A. *Right-Hand Man: The Life of George W. Perkins.* Harper and Bros., 1957.

Reports

PIPC Minutes, 1910–1912.

Pamphlets and Periodicals

"Frank Eugene Lutz, 1873–1943." *Journal of the New York Entomological Society,* vol. LII (March 1944).
"Frank E. Lutz." *American Museum of Natural History,* vol. 22, no. 5 (May 1997).
Garraty, John A. "Millionaire Reformer: The Progressives Part II." *American Heritage* (February 1962).

Correspondence

PIPC Correspondence, 1910–1912.
From J. DuPratt White, in *New York Times* (March 19, 1911).
Samuel Broadbent to George W. Perkins, October 16, 1911.
George W. Perkins to John D. Rockefeller Jr., November 9, 1912.

Newspapers

"Perkins Retiring at End of Year from J. P. Morgan and Company." *New York Times* (December 11, 1910).
"Commission Buys the Hook." *Rockland Journal News* (March 11, 1911).
"South Mountain to the Public." *Rockland Journal News* (October 28, 1911).
"George W. Perkins Dies in 58th Year." *New York Times* (June 19, 1920).
"Welch to Get Gold Medal." *Rockland Journal News* (December 16, 1935).
"Major Welch Dies, Builder of Parks." *New York Times* (May 5, 1941).

Chapter 7: Bear Mountain

Books

Abramson, Rudy. *Spanning the Century: The Life of W. Averell Harriman 1891–1986.* William Morrow, 1992.

Pamphlets and Periodicals

"Maintenance of Order." PIPC Archives, PIPC.
"Financial Data Prepared for Inspection Trip: June 11th, 1914." PIPC.
"Palisades Interstate Park." *American Scenic and Historic Preservation Society* (April 19, 1915).

"Proceedings of the National Parks Conference." Government Printing Office, 1917.

"In Memoriam: Frederick Charles Sutro." *Your Parks: New Jersey Parks and Recreation Association* (December 1964).

Correspondence

Of Leonard Smith, 1913.

George W. Perkins to John D. Rockefeller, September 18, 1913.

William Welch to the Superintendent of Yellowstone National Park, July 27, 1914.

H. Percy Silver, Chaplain, U.S. Military Academy, West Point, to J. DuPratt White, October 3, 1914.

Frederick C. Sutro to George W. Perkins, December 5, 1914.

George W. Perkins to Frederick C. Sutro, December 11, 1914.

George W. Perkins to Abby Barstow Bates, Secretary, Appalachian Mountain Club, December 11, 1914.

George W. Perkins to John D. Rockefeller, June 2, 1915.

John D. Rockefeller Jr. to George W. Perkins, June 18, 1915.

William Welch Files, 1916–1918, PIPC Archives.

Of Elbert W. King, 1916–1918.

Nathan F. Barrett to George W. Perkins, March 1, 1916.

Between Reverend Lee W. Beattie and George W. Perkins, March 24, April 19, 1917.

John D. Rockefeller Jr. to John D. Rockefeller, April 17, 1917. Rockefeller Family Archives.

Col. A. C. Dalton to George W. Perkins, April 30, 1918.

Theodore Horton to George W. Perkins and Dr. Hermann M. Biggs, September 17 and October 21, 1918.

Reports

PIPC Minutes, 1913–1918.

Newspapers

"Palisades Park Commission to Be Sulerized Next." *Newburgh Daily News* (June 12, 1913).

"Perkins Denies Palisades Charges." *New York Times* (June 13, 1913).

"Inquiry into Interstate Park." *Rockland County Journal* (June 21, 1913).

"No Road for Palisades." December 1914 newspaper article found in PIPC Archives.

"Governor Dedicates Park." *Rockland County Times* (June 3, 1915).

"Approves $2,650,000 Award." *New York Times* (May 3, 1916).

"Perkins to Be Made Food Commissioner." *New York Times* (August 5, 1917).

"Hatch & Perkins Clash on Bathing." *New York Times* (August 21, 1918).

"R. V. Lindabury Dies on Horseback Ride." *New York Times* (July 16, 1925).

"Charles W. Baker, Engineer 55 Years." *New York Times* (June 7, 1941).

"Ex-Gov. Averell Harriman, Adviser to 4 Presidents, Dies." *New York Times* (July 27, 1986).

Chapter 8: Perkins

Books

Garraty, John A. *Right-Hand Man: The Life of George W. Perkins*. Harper and Bros., 1960.

Reports

Adams, Dr. Charles C. "Natural History Survey of Palisades Interstate Park." *Empire Forester*, vol. 5. New York State College of Forestry (1919).

Perkins, George W. "Sight Seeing Buses on the Palisades." State Service. Governor of the State of New York, 1919.

"Directory of Encampments–1919." PIPC Archives.

"Appropriations Made to Palisades Interstate Park Commission." Rockefeller Family Archives.

Senate Act Nos. 1764, 1905 Rec. 425, April 8, 1919, State of New York.

PIPC Minutes, 1919, 1920.

Clyne, Patricia Edwards. "Iona Island." Hudson Valley (April 1985).

Correspondence

PIPC Correspondence 1919, 1920.

George W. Perkins to Richard V. Lindabury, January 28, 1919.

George W. Perkins to William Welch, January 28, 1919.

Brig. Gen. Brice P. Disque, U.S. Army, to Director, Bureau of Aircraft Production, Washington, D.C., February 11, 1919.

William Welch to George W. Perkins, February 22, 27, 1919.

Edward F. Brown to William Welch, March 5, 1919.

William Welch to Edward F. Brown, March 7, 1919.

Elbert W. King to J. DuPratt White, March 5, 1919.

Edward F. Brown to Franklin W. Hopkins, March 20, 1919.

Edward F. Brown to Edward L. Partridge, M.D., March 20, 1919.

William Welch to Abram Deronde, March 27, 1919.

From William Welch, March 27, 1919.

R. B. Potter to J. DuPratt White, March 26, 1919.

Elbert W. King to William Welch, April 2, 1919.

William Welch to Dr. Hugh Baker, October 10, 1919.

William Welch to Miss Marjorie B. Jones, April 10, 1919.

William Welch to A. M. Herbert, April 19, 1919.

Elbert W. King to A. M. Herbert, April 29, 1919.

J. DuPratt White to Elbert W. King, May 1, 1919.

William Welch to Files, May 1919. Welch Files.

Edward F. Brown to William Gee, May 26, 1919.

George W. Perkins to William Welch, June 5, 1919.

William Welch to F. Kingsbury Curtis, June 12, 1919.

Frederick Law Olmsted Jr. to William Welch, June 18, 1919.

Maurice M. Lefkowitz to William Welch, June 24, 1919.

Mrs. Tannebaum to William Welch, July 9, 1919.

Edward F. Brown to George W. Perkins, July 9, 1919.

Smith Riley to Edward F. Brown, July 12, 1919.

Charles B. Webb to PIPC, no date (mid-August 1919).

Fred K. Stuart Greene to William Welch, August 27, 1919.

Edward F. Brown to George W. Perkins, August 30, 1919.

Enos A. Mills to William Welch, October 14, 1919.

William Welch to George W. Perkins, November 26, 1919.
F. L. Fisher to William Welch, December 5, 1919.
John D. Rockefeller Jr. to George W. Perkins, January 28, 1920.
Between John D. Rockefeller Jr. and George W. Perkins, February and March 1920.
Stephen T. Mather to William Welch, February 16, 1920.
George W. Perkins to Commissioner Lindabury, March 18, 1920.
William Welch to Jay Downer, March 24, 1920.
From George W. Perkins's secretary, May, June 1920.
H. W. Jenkins (mayor, Englewood Cliffs, New Jersey) to PIPC, June 2, 1920.
William Welch to Dr. George F. Kunz, June 19, 1920.

Newspapers

"Bath Houses and Lunch at Hook Mountain This Year." *Nyack Evening Journal* (April 4, 1919).
"Provide Mothers' Rest Stations." *Messenger*, Haverstraw, N.Y. (August 7, 1919).
"Making Others Happy." *Rockland County Times* (August 1919 [exact date unknown]). PIPC Archives.
Benson, Captain William O. "Memories of a Visit to My Brother on the 'Onteora.'" *Sunday Freeman* (August 27, 1972). PIPC Archives.

Chapter 9: Jolliffe

Reports

"Bear Mountain—Harriman Attendance Reports for Entire Season."
"New Jersey Attendance Reports for Entire Season 1920." PIPC Archives.
"1920 Values (Purchased)." PIPC Archives.
PIPC Minutes. 1920–1921.
"PIPC Annual Report." 1921.

Correspondence

George W. Perkins's secretary to Richard Lindabury, June 24, 1920.
Between W. H. Radcliffe and William Welch, June 23, 25, 1920.
Fred Schiebelhuth to William Welch, August 3, 1920.
Elbert W. King to Peoples Bank of Haverstraw, August 14, 1920.
William Welch to Enos Mills, August 30, 1920.
Elbert W. King to Dr. Edward L. Partridge, September 2, 1920.
William Welch to PIPC, September 9, 1920.
Boy Scouts of America to William Welch, September 17, 1920.
Elbert W. King to William Welch, September 18, 1920.
Between William Welch and Dr. Edward L. Partridge, September 20, October 3, 1920, and January 25, 1921.
William Welch to Horace M. Albright, September 28, 1920.
Between James M. Lynch, New York State Department of Labor, and PIPC/William Welch, September 28, 30, October 4, November 8, 1920.
William Welch to Mary Ethel McAuley, September 30, 1920.

William Welch to Frederick C. Sutro, October 11, 1920.

Edward F. Brown to Dr. Edward L. Partridge, October 13, 1920.

George A. Blauvelt to William Welch, October 13, 1920.

F. Martin Brown to William Welch, October 22, 1920.

Elbert W. King to William Welch, October 23, 1920.

Between Stephen Mather and William Welch, October 23, 30, December 4, 1920.

Elbert W. King to Charles Whiting Baker, December 1, 1920.

Elbert W. King to P. V. D. Gott, December 3, 1920.

Everett G. Griggs to William Welch, December 27, 1920.

PIPC Correspondence, 1920–1921.

William Welch to A. J. James, February 21, 1921.

Elbert W. King to Committee on the Advisability of Selling the Dana Property, February 25, 1921.

Elbert W. King to Capt. James Conway, March 3, 1921.

George D. Pratt, Commissioner New York State Conservation Commission, to William Welch, March 4, 1921.

Between Mrs. Henry H. Dawson and William Welch, March 1921.

Elbert W. King to Mr. Deyo, March 28, April 18, 1921.

Elbert W. King to Richard V. Lindabury, April 14, 1921.

Elbert W. King to William Welch, April 18, 1921.

Elbert W. King to William Welch, April 26, 1921.

Elbert W. King to J. DuPratt White, April 27, 1921.

Elbert W. King to Dr. Edward L. Partridge, April 28, 1921.

Elbert W. King to Messrs. Blauvelt and Warren, April 30, 1921.

Elbert W. King to John F. Daschner, May 5, 1921.

William Welch to J. L. Ryan, May 17, 1921.

Elbert W. King to William Welch, May 20, 1921.

Elbert W. King to Charles W. Baker, May 21, 1921.

William Welch to Miss Jolliffe, May 24, 1921.

Elbert W. King to William Welch, July 7, 1921.

Elbert W. King to William Welch, October 6, 1921.

William Welch to Averell Harriman, October 7, 1921.

William Welch to Otis H. Cutler, December 19, 1921.

Newspapers

"Interstate Park Commissioners' Map Opens Forest Roads for Tourists." *New York Evening Post* (September 10, 1920).

"May Drill for Oil Under Palisades." *New York Times* (May 17, 1921).

Chapter 10: Trail and Bridge

Books

Myles, William J. *Harriman Trails: A Guide and History*. New York–New Jersey Trail Conference, 1992.

Waterman, Laura and Guy. *Forest and Crag: A History of Hiking, Trail Blazing and Adventure in the Northeast Mountains*. Appalachian Mountain Club Books, no date.

Pamphlets and Periodicals

"The Growth of the Appalachian Trail." PIPC press release, no date. PIPC Archives.

Caro, Robert. "Annals of Biography: The City-Shaper." *New Yorker* (January 5, 1998).

Dickinson, H. V. "The Bear Mountain Bridge." *Parks and Recreation* (March/April 1927).

MacKaye, Benton. "Progress Toward the Appalachian Trail." *Appalachia*, vol. XV, no. 3 (December 1922).

———. "The Appalachian Trail.: A Guide to the Study of Nature." *Scientific Monthly*, vol. XXXIV (April 1932).

Place, Frank. "Raymond H. Torrey." *Yearbook of the Appalachian Club* (January 1939).

Scherer, Glenn. "A Celebration of Vision, Planning, and Grass-roots Mobilization: Paying Homage to Benton Mackaye." *Appalachian Trailway News* (March/April 1997).

Reports

"Proceedings of the Second National Conference on State Parks." May 22–25, 1922.

28th Annual Report American Scenic and Historic Preservation Society. April 12, 1923.

DeWan, George. "The Master Builder: How Planner Robert Moses Transformed Long Island for the 20th Century and Beyond." http://www.lihistory.com/7/hs722a.htm.

PIPC Minutes 1921, 1922.

Annual Report 1923. PIPC.

PIPC Minutes April 12, November 20, December 28, 1923.

Correspondence

Elbert W. King to William Welch, January 3, 1922.

Elbert W. King to William Welch, January 10, 1922.

William Welch to Dr. E. L. Partridge, January 25, 1922.

Elbert W. King to Otis H. Cutler, February 24, 27, 1922.

Elbert W. King to Leib Deyo, March 15, 1922.

Elbert W. King to John E. Robinson, editor, *New York Amsterdam News*, March 20, 1922.

Elbert W. King to Dr. Edward L. Partridge, April 8, 1922.

Elbert W. King to William Welch, August 9, 1922.

Elbert W. King to J. Finley Bell, M.D., August 25, 1922.

Elbert W. King to Richard V. Lindabury, September 12, 1922.

Elbert W. King to Carl Bannwort, October 5, 1922.

Elbert W. King to William Welch, January 8, 1923.

Elbert W. King to J. DuPratt White, February 24, 1923.

To PIPC, February 27, 1923.

Elbert W. King to William H. Porter, March 26, 1923.

Edward L. Partridge and Franklin W. Hopkins to PIPC, April 10, 1923.

Elbert W. King to J. DuPratt White, April 11, 1923.

J. DuPratt White to Dr. Ernest Stillman, April 12, 1923.

Elbert W. King to Senator Blauvelt, April 13, 1923.

Raymond H. Torrey to Frederick C. Sutro, January 26, 1932.

Newspapers

"Would Span Hudson at Bear Mountain." *New York Times* (February 9, 1922).

"Motorists Urge New Bridge to Bear Mountain." *New York City American* (March 4, 1922).

"Winter at Bear Mountain." *Peekskill Union* (February 8, 1922).
"Bill Authorizes Huge Suspension Bridge Over Hudson." *Brooklyn New York Eagle* (February 8, 1922).
"Harrimans Back New Park System." *New York City World* (February 13, 1922).
"New Bridge Plan Endorsed by Westchester Interests." *New York City Telegram* (February 16, 1922).
"Mrs. Harriman Backs Hudson Bridge Plan." *Newburgh News* (February 18, 1922).
"Westchester Park Plan." *New York Times* (February 19, 1922).
"Both Farmers and City Dwellers Gain by New Bridge Across the Hudson." *Brooklyn New York Citizen* (February 26, 1922).

Chapter 11: Uncle Bennie

Pamphlets and Periodicals

Van Ingen, W. B. "A Wilderness Transformed." *New York Times Magazine* (September 7, 1924).

Reports

Adolph, Eleanor Bazzoni. "The Man Who Is Uncle to All the Boys of America." No date. PIPC Archives.
PIPC Annual Reports, 1924–1927.
Snead, J. E. "That Fair and Ancient Land: Archaeology and Society in the American Southwest, 1890–1915."
PIPC Minutes, April 28, May 20, September 23, 1924.
PIPC Minutes, March 11, 17, and October 20, 1925.

Correspondence

Robert Moses to J. DuPratt White, May 21, 1924.
Robert Sterling Yard, executive secretary, National Parks Association, to Maj. W. A. Welch, July 15, 1924.
Press Release from PIPC, July 31, 1924.
Elbert W. King to Frederick C. Sutro, August 19, 1924.
Elbert W. King to Miss Marvin, September 3, 1924.
Elbert W. King to George W. Perkins, Jr., September 11, 1924.
Elbert W. King to Hamilton Ward, September 12, 1924.
Frederick C. Sutro to PIPC, September 15, 1924.
Elbert W. King to PIPC, October 16, 1924.
W. A. Welch to Laura Spelman Rockefeller Foundation, November 3, 1924.
Kenneth Chorley to Col. Arthur Woods, November 11, 26, and December 20, 1924.
Internal Rockefeller, dictated by Mr. Chorley, December 13, 1924.
W. A. Welch to Col. Glenn H. Smith, December 4, 1924.
W. A. Welch to Secretary of the Interior Hubert Work, December 12, 1924.
W. A. Welch to Col. W. B. Greeley, February 7, 1925.
Harold Allen to Maj. W. A. Welch, February 25, 1925.
Elbert W. King to Paul H. Tichnor, March 12, 1925.
Elbert W. King to Major Welch, March 14, 1925.
Telegram of Elbert W. King to Vincent B. Murphy, comptroller, March 20, 1925.

Elbert W. King to Mr. Knowles, April 10, 1925.
Allegheny State Park Commission to Maj. W. A. Welch, April 11, 1925.
Arthur Woods to W. A. Harriman, April 17, 1925.
Elbert W. King to PIPC, May 13, 1925.
Elbert W. King to Scott R. Knowles, May 15, 1925.
Arthur Woods to Mr. Fosdick, May 22, 1925.
W. A. Welch to P. H. Elwood, Jr., June 4, 1925.
Elbert W. King to Scott R. Knowles, August 12, 1925.
Elbert W. King to Major Welch, October 28, 1925.
Elbert W. King to William Shepherd Dana, December 24, 1925.

Newspapers

"Trail Typhoid Peril to Palisades Brook." *New York Times* (July 16, 1924).
"Bond Issue May Go into Courts." *New York City Sun* (October 20, 1924).
"Cardinal Hayes Endorses Park as Aid to Children." *New York City Evening World*
    (October 20, 1924).
"Battle Over Parks at Albany Hearing." *New York Times* (February 12, 1925).
"Big Engineering Feat Now Going on in This County." *Nyack Evening Journal* (April 21,
    1925).
"R. V. Lindabury Dies on Horseback Ride." *New York Times* (July 15, 1925).
"City Migrates to Camps." *New York Times* (August 9, 1925).
"Mather And Welch Enroute to Park Are Fresh from Dynamiting Big Saw Mill in Glacier
    National Park; Inspect Tetons." *Livingston (Montana) Enterprise* (August 12, 1925).
"Welch to Get Gold Medal." *Journal-News* (December 16, 1935).
"'Uncle Benny' Hyde Dies of Injuries; In Crash." *Santa Fe New Mexican* (July 17, 1933).
"Dr. Stillman Gives 600 Acres to Park." *New York Times* (May 26, 1922).
"Call Storm King Motor Highway Triumph of Road Construction." *New York Times* (June
    10, 1923).
"To Move a Mountain for a Great Park." *New York Times* (July 11, 1923).
"It Seems to Be All Right." *New York Times* (July 12, 1923).
"Commission Will Inspect New Bear Mountain Span Wednesday." *New York City Herald*
    (October 5, 1924).
"Novel Honeymoon Trip." *New York Times* (November 15, 1924).
"Mrs. E. H. Harriman Opens Bear Mountain Span." *New York Times* (November 26, 1924).
"Bear Mountain Bridge Formally Opened Today for Thanksgiving." *Middletown Daily
    Times Press* (November 26, 1924).

Chapter 12: Black Thursday

Reports

Bradley, Stanley W. *Crossroads of the Hudson: The Story of Alpine, New Jersey*. Alpine
    Bicentennial Committee.
Haugland, Gary. "Thomas A. Edison." Trailside Museum and Zoo, Historical Papers no.
    H-3/93. PIPC archives.
Minutes of Southern Appalachian National Park Commission, January 8, 1926.
Jones, Mark M. "Palisades Interstate Park" (February 15, 1926).
PIPC Annual Reports, 1926–1928.

PIPC Minutes, April 19, June 21, 1927.

Program: "Outdoor Speed Skating Handicap Meet, Conducted by Bear Mountain Sports Association at Bear Mountain Park Sunday February 6th, 1927."

Program: "New York State Ski Jumping Championship Tournament Sanctioned by the Eastern Amateur Ski Association Under Auspices of Swedish Ski Club of New York At Bear Mountain Palisades Interstate Park Sunday, February 13th, 1927."

PIPC Minutes, June 6, August 21, October 9, October 11, December 28, 1928.

PIPC Minutes, March 19, May 21, 1929.

Correspondence

Address by Frederick Sutro on the "Occasion of the Dedication of the Memorial Women's Federation Park, April 30, 1929."

Proclamation by Franklin Roosevelt, August 30, 1929.

Mark Jones to Rockefeller Staff, January 7, 1926.

William Welch to Mark Squires, January 19, 1926.

DeHart H. Ames to William Welch, January 20, 1926.

Elbert W. King to Miss Young, January 21, 1926.

Frederick C. Sutro to Governor Silzer, January 27, 1926.

Raymond H. Torrey to Franklin D. Roosevelt, February 14, 1926.

Elbert W. King to Mr. E. C. Wallin, International Newsreel Corporation, February 19, 1926.

Elbert W. King to Franklin W. Hopkins, March 2, 1926.

Elbert W. King to Major Welch, March 13, 1926.

Elbert W. King to Miss Marvin, March 16, 1926.

Elbert W. King to Oscar R. Ewing of Messrs. Hughes, Rounds, Schurman and Dwight, March 16, 1926.

Elbert W. King to Mr. Hopkins, April 1, 1926.

Elbert W. King to Mr. L. O. Rothschild, April 5, 1926.

Elbert W. King to J. DuPratt White, April 5, 1926.

Elbert W. King to Miss J. A. Marvin, May 4, 1926.

W. P. Davis to Maj. William Welch, May 14, 1926.

From Kenneth Chorley, May 7, 1926.

Elbert W. King to Maj. W. A. Welch, October 25, 1926.

Elbert W. King to J. DuPratt White, December 1, 1926.

A. J. Joseph to PIPC, January 5, 1927.

Mrs. H. Pleus to William Welch, undated.

Elbert W. King to Franklin Hopkins, February 2, 1927.

William A. Welch to Mr. P. H. Elwood Jr., professor of Landscape Architecture, Iowa State College, February 23, 1927.

J. DuPratt White to Governor A. Harry Moore, March 16, 1927.

Elbert W. King to Commissioner George T. Smith, March 24, 1927.

Elbert W. King to Frederick Osborn, March 25, 1927.

Elbert W. King to Mr. Richards of Messrs. Hughes, Rounds, Schurman and Dwight, August 2, 1927.

Elbert W. King to Maurice Marmer, June 24, 1927.

Willis G. Corbitt to Mrs. George Leviston, September 14, 1927.

John D. Rockefeller, Jr., to Major Welch, September 26, 1927.

William A. Welch to Edsel B. Ford, November 21, 1927.

William A. Welch to Daniel P. Wine, January 12, 1928.
John D. Rockefeller Jr. to Mr. Cammerer, January 23, 1928.
William A. Welch to George Leviston, February 29, 1928.
E. Childs to Maj. W. A. Welch, March 6, 1928.
Howard J. Benchoff to Maj. W. A. Welch, March 8, 1928.
Raymond H. Torrey to Maj. W. A. Welch, March 12, 1928.
Howard H. Parsons to Maj. Wm. A. Welch, March 14, 1928.
Harold M. Lewis to Maj. Wm. A. Welch, March 15, 1928.
Elbert W. King to Mrs. William H. Osborne, April 18, 1928.
W. S. Richardson to John D. Rockefeller Jr., May 24, 1928.
Elbert W. King to L. H. Harrison, June 2, 1928.
Elbert W. King to Mr. J. F. Verrips, June 22, 1928.
Between Major Welch and Mrs. S. C. Eristoff, July 1, 20, 1928.
Elbert W. King to Major Welch, July 16, 1928.
John D. Rockefeller Jr. to Arthur Woods, August 22, 1928.
Thomas W. Lamont to Mr. Debevoise, August 24, 1928.
Arthur Woods to John D. Rockefeller Jr., August 31, 1928.
Elbert W. King to Major Welch, October 3, 1928.
Charles Heydt to John D. Rockefeller Jr., November 8, 1928.
Elbert W. King to William Welch, March 11, 1929.
Elbert W. King to Department Superintendents, March 28, 1929.
W. A. Welch to the International Olympic Committee, March 28, 1929.
W. A. Welch to Christine C. Heis, April 10, 1929.
Elbert W. King to Frederick C. Sutro, July 31, 1929.
John Hays Hammond to W. A. Welch, September 1, 1929.
W. A. Welch to Mrs. Henry Fairfield Osborne, September 10, 1929.

Newspapers

"Wm Childs Dead; Restaurant Man." *New York Times* (November 26, 1926).
"Parks and Nature Study." *New York Times* (October 31, 1926).
"Hook Mountain Park to Be Picnic Ground." *New York Times* (November 26, 1926).
"W. H. Porter Dead; Stricken in Street." *New York Times* (December 1, 1926).
"Park Museum Under Way." *New York Times* (May 5, 1927).
"Frederick Osborn, A General, 91, Dies." *New York Times* (January 7, 1981).
"Bear Mountain Case Heard." *New York Times* (July 30, 1927).
"Bear Mountain Ban Upset." Newspaper unknown (July 31, 1927).
"Trail Hut Now Open for Hikers." *New York Times* (October 23, 1927).
"Park Invites Fox Hunters to Shoot 500 Which Kill Game." *New York Times* (March 5, 1928).
"Foreword to A Plan." *New York Times* (May 19, 1928).
"Consider Parkway to Save Palisades." *New York Times* (May 21, 1928).
"Regional Plan Sees Big Jersey Growth." *New York Times* (May 30, 1929).
"Jury Lists Palisades High in Artistic Parks." *Herald Tribune* (January 10, 1929).
"Park Commission Sued." *New York Times* (March 26, 1929).
"Palisades Board Loses Land Ruling." *New York Times* (June 8, 1929).

"Governors Praise Park Cooperation." *New York Times* (August 7, 1929).
"High Court Reverses Palisades Case." *New York Times* (February 25, 1930).

Chapter 13: The Compact

Reports

"History of the New Jersey State Federation of Women's Clubs 1927–1947." New Jersey
   State Federation of Women's Clubs.
PIPC Minutes, March 6, November 6, 1930.
PIPC Minutes, July 9, December 3, 1931.
PIPC Annual Report, 1932.
PIPC Minutes, October 5, November 2, 1933.
"July Excursions 1934." Report in PIPC files.
"Business Booked for Summer Season—1934 (Up to June 27th, 1934)." Report in PIPC
   files.
PIPC Minutes, September 13, 1934.
PIPC Minutes, September 10, 1935.
PIPC Minutes, January 8, May 14, December 10, 1936.
New York State Council of Parks Minutes, March 6, 1936.
White, J. DuPratt. "Statement of J. DuPratt White, President, Commissioners of the
   Palisades Interstate Park, at hearing on November 21, 1936 before Committee on
   Interstate Cooperation regarding Proposed Compact between New York and New
   Jersey to create Palisades Interstate Park Commission."
PIPC Minutes, 1937.
Resume of Te Ata (undated).
"60 Years of Park Cooperation." Palisades Interstate Park Commission, 1960.

Pamphlets and Periodicals

Delavan, D. Bryson, M.D. "Long a Valued Member of Scenic Society—Originated
   Movement for Conservation of Hudson Highlands." Scenic and Historic America, vol.
   II, no.2 (June 1930).
"To the Ladies." *Liberty Magazine* (August 17, 1935).

Legislation

"An Act to Provide for the Creation by Interstate Compact of the Palisades Interstate
   Park Commission as a Joint Corporate Municipal Instrumentality of the States of
   New York and New Jersey with Appropriate Powers and Thereby to Continue the
   Palisades Interstate Park." State of New York No. 828 Int. 791 In Assembly January
   30, 1936.
"An Act to Provide for the Creation by Interstate Compact of the Palisades Interstate
   Park Commission as a Joint Corporate Municipal Instrumentality of the States of New
   York and New Jersey with Appropriate Rights, Powers, Duties and Immunities, for
   the Transfer to Said Commission of Certain Functions, Jurisdiction, Rights, Powers
   and Duties Together With the Properties of the Bodies Politic Now Existing in Each
   State Known as 'Commissioners Of the Palisades Interstate Park' and for the

Continuance of the Palisades Interstate Park." State of New York No. 1973 Int. 1758 In Assembly March 10, 1937.

Correspondence

Charles O. Heydt to John D. Rockefeller Jr., February 14, 1930.
Charles O. Heydt to Mr. Rockefeller, February 10, 1930.
Elbert W. King to Frederick C. Sutro, May 5, 1930.
E. W. King to Frederick C. Sutro, June 13, 1930.
W. A. Welch to Robert Moses, November 21, 1930.
John D. Rockefeller Jr., to Mr. Heydt, December 31, 1930.
Elbert W. King to Mrs. Brobst, February 2, 1931.
E. W. King to Mrs. J. F. Dashner, April 22, 1931.
Chief Clerk to George W. Perkins, May 25, 1931.
Chief Clerk to Mr. Vincent J. Kennedy, June 10, 1931.
Ruby Jolliffe to Upper Cohasset Camps, August 25, 1931.
J. DuPratt White to Winfield Scott, September 28, 1931.
Edmund W. Wakelee to Governor Morgan F. Larson, December 3, 1931.
Frederick C. Sutro to Charles Whiting Baker, January 21, 1932.
Frederick C. Sutro to Edmund W. Wakelee, June 8, 1933.
J. DuPratt White to John D. Rockefeller Jr., June 30, 1933.
John D. Rockefeller Jr., to the commissioners of the Palisades Park, July 7, 1933.
Raymond H. Torrey to Frederick C. Sutro, October 21, 1933.
W. A. Welch to President Franklin D. Roosevelt, October 24, 1933.
Carola Lehrenkrauss to Mr. Sutro, April 16, June 19, October 2, December 19, 1934.
Frederick C. Sutro to Dan F. McAllister, August 14, 1935.
Robert G. Mead to Thomas M. Debevoise, August 23, 1935.
Thomas M. Debevoise to PIPC, November 29, 1935.
W. A. Welch to Frank Storer Wheeler, December 7, 1935.
Robert Moses to Jay Downer, November 16, 1936.
J. DuPratt White to Thomas M. Debevoise, March 30, 1937.
John D. Rockefeller Jr. to J. DuPratt White, April 6, 1937.
J. J. Tamsen to K. T. Ross, June 18, 1937.

Press Releases

Torrey, Raymond H. "Beware Picking Flowers in the Bear Mt. Park." Book 1 (1939).
———. "Naturalization of Artificial Lakes in Harriman State Park." Book 2 (August 5, 1931).
———. "A Promising Experiment in Rehabilitation of Down and Outs." Book 4 (March 1, 1933).
———. "Civilian Conservation Corps Work in Palisades Interstate Park." Book 4 (July 5, 1933).
———. "Delay in Re-Opening Dyckman Street Ferry." Book 5 (April 24, 1935).
———. "Storm King Mountain Now Preserved in Interstate Park." Book 6 (April 1, 1936).
———. "Bronze Elk Head Dedicated in Bear Mountain Park." Book 6 (May 7, 1936).

Newspapers

"Ask State to Bar Dam as a Peril." *New York Times* (January 20, 1930).

"Alfred E. Smith Is Appointed a Member of Palisades Interstate Park Commission." *New York Times* (May 26, 1930).

"Jersey Confirms Smith." *New York Times* (May 28, 1930).

"A Mountain Folk in New York's Shadow." *New York Times* (July 27, 1930).

"Better Vegetation Along the Hudson." *New York Times* (July 26, 1931).

"Jobless Build Road on Bear Mountain." *New York Times* (December 2, 1932).

"State Relief Programs Stress Work." *New York Times* (December 25, 1932).

"Mrs. Roosevelt to Unveil Plaque at Bear Mt. Saturday Afternoon." (publication unknown) (May 25, 1933).

"Mather Is Honored by Park Leaders." *New York Times* (May 28, 1933).

"Big Tract Donated by Rockefeller Jr. to Save Palisades." *New York Times* (July 12, 1933).

"Beer by Nearly 2-1 Voted in Oklahoma." *New York Times* (July 12, 1933).

"A New Move to Preserve the Palisades." *New York Times* (July 16, 1933).

"Treasure Hunters Blast Mountain in Mysterious Quest in Palisades." *New York Times* (November 9, 1934).

"Log-Cabin Pioneers of Ramapos Driven Out by Man-Made Lake." *New York Times* (February 18, 1935).

"Rockefeller Gives Land on Palisades." *New York Times* (June 26, 1935).

"Privations of WPA Workers at Bear Mountain Revealed." *Citizen-Register*, Ossining (December 12, 1935).

"Joint Board Urged on Palisades Park." *New York Times* (January 27, 1936).

"White Urges Interstate Park Treaty." *Rockland Journal-News* (March 18, 1936).

"New Jersey Assembly Votes a Luxury Tax; Approval of Senate Is Expected Tonight." *New York Times* (April 14, 1936).

"Assembly Rejects 2 Palisades Bills." *New York Times* (May 15, 1936).

"Moses & Smith Oppose Park Plan." *New York Times* (November 22, 1936).

"Palisades Park Bill Passed at Albany Setting up a Corporate Bi-State Board." *New York Times* (March 24, 1937).

Chapter 14: The Palisades Parkway

Reports

PIPC Minutes, 1938–1945.

Fischer, Louis. "Report of Louis Fischer to the Palisades Interstate Park Commission." March 31, 1941.

PIPC Minutes, January 1, 1941.

PIPC Minutes, April 21, 1943.

PIPC Minutes, December 13, 1944.

PIPC Minutes, June 16, 1945.

Ward, Gene. "Man of the Mountain: Big Jack Martin." 1946.

PIPC Minutes, January 10, November 9, 1948.

PIPC Minutes, September 11, 1951.

PIPC Minutes, April 21, 1952.

PIPC Minutes, January 12, December 21, 1953.
PIPC Minutes, January 17, May 26, July 9, 1956.
PIPC Minutes, August 19, 1957.
PIPC Minutes, May 24, June 28, 1958.
PIPC Minutes, March 23, November 11, 1959.
PIPC Minutes, February 2, 1960.

Pamphlets and Periodicals

Lamont, Corliss. "The Palisades—3d Call." *Survey Graphic* (July 1945).

Correspondence

Robert Moses to Laurance Rockefeller, August 9, 1939.
George W. Perkins to A. K. Morgan, November 14, 1939.
To Palisades Interstate Park Commission, February 1, 1940.
Frederick C. Sutro to A. K. Morgan, May 19, 1941.
Edmund W. Wakelee to A. K. Morgan, October 11, 1943.
Robert Moses to George W. Perkins, February 6, 1945.
Bill Carr to A. K. Morgan, April 15, 1945.
A. K. Morgan to General Osborn, May 3, 1945.
John D. Rockefeller Jr. to Governor Edge, July 20, 1945.
Robert Moses to A. K. Morgan, August 10, 1945.
A. K. Morgan to Col. George W. Perkins, August 25, 1945.
John D. Rockefeller Jr. to A. K. Morgan, November 27, 1945.
Ellen W. Rionda to Senator Van Alstyne Jr., January 1, 1946.
Emma G. Foye to Colonel Perkins, January 18, 1946.
Governor Edge to John D. Rockefeller Jr., February 8, 1946.
Ridsdale Ellis to John D. Rockefeller Jr., October 15, 1951.
Ida W. Certo to John D. Rockefeller Jr., May 5, 1952.
A. K. Morgan to Laurance S. Rockefeller, July 1, 1952.
Josiah M. Hewitt to Francis V. D. Lloyd, September 23, 1952.
A. K. Morgan to George W. Perkins, February 19, 1953.
Transcript of E. W. Kilpatrick to A. K. Morgan, February 16, 1953.
A. K. Morgan to George W. Perkins, March 8, 1954.
A. K. Morgan to John D. Rockefeller Jr., March 23, 1954.
A. K. Morgan to George W. Perkins, November 24, 1954.
William Kean to Governor Harriman, February 7, 1955.
Averell Harriman to George Perkins, March 2, 1955.
Donald G. Borg to Ambassador George W. Perkins, May 4, 1955.
A. K. Morgan to Laurance S. Rockefeller, November 10, 1955.
A. K. Morgan to George Perkins, January 10, 1958.
Dana S. Creel to George W. Perkins, April 23, 1958.
Dana Creel to George W. Perkins, December 5, 1958.
Lt. Col. J. B. Meanor Jr. to A. K. Morgan, June 1, 1959.
A. K. Morgan to Robert Moses, June 2, 1959.
Col. Miles H. Thompson to George W. Perkins, June 10, 1959.
George W. Perkins to A. K. Morgan, June 25, 1959.

Newspapers

"High Tor in the Hudson Being Sought for a Park." *New York Times* (November 15, 1942).

"On Palisades Park Board." *New York Herald Tribune* (March 23, 1945).

"Bird's-Eye View of Palisades Parkway Shows This Rolling Bomb Gathered Moss 10 Years." *Bergen Evening Record* (November 8, 1945).

"Rockefeller Gives Views on Parkway." *New York Times* (December 7, 1945).

"Palisades Parkway Project Defended at Public Forum." Newspaper unknown (probably the end of 1945).

"Expert Cites Many Reasons for Parkway." *Englewood Press* (January 3, 1946).

"Good Faith and Bad on Parkway." *Bergen Evening Record* (August 8, 1946).

"New Nature Unit Gets Welcome from Commission." *Englewood Press* (November 7, 1946).

"Driscoll Pens Name to Bill for Parkway." *Bergen Evening Record* (April 22, 1947).

"Tamed Josephine Returns to Museum; Pilot Black Snake Toured for 10 Years." *New York Times* (August 10, 1947).

"Laurance S. Rockefeller Gives Dunderberg to Palisades Park." *New York Herald Tribune* (October 17, 1951).

"'Save Palisades' Campaign Opens in Fort Lee as Boro Views Offer for Cliff-Top." *Fort Lee Sentinel* (April 17, 1952).

"May Raze Riviera." *Newark Sunday News* (June 8, 1952).

"Bear Mountain Uranium Hunters Put Ski Jump Among Filed Claims." *New York Times* (September 8, 1954).

"Cops to Guard Palisades Zone Hearing." *New York Daily News* (March 13, 1955).

"Fantastic Jersey Estate." *Newark Sunday News* (April 24, 1955).

"Fort Lee Ordered to Sell 1776 Site." *New York Times* (January 30, 1956).

"Fort Lee to Fight Sale Order." *New York Times* (February 1, 1956).

"Fort Lee Sale Stayed." *New York Times* (February 7, 1956).

"Rockefellers, Harrimans Get Praise at Parkway Dedication." *Newburgh News* (August 29, 1958).

"A F 'Invaded' Area in '59." *Cornwall Local* (September 30, 1992).

Chapter 15: Storm King

Reports

PIPC Minutes, February 2, March 21, June 18, August 15, 1960.

McKeon, Warren H. "Recommendations for a Deer Management Program in the Bear Mountain State Park Area." June 1960.

PIPC Minutes, February 27, March 20, November 27, 1961.

PIPC Minutes, May 19, 1962.

Brochure: "Restoration and Power: Creative Conservation at Cornwall-On-Hudson." Con Edison, probably 1962.

Legal Brief: Doty, E. Dale. "Brief Opposing Exceptions of the Scenic Hudson Preservation Conference." October 8, 1964.

"PIPC A History 1900–1973." PIPC.

PIPC Minutes, April 15, May 18, October 21, November 18, 1963.

PIPC Minutes, February 17, 1964.

Report to the Members: John D. Dale to the Hudson River Conservation Society, May 10, 1964.

Tamsen, John J. "Testimony Before Joint Legislative Committee." November 20, 1964.

PIPC Minutes, March 23, May 15, 1965.

PIPC Minutes, February 17, 1966.

PIPC Minutes, April 17, December 18, 1967.

Memorial Tribute: Lembo, Margaret. 44th Annual Camp Director's Conference Bear Mountain Park, New York. March 30, 1968.

PIPC Minutes, September 16, 1968.

PIPC Minutes, December 16, 1968.

Prepared testimony of Conrad L. Wirth, March 24, 25, 1969.

PIPC Minutes, May 17, June 14, November 17, 1969.

PIPC Minutes, June 20, October 19, November 16, 1970.

PIPC Minutes, February 16, March 15, June 5, 1971.

PIPC Minutes, January 17, June 26, September 25, 1972.

Correspondence

Telegram of Robert Moses to Hon. Nelson A. Rockefeller, January 3, 1960.

Robert Moses to Hon. Nelson A. Rockefeller, January 14, 1960.

Laurance S. Rockefeller to Hon. Nelson A. Rockefeller, February 6, 1960.

Telegram of A. K. Morgan to Mrs. George W. Perkins, March 23, 1960.

A. K. Morgan to G. W. Perkins, March 15, 1960.

A. K. Morgan to Dr. Carl O. Gustafson, February 6, 1961.

Albert R. Jube to Laurance S. Rockefeller, September 1, 1961.

J. Willard Marriott Jr. to Donald Borg, November 27, 1961.

A. K. Morgan to files, September 24, 1962.

A. K. Morgan to Mr. L. L. Huttleston, October 11, 1962.

J. O. I. Williams to A. K. Morgan, January 2, 1963.

Frederick Osborn to A. K. Morgan, January 15, 1963.

A. K. Morgan to J. K. McManus, C. A. Marks, T. LeNoir, February 4, 1963.

Robert Moses to Mr. H. Philip Arras, April 12, 1963.

L. O. Rothschild to Hon. William J. Ronan, April 23, 1963.

Harry J. Gommoll to Hon. Richard J. Hughes, May 23, 1963.

A. K. Morgan to J. O. I. Williams, May 24, 1963.

A. K. Morgan to Hon. Robert Moses, May 31, 1963.

L. O. Rothschild to Hon. Harold G. Wilm, Commissioner of Conservation, May 31, 1963.

Laurance S. Rockefeller to Lelan F. Sillin Jr., July 17, 1963.

Laurance S. Rockefeller to Mr. Harland C. Forbes, July 17, 1963.

Carl O. Gustafson to A. K. Morgan, July 30, 1963.

Leo Rothschild to Hon. William J. Ronan, July 19, 1963.

Carl O. Gustafson to A. K. Morgan, October 8, 1963.

A. K. Morgan to Mr. Carl O. Gustafson, October 10, 1963.

A. K. Morgan to Laurance S. Rockefeller, November 27, 1963.

A. K. Morgan to J. K. McManus and E. Van Houtin, January 31, 1964.

William J. Ronan to A. K. Morgan, March 26, 1964.

Between Mrs. Alexander Saunders and Robert Moses, April 3, 13, 1964.

Randall J. LeBoeuf Jr. to A. K. Morgan, April 10, 1964.

Dale E. Doty to A. K. Morgan, April 10, 1964.

A. K. Morgan to Randall J. LeBoeuf Jr., April 14, 1964.

A. K. Morgan to Laurance S. Rockefeller, May 5, 1964.

Form from A. K. Morgan, May, 1964.

Mrs. LeRoy Clark to A. K. Morgan, May 17, 1964.

A. K. Morgan to Laurance S. Rockefeller, June 23, 1964.

A. K. Morgan to the Files, July 9, 1964.

R. Watson Pomeroy to Hon. Albert R. Jube, November 30, 1964.

Albert Wilson to A. K. Morgan, August 16, 1965.

Action from Scenic Hudson Preservation Conference, New York–New Jersey Trail
    Conference, Sierra Club, Atlantic Chapter, North Jersey Conservation Foundation, 1968.

Albert R. Jube to "People Objecting to PIPC Land for Con Ed Project." November 1968.

Henry L. Diamond to A. K. Morgan, December 4, 1968.

A. K. Morgan to Laurance S. Rockefeller, January 9, 1969.

J. O. I. Williams to Henry Diamond, February 28, 1969.

Letter and resume of Nash Castro to Robert Binnewies, November 23, 1999.

Newspapers and Periodicals

"Jube Elected President of N.Y. Park Unit." *Bergen Evening Record* (February 2, 1960).

"Palisades Tract Marks 60th Year." *New York Times* (May 13, 1960).

"New State Beach Gets a Wet Start." *New York Times* (June 16, 1962).

"Huge Power Plant Planned on Hudson." *New York Times* (September 27, 1962).

"Hudson Day Line Bought by Circle." *New York Times* (October 6, 1962).

"Businessmen Get Outline of Plans by Con Ed Aide." *Evening News* (November 29, 1962).

"Opposition to Power Cable Registered by Garden Club." *Evening News* (December 19,
    1962).

"Power Plan Stirs Battle on Hudson." *New York Times* (May 22, 1963).

"Court Approves Palisades Motel." *New York Times* (May 27, 1963).

"Defacing the Hudson." *New York Times* (May 29, 1963).

"Conservation Aid, Wife and Daughter Die in Plane Crash." *New York Times* (October 18,
    1963).

"Con Ed on Hudson Opposed." *New York Times* (February 15, 1964).

"F.P.C. Report Backs Hudson Power Plan." *New York Times* (March 11, 1964).

"Con Edison Hearings Expected." *Evening News* (March 25, 1964).

"State Pushing Park Program." *Middletown Record* (May 16, 1964).

"Preserving the Hudson Highlands." *New York Times* (May 23, 1964).

"Con Edison Plant on Hudson Backed." *New York Times* (June 16, 1964).

"Cornwall Con Ed Plant Opposed." *New York Times* (June 16, 1964).

"Hudson Group Plans Con Ed Opposition." *New York Times* (June 23, 1964).

"Must God's Junkyard Grow?" *Life Magazine* (July 31, 1964).

"Armada of Foes Invades Site of Con Ed Project on Hudson." *New York Times* (September
    7, 1964).

"Protecting the Highlands." *New York Times* (September 8, 1964).

"Saving the Hudson Highlands." *New York Times* (November 17, 1964).

"Con Ed-on-Hudson Hearings On." *New York Times* (November 20, 1964).

"Retreat of 1776 Noted in Jersey." *New York Times* (November 21, 1964).

"Governor Backs Storm King Plant." *New York Times* (December 11, 1964).

"Mr. Rockefeller's Wrong Move." *New York Times* (December 14, 1964).
"Con Edison Plan Is Called Fatal to 2 Fish Industries." *New York Times* (December 29, 1964).
"Housewives Picket to Preserve Manor." *New York Times* (January 17, 1966).
"Joseph Kearns McManus, 59; Interstate Park Superintendent." *Middletown Record* (September 5, 1967).
Gill, Bo. "Stray Boots." *Newburgh Evening News* (September 27, 1967).
Dobbin, William J. "Palisades Puts Play Above Par." *New York Times* (September 8, 1968).
"City Asks F.P.C. to Block Con Ed at Storm King." *New York Times* (October 28, 1968).
"Hearings Reopened on Plan by Con Ed to Build in Cornwall." *New York Times* (November 20, 1968).

Chapter 16: Minnewaska

Reports

Death Certificate: A. K. Morgan, October 16, 1969.
Memorial Service Program: Mamie Phipps Clark, 1917–1983.
"PIPC A History 1900–1973." PIPC.
PIPC Minutes, October 29, 1973.
PIPC Minutes, April 15, May 20, December 16, 1974.
PIPC Minutes, January 27, February 24, April 21, May 19, October 20, 1975.
PIPC Minutes, January 19, 1976.
PIPC Minutes, June 20, September 19, 1977.
PIPC Minutes, July 31, December 18, 1978.
PIPC Minutes, May 21, July 5, September 17, 1979.
Statement: New York Office of Parks and Recreation and the Palisades Interstate Park Commission on DEC Project No. 356-15-0080 Presented at Public Hearings Conducted in Rochester, New York, July 14, 1980.
Record of above hearing, July 18, 1980.
Statement: Findings and Closing Brief of the New York State Office of Parks and Recreation and the Palisades Interstate Park Commission on DEC Project No. 356-15-0080, March 13, 1981.
Proposed Findings of Fact, Issues and Recommendations: DEC Project No.356-15-0080, New York State Office of Parks and Recreation and Palisades Interstate Park Commission, March 31, 1981.
Reply Brief of Applicant: DEC Project No. 356-15-0080, undated.
Decision: DEC Project No. 356-15-0080, June 2, 1981.
PIPC Minutes, July 20, 1981.
PIPC Minutes, March 25, November 25, 1985.
PIPC Minutes, January 13, 1986.
PIPC Minutes, June 22, 1987.
"Minnewaska and Surrounds Historical Time Line Revised 10/06/96." PIPC.

Correspondence

Press Release: PIPC, July 18, 1969.
J. O. I. Williams to A. K. Morgan, August 28, 1969.
Albert R. Jube to Hon. William T. Cahill, June 30, 1970.

Nash Castro to Frederick Osborn, November 29, 1971.

John F. Haggerty to Laurance S. Rockefeller, September 7, 1972.

John P. Keith to Nash Castro, February 4, 1972.

Conrad L. Wirth to Nash Castro, January 13, 1972.

Christopher J. Schuberth to Donald B. Stewart, October 24, 1972.

Linn M. Perkins to Laurance [Rockefeller], September 18, 1973.

Press Release from State of New York, Executive Chamber, Nelson A. Rockefeller, Governor, September 21, 1973.

Nash Castro to Mr. Jeffrey Ketterson, June 12, 1974.

George R. Cooley to Nash Castro, November 14, 1974.

Kenneth B. Phillips to Nash Castro, August 8, 1975.

Adrienne K. Wiese to Nash Castro, August 30, 1975.

Assemblyman Maurice D. Hinchey to Commissioner Orin Lehman, September 6, 1975.

Commissioner Orin Lehman to Peter Goldmark, September 10, 1975.

Arthur Schleifer to Nash Castro, October 6, 1975.

Bradford C. Northrup to Commissioner Orin Lehman, January 5, 1976.

Richard L. Erdmann to C. Mark Lawton, January 27, 1976.

Roger W. Tubby to Richard L. Erdmann, February 11, 1976.

L. Sisk to M. O'Loughlin, February 11, 1976.

Nash Castro to Roger Tubby, February 17, 1976.

Nash Castro to Bradford C. Northrup, April 15, 1976.

R. R. Paige to Nash Castro, April 20, 1976.

Nash Castro to Bradford C. Northrup, April 27, April 28, May 4, May 10, May 25, August 6, 1976.

Message from Joan to Mr. Castro, May 11, 1976.

Orin Lehman to Mr. Northrup, June 14, 1976.

David W. Burke to Commissioner Orin Lehman, August 2, 1976.

Roland R. Page to Bradford C. Northrup, September 9, 1976.

R. R. Page to Nash Castro, October 14, 1976.

Albert E. Caccese to Hon. R. Lewis Townsend, October 18, 1976.

Roland R. Page to Bradford C. Northrup, October 29, 1976.

John Crutcher to Orin Lehman, January 19, 1977.

Lucille Phillips to Nash Castro, March 1, 1977.

Ivan P. Vamos to C. Mark Lawton, September 16, 1977.

To Anthony Corbisiero, September 16, 1977.

Robert M. Watkins to William Averell Harriman, October 5, 1977.

Orin Lehman to Mr. Phillips, October 14, 1977.

Press Release: "State Parks Agency to Acquire 1,300 Acres at Minnewaska." October 20, 1977.

W. J. Kiely to Pete Lynch, November 3, 1977.

Laurance S. Rockefeller to Governor Hugh L. Carey, July 12, 1978.

Ron Karner to William Kiely, April 3, 1979.

James L. Stapleton to Nash Castro, July 20, 1979.

Edward L. Bednarz to Nash Castro, June 18, 1980.

Nash Castro to the commissioners, August 13, 1980.

Nash Castro to Kenneth Phillips, Jr., March 26, 1981.

Nash Castro to Edward L. Bednarz, June 23, 1981.

Nash Castro to Orin Lehman, July 31, 1981.

Newspapers

"$18 Million Project Proposed for Minnewaska." *Times Herald Sunday Record* (January 12, 1975).

"Planners Advise against Minnewaska Condominium." *Times Herald Record* (March 7, 1975).

"Equity Partner Sought" (advertisement). *New York Times* (June 6, 1975).

"Foreclosure Action Is Filed Against Lake Minnewaska." *Daily Freeman* (July 25, 1975).

"Group Asks State to Buy Lake Minnewaska Park Land." *Times Herald Record* (August 19, 1975).

"In the Shawangunks." *New York Times* (October 25, 1975).

"Real Property Foreclosure Sale" (advertisement). *New York Times* (October 6, 1977).

"State Makes New Bid for Minnewaska Land." *Times Herald Record* (October 19, 1977).

"State Agrees to Buy Minnewaska Land." *Times Herald Record* (October 21, 1977).

"Minnewaska Saved." *Newburgh Evening News, Beacon Edition* (October 22, 1977).

"29 Acre Historical Park Is Dedicated in Fort Lee." *New York Times* (May 15, 1976).

"Fire Levels Minnewaska's Cliff House." *Times Herald Record* (January 3, 1978).

Martin, Neil S. "The View Today." *Westchester Rockland Newspaper Co.* (January 7, 1979).

"Rochester Approves Plan for Minnewaska Building." *Weekly Freeman* (May 1979).

"Marriott Plan Termed 'Environmentally unsound.'" *Times Herald Record* (August 7, 1979).

"At N.J.'s Edge, a Park Is Dying." *Sunday Record* (September 9, 1979).

"Neglect on the Palisades." *Record* (September 16, 1979).

"Marriott Plans Spring Startup." *Times Herald Record* (November 16, 1979).

"Tests Polluting Stream, Minnewaska Opponents Say." *Times Herald Record* (January 18, 1980).

"Marriott, Ulster Leaders Laud Resort Plans." *Times Herald Record* (January 18, 1980).

"Minnewaska Sale Near, Marriott Says." *Times Herald Record* (January 18, 1980).

"Archeologist Urges Study of Minnewaska Hotel Site." *Times Herald Record* (February 1, 1980).

Haddad, Ron. "Will Marriott Mar Lake Minnewaska?" *Sierra Atlantic* (February 1980).

"Court Oks Marriott Land Purchase." *Times Herald Record* (April 24, 1980).

"Marriott Big Issue in New Paltz School Vote." *Times Herald Record* (May 3, 1980).

"'Casinos Not Marriott Issue.'" *Times Herald Record* (May 20, 1980).

"Marriott Studies Disputed." *Daily Freeman* (May 25, 1980).

"Citizens' Group Raps Marriott Impact Report." *Times Herald Record* (June 7, 1980).

"DEC Rejects Delay on Marriott Hearing." *Times Herald Record* (July 2, 1980).

"PIPC Talks Result in Suggested Marriott Trade-Offs." *Huguenot Herald* (August 6, 1980).

"Marriott Hearings Turn to Effect on Lake." *Times Herald Record* (August 7, 1980).

"Lawyer Threatens Legal Action over Minnewaska Records." *Poughkeepsie Journal* (August 16, 1980).

"Owner: Minnewaska Interests Moon." *Times Herald Record* (August 19, 1980).

"A Peace Treaty for the Hudson." *New York Times* (December 20, 1980).

"Marriott Ready to Spring Water Study." *Times Herald Record* (January 7, 1981).

"Catskill Center Endorses Proposed Marriott Complex." *Times Herald Record* (April 1, 1981).

"Fate of Scenic Upstate Lake Resort Hinges on a Judge's Opinion." *New York Times* (April 11, 1981).

"Marriott Challenges State Report on Minnewaska Plans." *Poughkeepsie Journal* (April 9, 1981).

"Marriott Resort Project's Opponents Shifting Focus to Commission." *Times Herald Record* (June 4, 1981).

"Marriott Pressed to Give Up Condominium Plan." *Daily Freeman* (June 5, 1981).

"Condominium Proposal in Jeopardy." *Times Herald Record* (June 5, 1981).

"Who Speaks for Us?" *Daily Freeman* (June 5, 1981).

"Heritage Pitted against Marriott." *Daily Freeman* (June 7, 1981).

"Key Opponent Explains Marriott Proposal Stand." *Times Herald Record* (June 9, 1981).

"Marriott Backers Plan Support Act, Organizer Says." *Times Herald Record* (June 12, 1981).

"Sides Argue at Marriott Meeting." *Times Herald Record* (June 16, 1981).

"Challenge Mounted against Marriott." *Times Herald Record* (August 4, 1981).

"Lawsuit Filed against Easement." *Times Herald Record* (October 6, 1981).

"Marriott Gets OK to Build Golf Course." *Times Herald Record* (November 6, 1981).

"Court Order Stalls Disputed Marriott Plan." *Times Herald Record* (November 25, 1981).

"Judge Lifts Fiat Blocking Marriott Deal." *Times Herald Record* (December 15, 1981).

"Obstacle Cleared to Marriott's Resort." *Times Herald Record* (May 5, 1982).

"Marriott Intentions Probed." *Times Herald Record* (January 21, 1983).

"Court Rejects DEC Approval for Marriott Condos." *Times Herald Record* (February 4, 1983).

"Marriott Pulls Backing in Condo Project." *Times Herald Record* (February 9, 1983).

"Making a Mountain of Protest." *Sunday Times Herald Record* (April 3, 1983).

"Lake Minnewaska Resort Plan Revived." *Times Herald Record* (May 25, 1983).

"Small-Scale Minnewaska plan Offered." *Times Herald Record* (June 1, 1983).

"Marriott Golf Site Gets Legal Green Light." *Times Herald Record* (February 3, 1984).

"Marriott Calls Off Plans to Revive Ulster Resort." *New York Times* (March 13, 1985).

"Battle Lines Drawn at Lake Minnewaska." *New York Times* (October 6, 1985).

"Albany Considering Minnewaska's Fate." *New York Times* (June 28, 1986).

"State to Acquire an Upstate Lake." *New York Times* (November 25, 1986).

"Lake to Be a State Park, Ending a 17-Year Battle." *New York Times* (June 3, 1987).

Chapter 17: Sterling Forest

Correspondences

JoAnn Dolan to Mr. Leon G. Billings, July 26, 1993.

Robert O. Binnewies to Assistant Secretary-Designate George T. Frampton Jr., April 7, 1993.

JoAnn Dolan to Robert O. Binnewies, February 26, 1993.

Bob Binnewies to Bob Thomson, July 9, 1993.

Klara B. Sauer, Robert Augello, Kim Elliman, and Olivia Millard to Hon. Mario M. Cuomo, January 19, 1994.

Bob Binnewies to the Public/Private Partnership to Save Sterling Forest, March 25, 1994.

Jim Tripp to Rob Pirani, April 20, 1994.

Bob Binnewies to the Public/Private Partnership to Save Sterling Forest, October 7, 1994.

Bob Binnewies to "Mac" [Commissioner Malcolm Borg], November 30, 1994.

Bob Binnewies to Rob Pirani, December 9, 1994.

JoAnn Dolan to Robert O. Binnewies, January 10, 1995.

Robert O. Binnewies to Dr. Edward Kubersky, January 31, 1995.

Governor Christine Todd Whitman to Hon. George Pataki, February 3, 1995.

Governor Christine Todd Whitman to Mr. Larry Rockefeller, February 23, 1995.

Robert O. Binnewies to Mr. Patrick F. Noonan, March 24, 1995.

Governor George Pataki to Hon. Christine Todd Whitman, April 6, 1995.
Barnabas McHenry to Commissioner Bernadette Castro, April 14, 1995.
Robert S. Davis to Hon. Bernadette Castro, June 15, 1995.
Bob Binnewies to the Public/Private Partnership to Save Sterling Forest, July 3, 1995.
Bob Binnewies to the Public/Private Partnership, April 19, 1996.
Bob Binnewies to the Public/Private Partnership, June 14, 1996.
Bob Binnewies to the Public/Private Partnership, July 10, 1996.
Bob Binnewies to the Public/Private Partnership, October 18, 1996.

Reports

"Sterling Forest Fund Receipts as of 2/1/96."
"Sterling Forest Fund Expenditures as of 2/1/96."
"Proposed Agreement for Acquisition of Sterling Forest." November 1, 1996.
"Sterling Forest: Impact of Development in Reservoir Watershed." North Jersey District Water Supply Commission.

Newspapers

"Would Protect a Portion of Sterling Forest." *Advertiser Photo News* (February 5, 1986).
"Future of Sterling Forest Stirs Imaginations." *Bergen Record* (February 23, 1986).
"Don't Spoil Sterling Forest." *Bergen Record* (March 7, 1986).
"Monroe-West Milford Green Belt Proposed." *Bergen Record* (March 8, 1986).
"Activist's Plan Borne Out of love for Forest." *Bergen Record* (April 13, 1986).
"$2M Grant to Help Buy Woodlands." *Bergen Record* (August 28, 1986).
"They're Mapping the 'Greenway.'" *West Milford Argus* (September 7, 1986).
"Sterling Forest Land Deal Appears Near." *Times Herald Record* (September 17, 1986).
"For Sale: Sterling Forest 19,990 Acres." *New York Times* (October 26, 1986).
"Sterling Forest Off Seller's Block." *Times Herald Record* (March 26, 1987).
"Sterling Forest Preservation Has Torricelli Support." *Green Lake and West Milford News* (May 18, 1988).
"Questions About Sterling Forest." *New York Times* (May 22, 1988).
"Sterling Forest: What's a Watershed Worth." *Bergen Record* (June 20, 1988).
"Battle for Sterling Forest." *Rockland Journal News* (August 21, 1988).
"Owner to Skip Talks on Sterling Forest's Future." *Times Herald Record* (September 23, 1988).
"Quick Seizure Approved for Forest Tract." *Bergen Record* (October 20, 1988).
"Passaic Takes Title to Sterling Forest." *Bergen Record* (November 1, 1988).
"Sterling Forest Chief to Stress Public Relations." *Times Herald Record* (February 6, 1989).
"Open Space Around Cities Is Shrinking." *Philadelphia Inquirer* (June 26, 1989).
"Exit 15A Plan Gets New Support." *Times Herald Record* (July 11, 1989).
"'Environmentally Sensitive' Firm to Decide Sterling Property's Fate." *Times Herald Record* (August 10, 1989).
"Sterling Forest: Feds Pushed to Intervene." *Times Herald Record* (October 3, 1989).
"Development of Sterling Forest Viewed as Threat to North Jersey." *Star-Ledger* (October 3, 1989).
"Forest Project Called Peril to New Jersey." *New York Times* (October 8, 1989).
"Heimbach Named Sterling President." *Times Herald Record* (October 11, 1989).

"Palisades Park Appoints Director." *Times Herald Record* (November 29, 1989).

"Sterling Forest Study Released." *Times Herald Record* (February 16, 1990).

"Sterling Forest Plan Would Preserve Much Wilderness." *Bergen Record* (April 23, 1990).

"Sterling Forest Principal Has Local Tie." *Times Herald Record* (August 16, 1990).

"British Firm Buys Sterling Forest Corp." *Rockland Journal-News* (August 16, 1990).

"Sterling Forest Sale a Done Deal." *Times Herald Record* (October 2, 1990).

"Albany Looks Longingly at Land It Can't Pay For." *New York Times* (November 25, 1990).

"Sterling Forest Land Plan Unveiled." *Bergen Record* (March 28, 1991).

"Environmentalists Criticize Plan for Sterling Forest Development." *New York Times* (March 31, 1991).

"Last Chance Effort." *Star Ledger* (May 12, 1991).

"Clock Is Running on Sterling Forest." *Times Herald Record* (May 15, 1991).

"You Can't See the Forest for the Buffers." *Times Herald Record* (May 28, 1991).

"Boom County, Bust Budget." *Wall Street Journal* (September 25, 1991).

"Sterling Forest Plan Errs on Costs: Report." *Times Herald Record* (October 5, 1991).

"Democrat Fights Sterling Forest Development." *Times Herald Record* (February 29, 1992).

"Selling Sterling Forest." *Times Herald Record* (March 30, 1992).

"Fund Match for Sterling Ruled Out." *Times Herald Record* (May 7, 1992).

"Lawmakers Clash Over Forest Land." *Times Herald Record* (June 24, 1992).

"Gilman Opposes Sterling Forest Purchase." *Rockland Journal-News* (June 24, 1992).

"House Panel OKs Sterling Forest Funds." *Bergen Record* (June 30, 1992).

"DEC to Oversee Sterling Forest Review." *Times Herald Record* (July 1, 1992).

"State Seeks Sterling Forest Deal." *Rockland Journal-News* (August 22, 1992).

"Preserve Forest, Citizen Group's Survey Finds." *Times Herald Record* (September 4, 1992).

"War of Woods." *Bergen Record* (February 14, 1993).

"Economy Aids a Compromise Over Developing Forest Land." *New York Times* (April 27, 1993).

"Sterling Forest, U.S. Dealing." *Bergen Record* (April 28, 1993).

"Protection for NJ Highlands." *Bergen Record* (May 15, 1993).

"Steps Taken for U.S. to Buy Sterling Tract." *Times Herald Record* (May 27, 1993).

"$35 Million Sought to Save Sterling Forest." *Bergen Record* (July 27, 1993).

"$25 Million Sought for Forest Purchase." *Green Lake and West Milford News* (August 22, 1993.

"NY 'New City' Project Threatens Jersey Water." *Star Ledger* (August 22, 1993).

"Babbitt Calls Sterling Land-Buy a First Step." *Times Herald Record* (September 29, 1993).

"Trygg-Hansa Recapitalizing Its Home Unit." *Wall Street Journal* (December 20, 1993).

"Park Panel Enlisted in Fight for Highlands." *Bergen Record* (April 26, 1994).

"National Park Service Dampens $35 Million Plan to Buy SF." *Star Ledger* (May 19, 1994).

"Setback in Plan to Safeguard Sterling Forest." *Bergen Record* (May 18, 1994).

"Congressmen Ask Babbitt to Rescue Sterling Forest Bill." *Bergen Record* (May 19, 1994).

"Saving Sterling Forest Must Be a Top Priority." *Bergen Record* (May 20, 1994).

"Park Service Does Grave Disservice to Highlands." *Bergen Record* (May 22, 1994).

"Park Service Now Backs Sterling Preservation." *Times Herald Record* (May 24, 1994).

"Sterling Forest Parkland Plan May Get $17M Federal Boost." *Rockland Journal-News* (May 24, 1994).

"Strings Tied Sterling Forest Aid." *Bergen Record* (May 26, 1994).

"Sterling Forest Can Be Preserved." *Rockland Journal-News* (May 26, 1994).

"Panel OK's NJ Funds to Acquire Woodland." *Bergen Record* (June 3, 1994).
"State Senate OK's $10M for Sterling Forest." *Bergen Record* (July 1, 1994).
"Sterling Forest Funding Backed." *Bergen Record* (August 4, 1994).
"Heimbach Moves Up." *Times Herald Record* (August 11, 1994).
"Sterling Forest Funding Moves Ahead." *Bergen Record* (August 22, 1994).
"SFC Goes Generic with Its EIS." *Green Lakes and West Milford News* (September 7, 1994).
"10M OK'd for Sterling Forest." *Bergen Record* (September 27, 1994).
"Senate Kills Quartet of Bills Dealing with New Jersey Issues." *Bergen Record* (October 11, 1994).
"Senate Backs Sterling Forest Proposal." *Bergen Record* (December 16, 1994).
"Sterling Talk Put to Rest." *Times Herald Record* (February 11, 1995).
"Forest Short Listed." *Green Lake and West Milford News* (March 1, 1995).
"Newt Favors Buying Forest." *Bergen Record* (March 10, 1995).
"DEC Clears Way for Sterling Hearings." *Times Herald Record* (April 12, 1995).
"SF: A Search for Answers." *Times Herald Record* (April 19, 1995).
"Park Chief Backs Forest Buy." *Times Herald Record* (May 5, 1995).
"Development Opposed." *Times Herald Record* (June 16, 1995).
"For Tuxedo, a Big Decision on Sterling Forest." *Bergen Record* (June 21, 1995).
"Editor's Corner: Reality Check." *Green Lakes and West Milford News* (June 21, 1995).
"Forest DGEIS Blasted in Tuxedo." *Green Lake and West Milford News* (June 21, 1995).
"House Panel Fires Away at Sterling Forest Plan." *Rockland Journal-News* (September 29, 1995).
"Gingrich Backs Buying Tract on Jersey-New York Border." *New York Times* (December 13, 1995).
"Gingrich's New Deal: Buy Sterling Forest." *Rockland Journal-News* (December 14, 1995).
"Another Victory for Sterling Forest." *Sterling Messenger* (February 2000).
"Saving Sterling Forest." *Bergen Record* (February 10, 2000).

Books

Abbott, Arthur P. *The Greatest Park in the World*. Historian Publishing, 1914.

Correspondence

Bob Binnewies to Henry DeCotis, Megan Lesser Levine, Steve Lewis, and Jim Economides.

Chapter 18: Honor and Electronics

Amicus Curiae Brief, Docket No. A-000259-13T1, Superior Court of N.J., Appellate Division, April 4, 2014

Correspondence

PIPC Jim Hall to Secretary, Englewood Cliffs Board of Adjustment, 6/30/2011.
PIPC Jim Hall to Secretary, Englewood Cliffs Board of Adjustment, 10/31/2011.
PIPC Jim Hall to Seog-Won Park, CEO, LG Electronics USA, 11/14/2012.
PIPC Jim Hall to William Cho, President, LG Electronics USA, 01/23/2014.
PIPC Meeting Minutes, LG Electronics Resolution, 2/24/2014.

Articles

Marist Environmental History Project, "The Scenic Hudson Decision," 12/9/2019.
PIPC National Purple Heart Hall of Honor History.
"Trail Conference Joins LG Electronics at Groundbreaking," 2/7/2017.
Tom Stable. BNP Media, 11/23/2018.
New York Magazine, "Awosting Reserve," 8/13/2001.

Newspapers

Hartford Courant, "Purple Heart Museum Threatened," Anne M. Hamilton, 8/12/1996.
Hartford Courant, "Purple Heart Museum Has Two Suitors," Anne M. Hamilton, 12/11/1996.
Times Herald Record, "Awosting Reserve," 11/8/2005.
Cyclists International, "Commission Rejects LG's Proposal to Ruin Palisades," 2/25/2014.
New York Times, "New Forces Join Lawsuit Fighting Palisades Tower," Robin Pogrebin, 1/22/2014.
New York Times, "LG to Reduce Height of Headquarters, Preserving Palisades Horizon," Jim Dwyer, 6/23/2015.
ENR New York, "LG Tops NJ Palisades after Rocky Road to Build New Headquarters," 10/21/2015.
NJ Advanced Media, "Palisades Commission Asks LG to Lower Tower Height of Planned Englewood Cliffs Headquarters," Myles Ma, 3/29/2019.

INDEX

Robert O. Binnewies was Executive Director of the Palisades Interstate Park Commission throughout the 1990s. During a conservation career that spanned nearly forty years, he also served as Superintendent of Yosemite National Park, Vice President of the National Audubon Society, and Executive Director of the Maine Coast Heritage Trust.

EMPIRE STATE EDITIONS SELECT TITLES FROM EMPIRE STATE EDITIONS

Britt Haas, *Fighting Authoritarianism: American Youth Activism in the 1930s*

David J. Goodwin, *Left Bank of the Hudson: Jersey City and the Artists of 111 1st Street.* Foreword by DW Gibson

Nandini Bagchee, *Counter Institution: Activist Estates of the Lower East Side*

Susan Opotow and Zachary Baron Shemtob (eds.), *New York after 9/11*

Andrew Feffer, *Bad Faith: Teachers, Liberalism, and the Origins of McCarthyism*

Colin Davey with Thomas A. Lesser, *The American Museum of Natural History and How It Got That Way.* Forewords by Neil deGrasse Tyson and Kermit Roosevelt III

Wendy Jean Katz, *Humbug! The Politics of Art Criticism in New York City's Penny Press*

Lolita Buckner Inniss, *The Princeton Fugitive Slave: The Trials of James Collins Johnson*

Mike Jaccarino, *America's Last Great Newspaper War: The Death of Print in a Two-Tabloid Town*

Angel Garcia, *The Kingdom Began in Puerto Rico: Neil Connolly's Priesthood in the South Bronx*

Jim Mackin, *Notable New Yorkers of Manhattan's Upper West Side: Bloomingdale–Morningside Heights*

Matthew Spady, *The Neighborhood Manhattan Forgot: Audubon Park and the Families Who Shaped It*

Marilyn S. Greenwald and Yun Li, *Eunice Hunton Carter: A Lifelong Fight for Social Justice*

Jeffrey A. Kroessler, *Sunnyside Gardens: Planning and Preservation in a Historic Garden Suburb*

Elizabeth Macaulay-Lewis, *Antiquity in Gotham: The Ancient Architecture of New York City*

For a complete list, visit www.fordhampress.com/empire-state-editions.